Dynamical Systems

This book offers a captivating exploration of the intersection between mathematics, chaos theory, and dynamical systems through the personal journeys of twelve renowned mathematicians and physicists from China, Europe, Russia, and the USA.

The first section of this book provides an intimate look into the formative experiences and early steps of these scientists. In these life stories, the names of other famous mathematicians arise, crisscrossing all the stories in unexpected ways. The second part of this book explores the practical applications of chaotic attractors in various fields. These include chaos-based encryption in cryptography, sensor and actuator placement in Chua circuits for control systems, and chaotic dynamics in remote sensing for crop modeling. It also highlights the role of chaos theory in the development of memristors following Leon Chua's 1971 discovery, leading to advances in nonlinear dynamics, hyperchaos, and memristor-based systems. The chapters further examine how chaos theory addresses modern challenges such as modeling COVID-19 spread using SEIR models and optimizing mobile network design, demonstrating the wide-reaching impact of chaotic systems in real-world applications.

This book will be of great value to students and researchers in mathematics, physics, engineering, and related disciplines seeking to deepen their understanding of chaotic dynamical systems and their applications.

This book includes a revised introduction and a new chapter. The remaining chapters were originally published in the *Journal of Difference Equations and Applications.*

René Lozi is Emeritus Professor at University Côte d'Azur, Nice, France, and Vice-President of the International Society of Difference Equations. His research areas include complexity and emergence theory, dynamical systems, bifurcations, control of chaos, cryptography based on chaos, and memristors.

Safwan El Assad is Professor at Polytech Nantes, France. Between 1988 and 2005, his research activities concerned radar imagery and digital communications. Nowadays his research largely focuses on chaos-based cryptography, encryption, crypto-compression, steganography, hash functions, authenticated encryption.

Mohammed-Salah Abdelouahab is Professor at Abdelhafid Boussouf University Center, Mila, Algeria. He is the head of the research team in fractional calculus and its applications at the laboratory of mathematics and their interactions and the editor-in-chief of the *Journal of Innovative Applied Mathematics and Computational Sciences.*

Dynamical Systems

Paths of Twelve Mathematicians and Physicists to Chaos and Its Applications

Edited by

René Lozi, Safwan El Assad and Mohammed-Salah Abdelouahab

CRC Press
Taylor & Francis Group
Boca Raton London New York

CRC Press is an imprint of the
Taylor & Francis Group, an **informa** business
AN AUERBACH BOOK

First edition published [2025]
by CRC Press
4 Park Square, Milton Park, Abingdon, Oxon, OX14 4RN

and by CRC Press
2385 NW Executive Center Drive, Suite 320, Boca Raton FL 33431

British Library Cataloguing-in-Publication Data
A catalogue record for this book is available from the British Library

ISBN: 978-1-032-94029-8 (hbk)
ISBN: 978-1-032-94030-4 (pbk)
ISBN: 978-1-003-56860-5 (ebk)

DOI: 10.1201/9781003568605

Typeset in Minion Pro
by codeMantra

Contents

Citation Information

The chapters in this book were originally published in the *Journal of Difference Equations and Applications*, volume 29, issue 9-12 (2023), except Chapter 1, which was published in volume 30, issue 1 (2024). When citing this material, please use the original page numbering for each article, as follows:

Chapter 1
The paths of nine mathematicians to the realm of dynamical systems
René Lozi, Vladimir Belykh, Jim Michael Cushing, Lyudmila Efremova, Saber Elaydi, Laura Gardini, Michał Misiurewicz, Eckehard Schöll and Galina Strelkova
Journal of Difference Equations and Applications, volume 30, issue 1 (2024), pp. 1–89

Chapter 3
Optimal placement of sensor and actuator for controlling the piecewise linear Chua circuit via a discretized controller
Christophe Letellier, Ludovico Minati and Jean-Pierre Barbot
Journal of Difference Equations and Applications, volume 29, issue 9–12 (2023), pp. 1341–1368

Chapter 4
Nonlinear dynamics and hyperchaos in a modified memristor-based Chua's circuit and its generalized discrete system
Nadjet Boudjerida, Mohammed-Salah Abdelouahab and René Lozi
Journal of Difference Equations and Applications, volume 29, issue 9–12 (2023), pp. 1369–1390

Chapter 5
Memristor-based Gauss chaotic maps with hidden/self-exited dynamics
Han Bao, Yang Gu, Jianjun Sun, Xi Zhang and Bocheng Bao
Journal of Difference Equations and Applications, volume 29, issue 9–12 (2023), pp. 1391–1407

Chapter 6
Enhanced design and hardware implementation of a chaos-based block cipher for image protection
Mahdi Madani, Safwan El Assad, Fethi Dridi and René Lozi
Journal of Difference Equations and Applications, volume 29, issue 9–12 (2023), pp. 1408–1428

Chapter 7

On secure communication scheme-based mixed discrete-time hyperchaotic systems
Narjes Khalifa and Mohamed Benrejeb
Journal of Difference Equations and Applications, volume 29, issue 9-12 (2023), pp. 1429–1448

Chapter 8

On the use of chaotic dynamics for mobile network design and analysis: towards a trace data generator
Martin Rosalie and Serge Chaumette
Journal of Difference Equations and Applications, volume 29, issue 9-12 (2023), pp. 1449–1466

Chapter 9

The SEIR Covid-19 model described by fractional-order difference equations: analysis and application with real data in Brazil
Abdullah Othman Almatroud, Noureddine Djenina, Adel Ouannas and Giuseppe Grassi
Journal of Difference Equations and Applications, volume 29, issue 9-12 (2023), pp. 1467–1479

Chapter 10

Chaotic attractors captured from remote sensing time series for the dynamics of cereal crops
Sylvain Mangiarotti and Flavie Le Jean
Journal of Difference Equations and Applications, volume 29, issue 9-12 (2023), pp. 1480–1502

Chapter 11

Nonlinear dynamic and chaos in a remanufacturing duopoly game with heterogeneous players and nonlinear inverse demand functions
Habiba Meskine, Mohammed-Salah Abdelouahab and René Lozi
Journal of Difference Equations and Applications, volume 29, issue 9-12 (2023), pp. 1503–1515

Notes on Contributors

Mohammed-Salah Abdelouahab, Laboratory of Mathematics and Their Interactions, Abdelhafid Boussouf University Center, Mila, Algeria.

Abdullah Othman Almatroud, Department of Mathematics, Faculty of Science, University of Ha'il, Saudi Arabia.

Bocheng Bao, School of Microelectronics and Control Engineering, Changzhou University, People's Republic of China.

Han Bao, School of Microelectronics and Control Engineering, Changzhou University, People's Republic of China.

Jean-Pierre Barbot, QUARTZ EA7393 Laboratory, ENSEA, France; LS2N, UMR 6004 CNRS, École Centrale de Nantes, France.

Vladimir Belykh, Volga State University of Water Transport, Nizhny Novgorod, Russia; Lobachevsky State University of Nizhny Novgorod, Russia.

Mohamed Benrejeb, National Engineering School of Tunis (ENIT), Laboratory of Research in Automatic Control (LARA), University of Tunis El Manar, Tunisia.

Nadjet Boudjerida, Department of Mathematics, Mentouri Brothers University, Constantine, Algeria; Laboratory of Mathematics and their interactions, Abdelhafid Boussouf University Center, Mila, Algeria.

Serge Chaumette, LaBRI, UMR 5800, Univ. Bordeaux, Talence, France.

Guanrong Chen, Centre for Complexity and Complex Networks, City University of Hong Kong, China.

Jim Michael Cushing, Department of Mathematics and Interdisciplinary Program in Applied Mathematics, University of Arizona, Tucson, USA.

Noureddine Djenina, Laboratory of Dynamical Systems and Control, University of Larbi Ben M'hidi, Algeria.

Fethi Dridi, Electronics and Microelectronics Laboratory (EmE), Faculty of Sciences of Monastir, University of Monastir, Tunisia.

Lyudmila Efremova, Department of Differential Equations, Mathematical and Numerical Analysis, National Research University of Nizhni Novgorod, Russia; Department of Fundamental Mathematics, Moscow Institute of Physics and Technology, Russia.

Saber Elaydi, Department of Mathematics, Trinity University, San Antonio, USA.

Safwan El Assad, Nantes Université, France.

Laura Gardini, DESP, University of Urbino Carlo Bo, Italy.

Giuseppe Grassi, Dipartimento Ingegneria Innovazione, università del Salento, Lecce, Italy.

Yang Gu, School of Microelectronics and Control Engineering, Changzhou University, People's Republic of China.

Narjes Khalifa, National Engineering School of Tunis (ENIT), Laboratory of Research in Automatic Control (LARA), University of Tunis El Manar, Tunisia.

Flavie Le Jean, Centre d'Études Spatiales de la Biosphère, CNRS-UPS-CNES-IRD, ObservatoireMidi-Pyrénées, Toulouse, France.

Christophe Letellier, Rouen Normandie University – CORIA, Saint-Etienne du Rouvray, France.

René Lozi, Université Côte d'Azur, Laboratoire J. A. Dieudonné, C.N.R.S., Nice, France.

Mahdi Madani, IETR UMR CNRS 6164, Université de Nantes/Polytech Nantes, France.

Sylvain Mangiarotti, Centre d'Études Spatiales de la Biosphère, CNRS-UPS-CNES-IRD, ObservatoireMidi-Pyrénées, Toulouse, France.

Habiba Meskine, Department of Mathematics, Mentouri Brothers University, Constantine, Algeria.

Ludovico Minati, Center for Mind/Brain Sciences (CIMeC), University of Trento, Italy; Institute of Innovative Research, Tokyo Institute of Technology, Yokohama, Japan.

Michał Misiurewicz, Department of Mathematical Sciences, Indiana University – Purdue University Indianapolis, USA.

Adel Ouannas, Laboratory of Dynamical Systems and Control, University of Larbi Ben M'hidi, Algeria.

Martin Rosalie, Laboratoire Génome et Développement des Plantes, UMR 5096, Centre National pour la Recherche Scientifique, Univ. Perpignan, Via Domitia, Perpignan, France.

Miguel Sanjúan, Nonlinear Dynamics, Chaos and Complex Systems Group, Departamento de Física, Universidad Rey Juan Carlos, Madrid, Spain.

Eckehard Schöll, Institut fur Theoretische Physik, Technische Universitat Berlin, Germany; Bernstein Center for Computational Neuroscience Berlin, Germany; Potsdam Institute for Climate Impact Research, Germany.

Galina Strelkova, Institute of Physics, Saratov State University, Russia.

Jianjun Sun, School of Microelectronics and Control Engineering, Changzhou University, People's Republic of China.

Xi Zhang, School of Microelectronics and Control Engineering, Changzhou University, People's Republic of China.

Bo Zheng, School of Mathematics and Information Science, Guangzhou University, China.

Introduction: Dynamical systems: paths of twelve mathematicians and physicists toward the chaos and its applications

René Lozi, Safwan El assad, Mohammed-Salah Abdelouahab

Introduction

The first part of this volume is dedicated to the first steps of 12 mathematicians and physicists from China, Europe, Russia, and the USA on their path to chaotic dynamical systems. These scientists during their career improved the theory of chaotic attractors which are now involved in many applications as presented in the second part of the book. We can find such applications in the domain of cryptography with, for example, enhanced design and hardware implementation of a chaos-based block cipher for image protection or secure communication scheme-based mixed discrete-time hyperchaotic systems. Chaos is also involved in sensors with the search of optimal placement of sensor and actuator for controlling the piecewise linear Chua circuit via a discretized controller, and chaotic attractors captured from remote sensing time series for the dynamics of cereal crops. Furthermore, the discovery by Leon Chua on the electric device called memristor in 1971 is nowadays strongly linked with chaos theory, as it is presented in nonlinear dynamics and hyperchaos in a modified memristor-based Chua's circuit and its generalized discrete system, memristor-based Gauss chaotic maps with hidden/self-exited dynamics. Chaos is also used for modelling the SEIR Covid-19 and for mobile network design. For three years (from Spring 2021 to Spring 2024), 36 authors from 10 countries (Algeria, China, France, Germany, Italy, Russia, Saudi Arabia, Spain, Tunisia, and the USA) worked jointly to publish new results in this field.

Chapter 1 describes the lives of nine mathematicians from France, Germany, Italy, Russia, and the USA and the paths they took to mathematics, chaos, and discrete dynamical systems, some from early childhood. In these life stories, the names of outstanding mathematicians arise, crisscrossing the nine stories in unexpected ways. These mathematicians also interacted with each other, forming an intriguing social network worldwide, across all borders of nationality and languages.

Chapter 2 is a continuation of Chapter 1 showing the lives of three more outstanding researchers in mathematics and physics from China and Spain. It reveals in a fascinating way their life stories, pointing out the steps they followed from early years to the present career, each professional trajectory having its challenging stages, struggles and inspiration, motivation and hard work, and finally international recognition and remarkable success. These three paths are completely distinct, but each one shows that the love for science, perseverance, and passion are the keys to outstanding results.

The Lorenz attractor published in 1963 [9] is the first example of a continuous chaotic attractor. This publication was followed in 1976 by the first example of discrete chaotic attractor, the Hénon attractor [5]. Since then, the study of chaotic dynamical systems has been developed along several directions in the second half of twentieth century. In the early 1990s, unexpected applications of these pure mathematical objects to many fields, including biology, cryptography, optimization, economics and finance, etc., were discovered. This book collects original research papers addressing such themes, in particular the theory of chaotic attractors.

The first paradigm shifts in the study of chaotic maps came in 1987, when instead of focusing on the theoretical study of their mathematical properties or the search for generalizations, some electrical engineers used them to build new devices, like Rodriguez-Vasquez et al. [17] who presented a special-purpose analog computer made of switched-capacitor circuit for analyzing chaos and bifurcation phenomena in nonlinear discrete dynamical systems modelled by discrete maps. They published results for four maps: the logistic map, a piece-wise linear map, and the Hénon and Lozi maps. Since then, and until today, these maps are continuously involved in the field of electronics or electrical devices like memristors, in secure communications, biological models, and even economy. Some of them are presented in this book.

In 1971, Leon Chua, then Professor at the University of California, Berkeley, predicted the existence of a missing fourth passive circuit element, in addition to the three classical ones: resistor, inductor, and capacitor [4]. He called this new element 'memristor' meaning it is a resistor with memory. It is characterized by a nonlinear constitutive relationship between the charge q and the flux φ. Such a physical device would not be reported until 2008, when a physical model of a two-terminal *hp* device behaving like a memristor was announced [20] sparkling intense research with thousands of papers published to date. A general Ohm's law for theorizing this device was published ten years ago [1].

In 1983, Chua invented a very simple electric circuit producing chaos [14]. The advantage of this circuit is that the variables of the mathematical equations correspond to voltage and current and can be viewed on the screen of an oscilloscope.

These two important improvements in the theory of electrical circuits are studied in three chapters of this book.

In Chapter 3, Letellier, Minati, and Barbot [8], considered the control of continuous-time and discretized Chua circuit, governed by differential and difference equations. They showed that selecting the measured variable and the actuator can be guided by considering the observability and controllability symbolic coefficients and applying the notion of flatness. The feedback linearization was compared to a homogeneous and a passivity-based control laws, the flat control laws being more efficient than the others. It was thus shown that the proposed flat control law by feedback linearization is very efficient. The authors suggested that, most likely, these results could be extendable to a large group of natural and experimental systems.

In Chapter 4, Boudjerida, Abdelouahab, and Lozi [3] reported a modified hyperchaotic memristor-based Chua's circuit and its generalized discrete model and explored their dynamics using different classical tools, showing that the proposed circuit model possesses a line of equilibrium and exhibits rich dynamics, including the coexistence of regular and chaotic attractors and mixed-mode oscillations. Using the Jury criterion, they constructed a useful stability test to localize some stable regions in three parameter planes of the discrete model. The presence of an exact periodic solution was numerically highlighted, and the occurrence of chaotic behavior was confirmed using the largest Lyapunov exponent and the 0-1 test.

In Chapter 5, Bao et al. [2] presented a unified memristor-based Gauss mapping model by coupling discrete memristors with Gauss map and then four two-dimensional (2-D) memristive Gauss maps using four different memristance functions were derived. These maps possess two specific cases of infinitely many line-fixed points and no fixed points, resulting in the appearance of hidden dynamics or self-exited dynamics. Their results demonstrated that the discrete memristor can make the four 2-D memristive Gauss maps produce complex dynamical behaviours and greatly improve the chaos complexity of the one-dimensional (1-D) Gauss map.

Nowadays, chaos-based cryptography systems are widely used to protect sensitive data. A variety of chaos-based cryptosystems with software implementation have been published in the literature. However, due to the rapid growth of connected devices, the development of hardware secure and low-resource cryptographic systems is necessary.

In Chapter 6 [11], Madani et al. enhanced the design of one of their already published software cryptosystems and then addressed its FPGA-based implementation. The improvement in the design consists essentially to introduce an Initial Vector (IV) entry to the Chaotic Pseudo Random Number Generator (CPRNG), thus allowing to generate new dynamic keys for the confusion layer (using a modified 2-D cat map) and diffusion layer (using a logistic map) with each new execution of the system. The proposed cryptosystem, which operates in Cipher Block Chaining (CBC) mode, is controlled by a Finite State Machine (FSM) to ensure that both the encryption and decryption are performed by the same reconfigurable architecture. The system was implemented on a Xilinx XC7Z020 ZYNQ FPGA platform.

In Chapter 7, Khalifa and Benrejeb [6] proposed a secure communication system based on mixed discrete-time chaotic systems. It is composed of two chaotic systems on the transmitter and receiver sides, to increase signal transmission security over the public channel. The case of electrocardiogram encryption and decryption is considered as an application to illustrate the effectiveness of the proposed chaotic encryption approach.

In Chapter 8, Rosalie and Chaumette [18] focussed on chaotic dynamics and reviewed their applications in the design of mobility models. They also provided a review of the nonlinear tools used to characterize mobility models, as it can be found in the literature. Finally, they proposed a method to generate traces for a given scenario involving moving people, using tools from the nonlinear analysis domain usually dedicated to topological analysis of chaotic attractors.

Chaotic attractors have also applications in life science and economy. Two chapters are related to the Covid-19 disease and the dynamics of cereal crops, and one in duopoly game.

In Chapter 9, Othman Almatroud et al. [16] made a contribution to the studies on epidemic mathematical models based on fractional-order operators by introducing a novel Covid-19 model described by noninteger-order difference equations. Their system has two fixed points at most, i.e. a disease-free fixed point which is proved globally stable, indicating that the pandemic will disappear when a simple condition on the system parameters is satisfied. Numerical simulations confirmed these results.

In Chapter 10, Mangiarotti and Le Jean [12] recalled that chaotic attractors were captured from remote sensing time series of the dynamics of cereal crops in semi-arid region in the early 2010s. They questioned the generality of this result by applying the same modelling techniques to four provinces in coastal and inland Morocco: Safi, El Jadida, Khourigba, and Khenifra. Several experimentations were considered, applying the analyses at the provinces scale, or to several provinces in association and in aggregation. The obtained models confirmed the results obtained in the early 2010s: the dynamics of cereal crops is chaotic and can be approximated by a weakly dissipative three-dimensional dynamical system.

In Chapter 11, Meskine, Abdelouahab and Lozi [15] investigated the dynamics of a remanufacturing duopoly game model with heterogeneous players and different competition strategies, assuming that the original equipment manufacturer has bounded rationality and aims to maximize profit, and the third-party remanufacturer has an adaptive nature and aims to maximize share in order to get a certain profit. The model behaviours were discussed mainly from the point of view of stability, bifurcations, and chaos. By utilizing Jury's stability criteria, the Cournot-Nash equilibrium's asymptotic stability was examined. Under certain conditions, when consumer willingness to pay was taken as a bifurcation parameter, the system undergoes a flip bifurcation and a Neimark-Sacker bifurcation. Lyapunov exponents were used to demonstrate that the system becomes chaotic through each of the previous bifurcations.

Research on applications of chaotic attractors is a hot topic today. Between April 2021 and October 2023, significant literature has been published. We give a brief overview of the contents of those articles that were not published in this book, although related to the same topic.

Recently, hidden attractors have also been discovered in some discrete memristors-based maps by Zhang et al. [23]. Wang et al. [22] included a discrete-time memristor to create a memristive Lozi map. This new 3-D memristor-based Lozi map was established by coupling a discrete memristor to the original 2-D Lozi map. Khennaoui et al. [7] using the Caputo-like delta difference obtained another memristor-based fractional Lozi map.

In [19], Sahnoune and Berkani presented a new method to design a deterministic interleaver with random-like behaviour based on Lozi map. Urrea et al. [21] compared the performances of such a chaotic interleaver with short block lengths, with a new chaotic interleaver based on Duffing map which showed slightly better performances.

Martinez-Vargas et al. [13] proposed a leader-follower optimization algorithm based on the combination of five different Population-based Algorithms (PAs) to tackle the high-dimensional optimization problem known as the base station (BS) deactivation problem in telephony. Among these five algorithms, chaotic binary particle swarm optimization (BPSO) has a Lozi map, embedded in.

Lozi [10] has writen a survey of recent applications of the chaotic Lozi map including 167 references.

Conclusion

Chaotic dynamical systems represent a revolution in mathematical analysis at the turn of the twenty-first century. Since the first example of a chaotic attractor 60 years ago, research on such objects and their applications is still ongoing. Twelve outstanding scientists describe their lives and the paths they have taken to explore this fascinating field. This book is just a glimpse of the last 60 years of this process that will continue for a long time to come.

References

[1] M. S., Abdelouahab, R., Lozi, and L. O. Chua, *Memfractance: A mathematical paradigm for circuit elements with memory*, Int. J. Bifurc. Chaos 24(9) (2014), p. 1430023. https://doi.org/10.1142/S0218127414300237

[2] H. Bao, Y. Gu, J. Sun, X. Zhang, and B. Bao, *Memristor-based Gauss chaotic maps with hidden/self-exited dynamics*, J. Differ. Equ. Appl. 29(9-12) (2023), pp. 1391–1407. https://doi.org/10.1080/10236198.2022.2144268

[3] N. Boudjerida, M.-S. Abdelouahab, and R. Lozi, *Nonlinear dynamics and hyperchaos in a modified memristor-based Chua's circuit and its generalized discrete system*, J. Differ. Equ. Appl. 29(9-12) (2023), pp. 1369–1390. https://doi.org/10.1080/10236198.2023.2172334

[4] L. O. Chua, *Memristor-the missing circuit element*, IEEE Trans. Circuit Th. 18(5) (1971), pp. 507–519. http://dx.doi.org/10.1109/TCT.1971.1083337

[5] M. Hénon, *A two-dimensional mapping with a strange attractor*, Commun. Math. Phys. 50(1) (1976), pp. 69–77. https://doi.org/10.1007/BF01608556.

[6] N. Khalifa and M. Benrejeb, *On secure communication scheme-based mixed discretetime hyperchaotic systems*, J. Differ. Equ. Appl. 29(9-12) (2023), pp. 1429–1448. https://doi.org/10.1080/10236198.2023.2222006

[7] A.-A. Khennaoui, V.-T., Pham, V. P. Thoai, A. Ouannas, G. Grassi, and S. Momani, *From Lozi map to fractional memristive Lozi map*, Eur. Phys. J. Spec. Top. (2023), pp. 1–9. https://doi.org/10.1140/epjs/s11734-023-00911-8

[8] C. Letellier, L. Minati, and J.-P. Barbot, *Optimal placement of sensor and actuator for controlling the piecewise linear Chua circuit via a discretized controller*, J. Differ. Equ. Appl. 29(9-12) (2023), pp. 1341–1368. https://doi.org/10.1080/10236198.2022.2161899

[9] E. N. Lorenz, *Deterministic nonperiodic flow*, J. Atmos. Sci., 20 (1963), pp. 130–141. https://doi.org/10.1175/1520-0469(1963)020⟨0130:DNF⟩2.0.CO;2.

[10] R. Lozi, *Survey of recent applications of the chaotic lozi map*, Algorithms 2023, 16(10), 491. https://doi.org/10.3390/a16100491

[11] M. Madani, S. El Assad, F. Dridi, and R. Lozi, *Enhanced design and hardware implementation of a chaos-based block cipher for image protection*, J. Differ. Equ. Appl. 29(9-12) (2023), pp. 1408–1428. https://doi.org/10.1080/10236198.2022.2069496

[12] S. Mangiarotti and F. Le Jean, *Chaotic attractors captured from remote sensing time series for the dynamics of cereal crops*, J. Differ. Equ. Appl. 29(9-12) (2023), pp. 1480–1502. https://doi.org/10.1080/10236198.2022.2152336

[13] A. Martínez-Vargas, J. A. Gomez-Aviles, M. A. Cosio-Leon, and A. G. Andrade, *Explaining the walking through of a team of algorithms*, Computer 56(2) (2023), pp. 67–81. https://doi.org/10.1109/MC.2022.3212998

[14] T. A. Matsumoto, *Chaotic attractor from Chua's circuit*, IEEE Trans. Circuits Syst. CAS 31(12) (1984), pp. 1055–1058. http://dx.doi.org/10.1109/TCS.1984.1085459

[15] H. Meskine, M.-S. Abdelouahab, and R. Lozi, *Nonlinear dynamic and chaos in a remanufacturing duopoly game with heterogeneous players and nonlinear inverse demand functions*, J. Differ. Equ. Appl. 29(9-12) (2023), pp. 1503–1515. https://doi.org/10.1080/10236198.2023.2228421

[16] A. Othman Almatroud, N. Djenina, A. Ouannas, and G. Grassi, *The SEIR Covid-19 model described by fractional-order difference equations: analysis and application with real data in Brazil*, J. Differ. Equ. Appl. 29(9-12) (2023), pp. 1467–1479. https://doi.org/10.1080/10236198.2023.2211168

[17] A. Rodriguez-Vazquez, J. L. Huertas, A. Rueda, B. Perez-Verdu, and L. O. Chua, *Chaos from switched-capacitor circuits: discrete maps*, Proc. IEEE, 75(8) (1987), pp. 1090–1106. https://doi.org/10.1109/PROC.1987.13852

[18] M. Rosalie and S. Chaumette, *On the use of chaotic dynamics for mobile network design and analysis: towards a trace data generator*, J. Differ. Equ. Appl. 29(9-12) (2023), pp. 1449–1466. https://doi.org/10.1080/10236198.2023.2194452

[19] A. Sahnoune and D. Berkani, *On the performance of chaotic interleaver for turbo codes*, SN Applied Sciences 3(1) (2021). https://doi.org/10.1007/s42452-021-04147-w

[20] D. B. Strukov, G. S. Snider, D. R. Stewart, and R. S. Williams, *The missing memristor found*, Nature 453(7191) (2008), pp. 80–83. https://doi.org/10.1038/nature06932

[21] U. Urrea, J. Kern, and R. Lopez-Escobar, *Design of chaotic interleaver based on Duffing map for turbo code*, Symmetry 14(12) (2022), p. 2529. https://doi.org/10.3390/sym14122529

[22] J. Wang, Y. Gu, K. Rong, Q. Xu,and X. Zhang, *Memristor-Based Lozi Map with Hidden Hyperchaos*, Mathematics 10(19) (2022), p. 3426. https://doi.org/10.3390/math10193426

[23] L. P. Zhang, Z. C. Wei, H. B. Jiang, W. P. Lyu, and Q. S. Bi, *Extremely hidden multistability in a class of a two dimensional maps with a cosine memristor*, Chin. Phys. B 31(10) (2022), p. 100503. https://doi.org/l10.1088l1674-10561ac7294

The paths of nine mathematicians to the realm of dynamical systems

René Lozi, Vladimir Belykh, Jim Michael Cushing, Lyudmila Efremova, Saber Elaydi, Laura Gardini, Michał Misiurewicz, Eckehard Schöll and Galina Strelkova

ABSTRACT
This article is devoted to the first steps of nine mathematicians from five countries on their path to mathematics, chaos and discrete dynamical systems, some from early childhood. In these life stories, the names of outstanding mathematicians arise, crisscrossing the nine stories in unexpected ways. These mathematicians also interacted with each other, forming an intriguing social network worldwide, across all borders of nationality and languages.

1. Introduction by René Lozi

This article is dedicated to the first steps of nine mathematicians from five countries on their path to mathematics, chaos and discrete dynamical systems, some beginning from their early childhood. In these life stories, the names of famous mathematicians arise, crisscrossing the nine stories in unexpected ways.

In March 2021, I was appointed associate editor of the *International Journal of Difference Equations*. Saber Elaydi the editor in chief asked me to edit a special issue dedicated to my area of research. He suggested the title 'Lozi, Hénon and other chaotic attractors, theory and applications'. I asked my long-time colleagues and friends Lyudmila Efremova, Mohammed-Salah Abdelouahab, Safwan El Assad and Michal Pluhacek to help me in this task, by agreeing to be co-editor.

For 2 years, we sent hundreds of emails to potential authors in more than 20 countries around the world to finally publish this special issue of 32 articles. During this period, inspired by the collective article 'Some elements for a history of dynamic systems theory' [110] in which around 15 authors, including me, described their early involvement in research on chaos, I asked several of the oldest authors of the special issue to describe their path to mathematics and discrete dynamical systems without setting a start date for their reminiscence. Some declined the invitation, others accepted after very long discussions. I addition I asked also three recipients of the prestigious Bernd Aulbach prize of the International Society of Difference Equations to contribute: Michał Misiurewicz who in 1979 coined the name 'Lozi map', Jim Cushing and Saber Elaydi (the fourth recipient Laura Gardini was already involved as author in the special issue). All of them were asked to focus on their early times without any other indication. Therefore, this article has to be considered as a companion of the special issue.

A history of dynamical systems and chaos theory was provided by Aubin and Dahan-Dalm'edico, focusing on three important contributors from the 1960s (Smale, Lorenz and Ruelle) [19]. However, their article is based on the publications of the scientists who constructed this domain of mathematics and linked sciences and does not highlight their motivations. From another point of view, the famous essay entitled 'On How I Got Started in Dynamical Systems' by Steve Smale [193] describes his encounters with the mathematicians who inspired him and made him want to study chaos and invented the 'horseshoe' (between other important discoveries he made). It is with this in mind that this recollection article was designed.

Mathematical research is often perceived as an exclusively male domain, despite the significant contributions made by women to the discipline. This misconception is partly due to their excessive modesty, which prevents them from promoting their discoveries in society, but also to the weight of prejudices that have affected them in society for centuries. This article is no exception to this imbalance, but I am delighted that a third of the contributions are by women.

This article is organized as follows: in Section 2, Michał Misiurewicz describes his love for mathematics since his early childhood (starting counting when he was 3 years old and reading math text book for the first grade at the age of 5 and solving 'problems' from that book). He emphasizes the importance of the International Mathematics Olympiads to him during his high school years (calling them a 'great adventure') and the brilliant idea of creating 'special groups' for gifted students at the university level. In 1966, he was admitted as a student to the Faculty of Mathematics and Physics of the University of Warsaw, field of mathematics, resisting to his Grandfather who tried to convince him that he should go to the Technical University instead, and become an engineer. When he was a second-year student, Włodzimierz Holsztyński, a topologist, started to mentor him. In particular, giving him a problem in lattice theory. He solved the problem which resulted in his first publication, in 1969. He describes the origin of dynamical systems in Warsaw around 50 years ago due to the influence of Stanisław Mazur, Andrey Kolmogorov and Yakov Sinai on Wiesław Szlenk and Karol Krzyżewski. Jan Maria Strelcyn was also member of the Warsaw group before leaving to France. He explains how he switched from topology to dynamical systems (but still working between both specialities), due to the move of Holsztyński to the USA, leaving Poland, and how he remained stuck with one-dimensional dynamics having about two-thirds of his papers about it. Eventually he gives a practical advice for young mathematicians.

Laura Gardini, in Section 3 recalls that her decision to study Mathematics was taken when she was 12 years old. There was not a specific event that led to her decision, she simply loved Mathematics and realized that she was happy studying it, and that it was easy for her. She was living in a very small village, and she knew that her goal was to convince her parents of her will to continue to study in a scientific lyceum (the closest was in Ravenna, 10 km, far from her village) and then to study Mathematics at the closest University which was in Bologna. When she graduated (with honours) in 1975, she found research position at ENI (Ente Nazionale Idrocarburi) group. After reading some books at the beginning of the 1980s, by Thom [197], Zeeman [209], Poston and Stewart [151], Collet and Eckmann [45], Guckenheimer and Holmes [79], she immediately loved dynamical systems and therefore decided not to change anymore. Since then it has been her area of research. She resigned from ENI in Summer 1987, studying to win some competition at the University, that happened the year after, and she started as Researcher at the University of Urbino in November 1988. She then came into contact with Christian Mira in Toulouse, and established a friendship and a fruitful collaboration with him and later with Ralph Abraham. She began other long collaboration with Yuri and Vladimir Maistrenko and Iryna Sushko. In Urbino (Italy), she invited many other well famous scholars, in particular, Leonid and Andrey Shilnikov. In Minneapolis in 1995, she collaborated with Christos Frouzakis and Ioannis Kevrekidis. Since then the number of her collaborators working on smooth and nonsmooth maps has constantly been increasing.

Vladimir Belykh says in Section 4 that from 1961 to 1965, he studied mathematics and physics in the Chair of Oscillation Theory and Automatic Control, named after its founding director, Alexander A. Andronov, at Gorky State University (nowadays Lobachevsky State University of Nizhny Novgorod). His professors were Dmitry A. Gudkov, Yuri I. Neimark, Nickolai A. Fufaev and Nickolai A. Zheleztsov, the brilliant team of Andronov's disciples. Later he was accepted at the Research Institute of Applied Mathematics & Cybernetics where he met Leonid P. Shilnikov and became friend with him. He met also Alexander Sharkovsky in 1965 who was later external reviewer of his Ph.D. thesis in 1972. During his long career not yet finished, he proved many important results and especially invented the 'Belykh attractor' (the name was coined by Valentin Afraimovich) in September 1976, which was remarked by Yakov Sinai. Dmitri Anosov who served as an opponent for his 'Doctor of Sciences' degree thesis in 1983, strongly recommended to include the description of this map and its detailed analysis in the thesis. The same year he met Vadim Anishchenko at a conference on the Oka river and both became quickly life-long friends. He collaborated also with Leon Chua and visited him at UC Berkeley and later with many other outstanding researchers like Martin Hasler at École Polytechnique Fédérale de Lausanne (EPFL).

The research of Eckehard Schöll encompasses mathematics and physics. He recalls in Section 5 that as a school boy in a small German town, Nürtingen near Stuttgart in Southwest Germany, he was interested in a very broad spectrum of subjects, ranging from Latin, English, French, German via History, Music, Art to Mathematics and Physics. But when he passed the final school exam (Abitur) with best grades in all subjects in 1970, it was clear that he wanted to study Physics at university. In a Physics Colloquium in 1974 in Tübingen, he heard a lecture by Werner Heisenberg and was fascinated by his personality. He attended a mathematically oriented conference 'Rencontre entre mathématiciens et physiciens théoriciens' in Strasbourg in 1976 when he was still an undergraduate student and was very impressed by a talk by David Ruelle. In summary, he studied a wide selection of basic and special courses in theoretical and experimental physics and pure and applied mathematics, and moreover also history of arts, musicology and philosophy. After finishing his Diploma degree in Physics in 1976, he started to work with Peter Landsberg in Southampton on combined generation and recombination processes in semiconductor with nonequilibrium thermodynamics and a nonlinear dynamical systems approach. There he met David Chillingworth who came from the Warwick dynamical systems group and put this research into the more mathematical framework of bifurcation theory and catastrophy theory. He completed a Ph.D. Thesis in Applied Mathematics in Southampton on 'Nonequilibrium Phase Transitions in Semiconductors' in 1978. After that he moved to Aachen and prepared another Doctoral Thesis (Dr. rer. nat.) in Physics, with Professor Friedrich Schlögl as mentor, and completed it in 1981. He describes then his outstanding career in details and reveals his secret in the life: since he has spent most time with young students, he has also stayed young!

In Section 6, Galina Strelkova says that during her high school years, her interests were quite broad. She liked mathematics and physics, but was very attracted to medicine, reading on surgery with great interest and sometimes being even allowed to attend operations in clinics (her mother was an anesthesiologist). In addition, over all her school years, she also seriously studied music (piano and vocals), composed music for poetry, took part in various competitions and became their laureate several times. However, by the time she graduated from high school, she definitely decided to become a school teacher in mathematics, physics and computer science. After graduating from high school, she went to Saratov (a big regional centre at the Volga river, 260 km far from her native place) to enter the Department of Radiophysics of the Faculty of Physics of Saratov State University. Since then her scientific life has been inextricably linked with this department. In the early 1990s, Professor Vadim Anishchenko director of the Nonlinear Dynamics Laboratory introduced her to dynamical systems and she became a member of this laboratory. After defending her Ph.D. thesis, she was invited by Professor Jürgen Kurths in his Nonlinear Dynamics Working Group in Potsdam University and published several joint papers with him. Thanks to the initiated collaboration and the activities of Professors E. Schöll and A. Zakharova since 2015, the Nonlinear Dynamics group was involved into a very interesting field of research devoted to the study of complex spatiotemporal structures, such as chimera states in networks of coupled nonlinear oscillators. This collaboration was very fruitful and successful and is still continuing and resulted in numerous joint publications, exchange research visits, plenary and invited talks at different international conferences and workshops, as well as participation in several international scientific grants and projects. After defending her Doctor of Sciences Thesis (Habilitation) on 1 October 2020 and the death of Vadim Anishchenko on 30 November 2020, G. Strelkova headed the Radiophysics and Nonlinear Dynamics Department of Saratov State University and mentored several Ph.D. students.

Section 7, is devoted to the particular history of Saber Elaydi, who from a young age, displayed an insatiable thirst for knowledge that transcended the confines of his circumstances. In a world plagued by uncertainty and limited resources, he sought solace and inspiration within the realm of mathematics. The numbers and theorems became his refuge, offering a sanctuary where possibilities knew no boundaries. He delved into the world of mathematics, voraciously consuming every piece of knowledge he could find. His perseverance caught the attention of mentors and benefactors who recognized his talent. This unwavering dedication earned him a scholarship that paved the way for formal education, propelling him beyond the confines of the refugee camp. In 1978, he received his Ph.D. from the University of Missouri under the guidance of Ping-Fun Lam and David Carlson. His dissertation, titled 'preferred sets in topological dynamics', was focused on transformation groups. After his graduation, he was hired as an assistant professor at Kuwait University where he worked with three different researchers H. Farran, F. Dannan and S. Kaul. In 1983, he moved to Case Western Reserve University, Cleveland, OH. He worked with Otomar Hajek on the study of dichotomy and trichotomy of nonlinear differential equations, two of the most important asymptotic properties of dynamical systems. However, he was eager to return closer to his original interest. And in 1987, he found it in difference equations and discrete dynamical systems. In 2001, together with Bernd Aulbach they created the International Society of Difference Equations (ISDE). This creation was announced during ICDEA 2001 at the University of Augsburg. Throughout his career, he has held several key positions, including Co-Editor-in-Chief of the *Journal of Biological Dynamics* (JBD) and Editor-in-Chief of the *Journal of Difference Equations and Applications* (JDEA). He mentored numerous Ph.D. students and authored various publications in discrete dynamical systems and Ecology including *Discrete Chaos: Applications in Science and Engineering, An Introduction to Difference Equations, Upside-Down: The Interplay between Life and Chaos*, and co-authored with Jim Cushing *Discrete Mathematical Models in Population Biology: Ecological, Epidemic, and Evolutionary Dynamics*. Two more people have had great influence on his research: Lord Robert May and Jim Yorke.

In Section 8, Jim Michael Cushing remembers that he was so enamoured with mathematics when he was student in the late 1950s in Cheyenne, Wyoming, that he would check out the text book from the local Carnegie library during the summer recess before he took a math course to teach himself, as best he could, the topic prior to the upcoming school year. He adds that, given the lack of activities available to young people in such a small town, he also spent a great deal of time studying piano performance, practicing sports (basketball, football and baseball) and reading philosophy (a lot of Aristotle). In College and university of Colorado in Boulder, he focused on courses in physics, chemistry and geology as well as astronomy and psychology. Eventually during his senior year, he enrolled in postgraduate mathematics classes and participated, as part of the University's team, in the preeminent national William Lowell Putnam Mathematical Competition for undergraduate students. Having obtained a National Defense Act Fellowship, he became in 1964–68 a graduate student in applied mathematics at College Park, MD, where he met Jim Yorke. In this section, he describes in details his outstanding career which followed a long path from his Ph.D. on the Levi–Civita's conjecture concerning deep water waves, with Monre H. Martin as advisor, to theoretical ecology. He explains in particular how was created the Beetle Team consisting of Bob Costantino, Bob Desharnais and him together Brian Dennis and later Shandelle Henson and Aaron King in 1989–90. This team gained the support of NSF for over a decade, during which they developed and parameterized a discrete time (Leslie) model using historical data, studied the predictions of the model using analysis and numerics, determined bifurcation sequences that included chaotic dynamics and experimentally feasible manipulations that place replicated cultures along that route-to-chaos, and statistically validated (without re-parameterization) the model predictions against the observation data obtained. He adds that during his travels on this long career path, he never stopped playing the piano.

In Section 9, Lyudmila Efremova native of Nizhny Novgoro (previously Gorky between 1932 and 1990) first highlights the well-known Andronov's school of nonlinear oscillations and dynamical systems. She recalls that the Andronov's great merit is the use and development of fruitful, but forgotten (by 20s–30s of twentieth century) Poincaré's ideas in the theory of differential equations for the needs of radiophysics and the theory of radio transmitters. In 1956, the Gorky University was named after the great Russian mathematician, a native of Nizhny Novgorod, N.I. Lobachevskii. In the struggle between her two serious hobbies, mathematics and music, mathematics won, and she entered the Faculty of Mechanics and Mathematics of the University. When she was a second-year student, she had a huge impression reading the Sharkovsky's paper [187] which was submitted to her by Dr. Rakhmankulov who give her next year the subject of her Ph.D. Now after many years, she can say that her first impression of Sharkovsky's Theorem determined her scientific biography. She says that her student years passed with a continuous feeling of happiness of learning new things, listening lectures given by major mathematicians and physicists, as well as famous musicologists at the Gorky Conservatory on music theory. During the second year of her Ph.D., she was invited by Sharkovsky to present her results at his seminar at the Institute of Mathematics of the Academy of Sciences of Ukraine and recommended her work to the Programme Committee of the IX International Conference on Nonlinear Oscillations (ICNO – IX, Kiev, 1981). It was the first major conference in her life and Professor Plykin was the chairman of the section where she presented her report. She later saw the poems written by him and was astonished by the fact that a mathematician could be a poet. Prof. Otrokov and Dr. Rakhmankulov were her scientific supervisors, and Sharkovsky and Belykh were the referees of her Ph.D. After that, another period was coming, in which the most interesting areas of activity seemed to her, first, the applications of one-dimensional dynamics to the study of discrete dynamical systems on manifolds of dimension at least two; and, second, the creation of dynamical systems theory on complicated one-dimensional ramified continua, which do not allow order topology. She worked on these topics under the guidance of Professors Anosov and Stepin. Eventually, Sinai chaired the jury of her Doctor of sciences thesis.

In Section 10, René Lozi indicates that like Misiurewicz, he was counting many things around him in his early childhood. He started his studies at the University of Nice in 1967. He was fascinated by particle physics. After being graduated in fields and particles, he changed his mind and became numerical analyst. The subject of his Ph.D. was related to bifurcation theory, which was near unknown in France in 1972, among mathematicians. In 1974, a group of researchers in bifurcation was constituted in the mathematics department in Nice, around Gérard Iooss. With this group, he attended a Conference in Roma where he heard David Ruelle speak about the Hénon map. He soon proposed a piecewise linear version of this map, known now as 'Lozi map' in 1978. After that, his interest shifted to research of realistic model of chaos, first studying slow-fast differential equations after his meeting with Claude Lobry and introducing the theory of confinors in place of attractors. René Thom was interested by his morphological modelling of the Lorenz equation and was in the jury of his doctoral thesis, along with Michel Hénon. He met in 1986 Leon Chua and studied the 'Chua circuit' for many years with Shigehiro Ushiki from Kyoto university. He worked also with many renowned scholars like Alexander Sharkovsky, Guangrong Chen, Leon Chua.

2. My road to dynamical systems by Michał Misiurewicz

To arrive to Dynamical Systems, of course I had first to arrive to Mathematics.

My first encounter with mathematics which I remember was counting the cars of a moving train. I do not remember how old I was then. At this moment, let me make two digressions.

The first one is about counting. The urge to count various things remained with me for my whole life. In particular, even now when I see a moving train I start counting cars.

The second digression is about my memory. It seems that it works in a not quite typical way. On one hand, people often are surprised that I remember so many songs and so many jokes. On the other hand, I remember only very small fragments of all years of my life. And problems with remembering things appear also in mathematics. For instance, I do not remember the direction of the inequality between the arithmetic and geometric means. I have to check it for 1 and 4, and only then I recover the direction. Similarly for the Fatou's Lemma [65], I have to go through an example to know which way the inequality goes. And from time to time funny things happen. Once I was listening to a conference talk and noted a very nice trick. After the talk I asked the speaker who is the author of this trick. And he answers 'I do not know, but I learned it from you.'

Fortunately, my Mother was keeping a dairy, so I know how I was progressing. I started counting when I was 3 years old. By the age of 5, when I learned to read, my favourite book was the math textbook for the first grade; I was solving the 'problems' from that book. Does this mean that it was then when I already become a mathematician?

However, during vacations I forgot almost all that I learned before. But later I either recalled quickly what I knew or I learned it again.

At a certain moment, I started to read the book *Lîlâvatî* by Szczepan Jeleński[1] and to solve problems from that book. I was fascinated by the book. In particular, I became interested in the story where somebody asks as a prize 1 grain on the first square of a chessboard, 2 on the second one, 4 on the third one, etc. (exponentially). I took a notebook and started to compute the consecutive powers of 2. The notebook survived (although I do not know where it is now), so much later I checked my computations with a calculator. It turned out that somewhere around 2^{30} I made an error, and of course all further powers was also incorrect.

After reading about the Fermat's Great Theorem, of course I decided to think about it. In such a way, I discovered by myself the formula $a^2 - b^2 = (a - b)(a + b)$. So it seems that I was already a mathematician by that time.

In the high school, I had a partner for mathematics. Péter Szeredi, a son of Hungarian diplomats, living at that time in Warsaw, was in the same class. He was subscribing the Hungarian journal *Matematikai Lapok*, with lot of interesting problems.

Finally, a great adventure – olympiad. When I was in the ninth grade (high school at that time consisted of grades 8–11) I took part in the Mathematical Olympiad. It consisted of three stages. I managed to qualify to the third stage. This already gave me the right to become a student of mathematics at any Polish university without the entrance exam; however, I had to graduate from the high school first. At the third stage competition I did not do well, in particular because I did not know trigonometry yet. I did not even know what sine and cosine were.

Next year (1964–1965), next olympiad, tenth grade. This time I became one of the laureates. What is more important, I qualified to the Polish team for the International

Figure 1. A photo from the 1965 International Mathematical Olympiad. Among people standing I am the rightmost one and Pavel Bleher is the leftmost one.

Mathematical Olympiad (IMO). There was a 'training camp' for the team. I still remember the number of the bus I had to take each day to attend it.

IMO took place in Berlin. I got a third degree diploma – now it is called a bronze medal, but at that time there were no medals. I got 29 points out of 40 possible. I still have an interesting photo, made on a boat (see Figure 1). A group of participants that include myself and Pavel Bleher. I could not guess at that time that in a distant future I will be working with him at the same Math Department in Indianapolis for 27 years. Another memory – a joke problem: evaluate the determinant of a big complicated matrix. The matrix was so big that it was not easily visible that it was a 13×14 one.

One more olympiad, 11th grade. Again I was a laureate, but this time without a diploma, because one can get only one diploma at the Polish Mathematical Olympiad. Years later I could see the grades I got. Each problem was graded by three people, and then the final grade was negotiated. From one problem I got (on the scale from 2 to 5) 2, 2+, and 5 with an exclamation mark (an extra praise). The final grade was not given, because nothing depended on it. I still do not know whether my solution was correct.

To graduate I have to pass school-leaving exams. The math exam was of course very easy for me, so the only thing I remember is that a sparrow flew into the room through an open window and left a 'present' on my table.

Again I was in the Polish team for the International Mathematical Olympiad (this was 1966). Again the training camp. I remember two things from it. One is that we tried to construct a theory of half-sets. The basic examples of half-sets were the intervals $(a, a]$ and $[a, a)$.

Surprisingly, much later this turned out to be not a pure joke; I used this idea in several papers (for instance [136] and [139]). The second thing I remember was a problem posed by Włodzimierz Kuperberg about a broken line and n points on the plane. The problem is relatively simple for $n = 1, 2, 3, 4$, but for larger ns it is still open.[2]

IMO that time was in Sofia. The problems were relatively simple, so I managed to get the maximal score, 40 points, and a diploma of the first degree (nowadays a gold medal). Not

Figure 2. The special group in Analysis I, 1966/67. I am sitting on the second row, third from left (in a sweater).

all of us were that lucky. One of my colleagues immediately after leaving the room (so he already could not change anything) realized that every stick has two ends. When counting the endpoints of segments in one of the problems he forgot about it.

In 1966, using one of the four documents (three for qualifying to the third stage of the olympiad and one diploma) I was admitted as a student to the Faculty of Mathematics and Physics of the University of Warsaw, field of mathematics. My Grandfather tried to convince me that I should go to the Technical University instead and become an engineer. Engineers are always needed. In 1917, he lived in Odessa, and after the revolution the bolsheviks not only spared him, but even were paying him in golden rubles for his work as an engineer. I recalled this many years later in Indianapolis, when I was working as a professor of mathematics with tenure, and a nearby factory producing auto parts was closed, flooding the job market with 600 unemployed engineers.

Switching from the school to the university was a great relief. At last I did not have to learn things that I absolutely did not care about. And at the university – mathematics, mathematics, and mathematics. Several additional, unimportant and not taking much time things, but basically only mathematics. And I did not have to be in the school at 8 in the morning.

An idea that some people consider is controversial, but from my perspective was absolutely great, was the creation of so called 'special groups.' The lectures were held for all students, but for recitations smaller groups were created. While during the lectures only the professor was talking, the recitations were devoted to solving problems by the students. After several first weeks, the students that wanted (and the professors agreed that they were able) to solve more difficult problems were moved to those special groups (see Figure 2). Usually it was not only about more difficult problems but also more theory.

The people that are against this idea claim that if the best students are kept in the same groups as the rest of the students, they will help those students. This may be partially true, but in such a way the potential of the best students will be wasted. One has to remember that the progress in any discipline is mainly achieved by the best minds in that discipline. And in my 47 years of teaching, I saw many times how difficult it is to teach a group of students whose knowledge and abilities are all over the spectrum.

Recitations from Analysis I were conducted by Witold Kołodziej. He was the best teacher among those I meet during my studies. He was demanding; sometimes my work was graded for 3 (in the scale from 2 to 5), but he really taught me Real Analysis (and at the end of the year I got a 5). The material I learned at the first year of studies was approximately the same that I much later taught in the USA at the 400 level course in Real Analysis.

A quick comment about the differences between Poland and the USA, when it comes to the teaching of the future mathematicians. In Poland, one is considered a fully educated mathematician after getting the MS degree; in the USA it is the Ph.D. Thus, compared to Poland, the education of a young person in the USA is shifted by about 3 years. And so is the choice of their career.

After each semester, we had to pass exams. I was usually preparing for the exams with my two colleagues: Michał Krych and Piotr Minc. To prepare for the theoretical part of an exam after a one-semester class, we needed 1 week of hard work. For a two-semester class, we needed 2 weeks. The exams were both written (solving problems) and oral (questions about the theory). However, if one got a 5 both from the recitations and from the written exam, usually the necessity of passing the oral exam was waived.

The next digression. A very important thing for a student (and also for anybody in many circumstances) is to make others to have a good opinion on yourself. This helps a lot. For example, one of my younger colleagues in Warsaw took the written part of the exam and to not to waste time immediately went for vacations. However, his exam got lost. What to do? Since the professor had good opinion on him, he gave him a 5.

During summer vacations two times (after the first and second years), I attended the Summer Schools in Algebraic Topology in Sopot (at the north of Poland). I remember a lecture by Peter Hilton, where he was talking (in English, of course) about von Neumann algebras and C^* algebras. I was wondering: I knew who von Neumann was, but who was this guy Sistar? The subject of the school was not only algebraic topology. Karol Krzyżewski and Jan Maria Strelcyn lectured on dynamical systems. That was my first encounter with this theory.

When I was a second-year student, Włodzimierz Holsztyński, a topologist, started to mentor me. In particular, he gave me a problem in lattice theory (it was really about the classes of open sets and of closed sets in a general topological space). I solved the problem and it resulted in my first publication, in 1969. I remember a funny incident connected with this mentoring. Holsztyński was one of the professors conducting the oral exam in topology. I enter the office, get several simple questions, and receive a 5. But then we start to talk about those lattices. When I am leaving the office, my colleagues waiting for the exam are already trembling with fright. What kind of an exam is it? Misiurewicz knows all the course material, but he is staying there already a full hour!

Holsztyński left Poland for America in 1969. If he didn't, I would probably be a topologist. However, even now I like the topological aspects of dynamical systems.

When I started my third year of the undergraduate studies, Wiesław Szlenk organized a seminar in mathematical analysis. In spite of the name, it was really a seminar in dynamical systems.

The origin of this discipline in Warsaw was quite interesting. At some moment two young Warsaw mathematicians, Wiesław Szlenk and Karol Krzyżewski, were sent by Stanisław Mazur to Moscow for a postdoctoral training. They were specialists in functional analysis and went to Andrey Kolmogorov for the advise of what they should do. Kolmogorov told them that the great days of functional analysis were already over, but there was a new interesting discipline, called dynamical systems (together with the ergodic theory). Then he sent them to Yakov Sinai (who had been his student) to learn it.

A small digression: according to the *Mathematical Genealogy Project*, Kolmogorov had 82 students, and 3901 descendants. Sinai has 39 students and 356 descendants (among them, two grandgrandgrandgrandsons).

Krzyżewski and Szlenk, after learning the basics of the dynamical systems, returned to Warsaw. They were followed by Sinai, who came and gave a cycle of lectures in ergodic theory. The next member of the Warsaw group was Jan Maria Strelcyn, who was taking notes from Sinai's lectures that were later available as a booklet. Around 1969 Strelcyn left Poland for France, but Szlenk, and later also Krzyżewski, started to build a group in this new area of mathematics. Krzyżewski concentrated on scientific supervision over new students, while Szlenk, with his extraordinary organizational skills, first organized a seminar (which, after many changes, still exists), and then was taking care of the development of the careers of the students. Having many contacts in the East and West, he was sending us to various conferences and trainings. First two students from the new generation were Michał Krych and myself. Next students that joined were Feliks Przytycki and Maciej Wojtkowski, and after them many others.

My next three papers, published in 1970, 1971, and 1973, were already on dynamical systems. According to the *Google Scholar*, they have together 198 citations and are still being cited. So that was definitely time when I could say that I arrived to Dynamical Systems.

And here I am after the next 50 years.

I started by investigating topological entropy. In 1972, at a so-called international course *Global Analysis and Its Applications* in Trieste, I met Karl Sigmund, who introduced me to the results of Rufus Bowen on entropy-expansive maps. I started to generalize this notion and in 1974 wrote a Ph.D. thesis on this generalization. I met Bowen only once, but we corresponded a little, and I still have his letters written by hand (explanation for young readers: this is what you did before the Internet era).

This stayed with me. I wrote 44 papers and 1 book with the word 'entropy' in the title.

In 1976, Michel Hénon published a paper about his attractor [85]. After spending about half a year on trying to prove that it is really an attractor, I decided that temporarily I will investigate interval maps and only when I gain substantial knowledge of them, I will return to the Hénon's attractor. As everybody knows, temporary solutions usually last very long, and so I remained stuck with one-dimensional dynamics. In fact, about two-third of my papers are about it. Return to the Hénon's attractor did not happen, although I managed to prove the existence of an attractor for similar Lozi maps [135].

However, my start in the theory of interval maps was quite successful. In 1977, I wrote with Szlenk a paper, where we proved some basic results [138].

Here a practical advice for young mathematicians. If you want, without studying a lot of technical tools, to write a paper that will be often cited, you should choose a subject which will soon become fashionable, but at a moment the basic theorems are not proved (or even basic definitions are missing). So you introduce definitions, prove basic theorems, and then sit down and observe how other people continue the research that you started. And from time to time you add some brick to the building that is being constructed.

And the only small problem is how to predict what will become fashionable. In fact, you cannot predict it. But you can try. Most of times you will fail. But maybe sometimes you will succeed?

3. My road to dynamics by Laura Gardini

My decision to study Mathematics was taken when I was 12 years old. There was not a specific event that lead to my decision, I just loved Mathematics, I realized that I was happy when studying it, and it was easy for me. But I was living in a very small village, and I knew that my goal was to convince my parents of my will to continue to study in a scientific lyceum (the closest was in Ravenna, 10 km, far from our village) and then to study Mathematics at the University (the closest University was in Bologna). For them it was a kind of 'investment' in my future. At the lyceum, I was always the first in Mathematics, and at the University in Bologna I was always among the best students in my class.

At those times, in Italy Ph.D. courses did not exist (they were introduced only a few years later, in 1983) and when I graduated (*cum laude*[3], in 1975) my supervisor told me that he could not have other collaborators, and he referred me to the ENI (Ente Nazionale Idrocarburi) group, which in those years had also a Department devoted to research work. I was employed immediately. However, the research subjects were decided by the Director, and these changed quite often in the years.

At the beginning of the 1980s, I was asked to study a few recent books by Thom [197], Zeeman [209], Postom and Stewart [151], Collet and Eckmann [45], Guckenheimer and Holmes [79]. In the meantime I also had 3 years of teaching experience at the Faculty of Engineering in Ancona (teaching rational mechanics in the academic years 1984/85/86/87 since as an ENI researcher, I was allowed to teach with a contract).

That was really a fascinating subject, I immediately loved dynamical systems so I decided not to change anymore. Since then it is my research field. However, to do so, I resigned from ENI in Summer 1987, studying to win some competition at the University, that happened the year after, and I started as researcher at the University of Urbino in November 1988.

As I started working for an Economics Department, my colleagues were proposing me to analyse several models, all systems in discrete time, that is, maps, and mainly noninvertible maps in high dimensions. It was a subject not so widespread in the literature of those years. Fortunately I have seen the books by Gumowski and Mira [80] and [81], in particular the one in French, and Mira [133], and several articles written by them, that pointed me in the right way. When I was preparing the paper [74], at the beginning of 1991, I wrote a letter to Professor Mira, since I thought that there were new dynamic behaviours, related to the saddles and their homoclinic bifurcations, that could be of interest also to him. He was very kind, inviting me to Toulouse, and I visited him in the spring time. That started my friendship and collaboration with Christian Mira.

He also invited me to participate in the *European Conference on Iteration Theory* held in Lisbon in September 1991, organized by Sousa Ramos. There, for the first time I met many researchers, besides Sousa Ramos, who were quite famous in Dynamical Systems, including scholars such as Alexander Sharkovsky, Jaroslav Smital, Sergei Kolyada, Ludomir Snoha, Francisco Balibrea, Jaume Llibre.

There I presented a work showing how homoclinic bifurcations that are not possible in invertible maps may occur in noninvertible ones involving the critical curves.[4]

In other words, the first homoclinic bifurcation of a fixed point or of a cycle which is a repelling focus or a repelling node occurs via critical points, i.e. via a contact of the absorbing area delimited by the critical segments and the stable set of the cycle (which in case of an expanding cycle consists in its pre-images). In the same way, the first homoclinic orbit

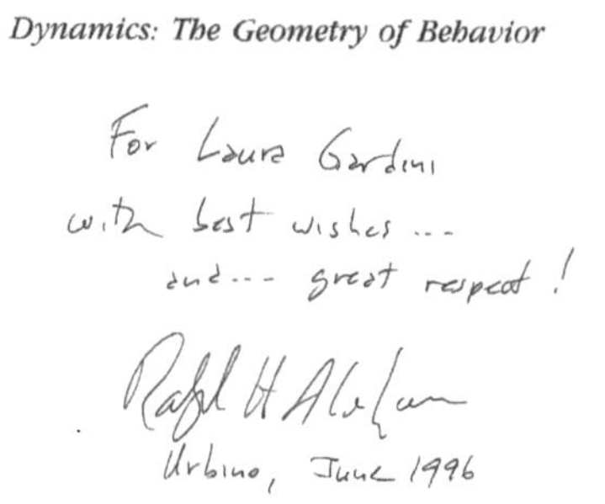
Dynamics: The Geometry of Behavior

For Laura Gardini
with best wishes ...
and... great respect!

Ralph H Abraham
Urbino, June 1996

Figure 3. Dedication of Ralph Abraham.

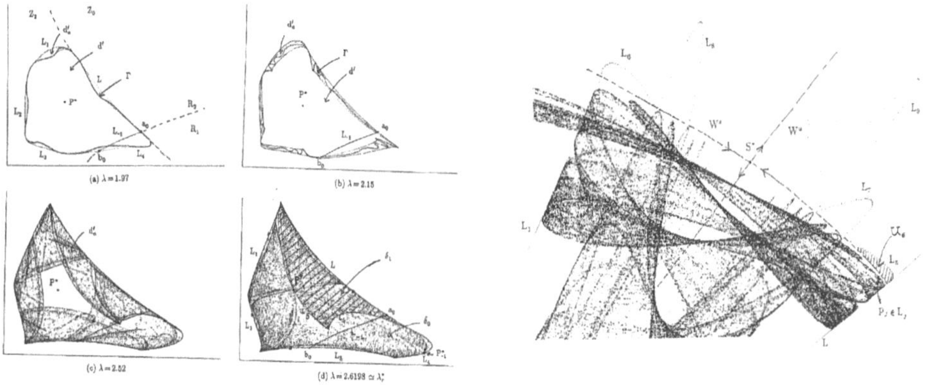

Figure 4. Critical curves and homoclinic tangency with the stable set of a saddle fixed point.

of a saddle cycle in a region of the phase plane with several preimages can occur via contact of the critical segments with the stable local saddle manifold [75].

This was my favourite subject for a couple of years. In 1992, I presented further results, at two other conferences, one in Germany and one in Italy where, in addition to Christian Mira, Ralph Abraham was also invited. Abraham, with whom I also began a long friendship and collaboration, encouraged my research, and was fascinated by the dynamics of non-invertible maps. He was a visiting professor in Urbino for several periods (six times from 1992 to 2000). I loved his book [4], which he gave me (with dedication, see scanned page Figure 3). We published together a textbook for beginners [5].

In the same year 1992, Christian Mira invited me to coauthor a book on noninvertible maps. There were so many open problems in two-dimensional noninvertible maps that we were fascinated and curious to understand the bifurcation mechanisms leading to different effects. Disconnected basins of attraction, loops on the critical curves, loops on a closed invariant curve after a Neimark–Sacker bifurcation (destroying its smoothness), contact bifurcations involving the critical curves, leading to chaotic repellers, or expansion and merging of attracting sets, and not only in smooth maps but also in nonsmooth ones (see Figures 4 and 5 from papers of the year 1994).

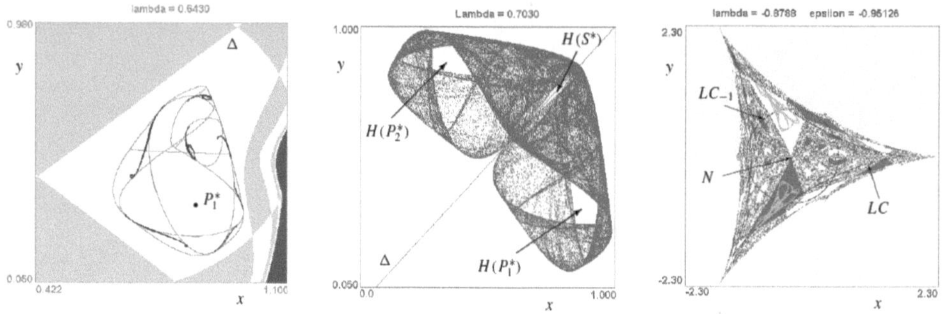

Figure 5. Examples of absorbing areas, snap-back repeller bifurcations and contact bifurcations.

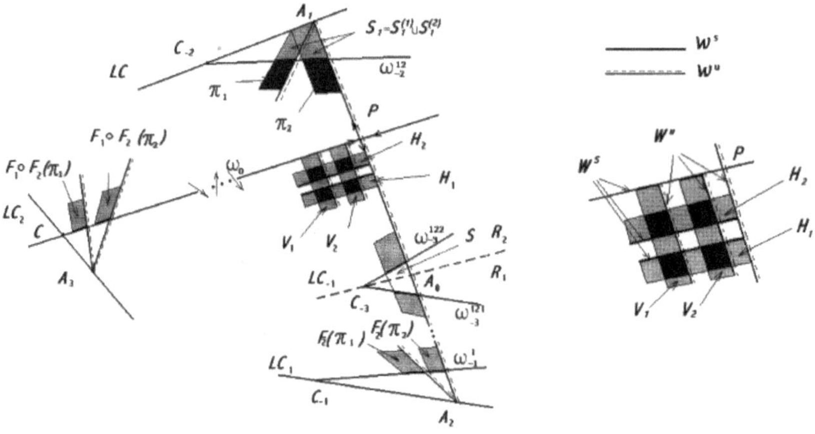

Figure 6. Details of the mechanism of the horseshoe in a noninvertible map.

All this, for the book, required a lot of work, and Mira and I met (in Toulouse or Urbino) several times, at least twice a year, up to its publication, in 1996 [134] (at that time we had exchanges only via fax and letters, we started with e-mails only in the fall of 1996).

The homoclinic bifurcations of expanding points and saddles were of interest also for other researchers, in particular Yuri and Vladimir Maistrenko, and the young Iryna Sushko (Ph.D. student of Professors Yuri Maistrenko and Alexander Sharkovsky, in Kyiv), with whom I started a long collaboration. They were experts on one-dimensional smooth and piecewise smooth maps, and were starting to investigate the dynamics of a two-dimensional noninvertible piecewise linear map. Iryna was preparing her Ph.D. thesis on that map, and they visited Urbino for some weeks every year, from 1993 to 1997. There is a paper that we liked a lot [128], in which it is also shown how the Smale-horseshoe mechanism works for the first homoclinic bifurcation of a saddle fixed point in a noninvertible map, involving the critical lines (see Figure 6).

In May 1994, Maistrenko and Sushko organized a conference 'Differential Equations, Bifurcations and Chaos', at Katsiveli, in Crimea, and there for the first time I met many other well-known scholars, in particular, Leonid and Andrey Shilnikov. The flight to Katsiveli from Kyiv deserves a movie. Although it was in May when we left Kyiv it was snowing, and the plane was an old model from the Second World War (I jokingly called it

Figure 7. Workshop 'Noninvertible Dynamical Systems', Minneapolis, 14–18 March 1995. First line from bottom, from the left, Andrey Shilnikov, Yuri Maistrenko, my son Martino, and Christian Mira. Second line from the bottom, from the left, Leonid Shilnikov, sitten behind him is Daniele Fournier Prunaret, I am behind Mira, close to my husband (with glasses) and behind him Ralph Abraham.

'the albatross'). Inside it was windy and cold, and to relax we had some vodka to drink, so at the end we were all happy. We have to recall that those years were soon after the Ukraine became independent, and we were conscious that everything was difficult. In particular, we had our social dinner without electricity, but the candles worked perfectly. The buildings sometimes had no heating, but the academic conferences were very impressive, and that was enough. We became friends also with Leonid and Andrey Shilnikov. At the time, Leonid Shilnikov was studying a three-dimensional map which, in a degenerate case, was reduced to the two-dimensional quadratic map that we and Mira have used in our book for many examples. So I prepared a kind of working paper collecting relevant bifurcations occurring in that map, and I showed it to Leonid at the workshop *Noninvertible Dynamical Systems* organized by Christos Frouzakis and Ioannis Kevrekidis in Minneapolis (in March 1995, see Figure 7).

Also at that conference (my first visit to the United States), I met many famous scholars. With Frouzakis and Kevrekidis, we started to work on a common paper (see [73] and Figure 8 from it) since they were also studying a discrete-time system in which noninvertibility led to particular dynamic behaviours.

Soon after, in 1995, Leonid and Andrey Shilnikov visited me in Urbino (see Figure 9), and I remember that I was so proud and happy when Leonid asked me to show the proof of my paper [76] on the homoclinic bifurcations of expanding cycles in $\mathbb{R}^n$, as he had heard about this work at another conference, mentioned by Ricardo Viana, and he was interested in those homoclinic bifurcations and orbits.

Moreover, the now famous *Blue sky catastrophe model* was presented (for the first time outside Nizhny Novgorod) by Leonid Shilnikov at the *European Conference on Iteration Theory* that I organized in Urbino in September 1996 (see Figure 10). At the same time, I worked closely with Christian Mira, not only on the book but also on numerous articles,

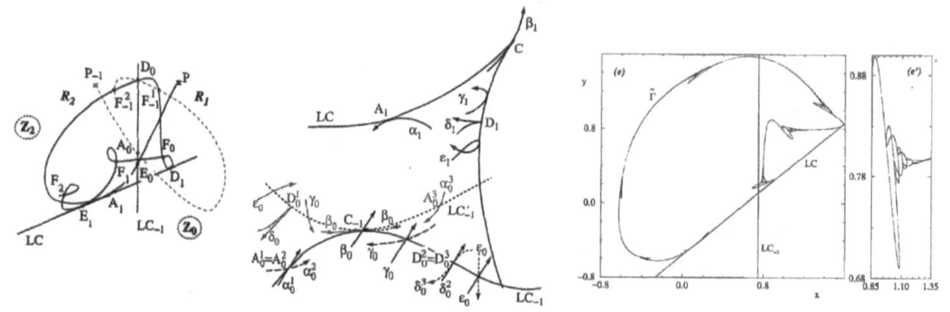

Figure 8. Examples of loops occurring on a closed invariant set of a noninvertible two-dimensional map.

Figure 9. Urbino, June 1995, with Andrey and Leonid Shilnikov.

Figure 10. Urbino, September 96, ECIT conference. From left to right, Leonid Shilnikov, Yuri Maistreno, me, Iryna Sushko, Vladimir Maistrenko, Alexander Kopansky.

until his retirement (in 2000). Since then, the number of my collaborators working on smooth and non-smooth maps has grown steadily.

4. My journey into the chaos theory: homoclinic bifurcations and Belykh attractor by Vladimir Belykh

Inspired by the famous essay titled 'On How I Got Started in Dynamical Systems' by Steve Smale [193], I would like to share my story on how I embarked on my journey into the Chaos theory. However, to arrive to Chaos, of course I had first to arrive to Mathematics. Like Misiurewicz, I have to say few words about my first school years.

After completing elementary school in the Gorky region, at 12, I was accepted into an elite boarding military school in Saint Petersburg. This was a one-of-a-kind college preparatory Suvorov's school for future military, foreign affairs, and intelligence careers. I studied there for 5 years between 1955 and 1960 and was destined to continue my path in the University of Saint Petersburg military branch (see Figure 11).

In 1960, Nikita Khrushchev (first secretary of the Communist Party of the Soviet Union from 1953 to 1964), initiated massive reorganizations of the Soviet Army and started significant cuts that led to the boarding military school's closure. As a high-school junior, I returned to Gorky. I graduated from high school in 1961 and was accepted into Gorky State University (nowadays Lobachevsky State University of Nizhny Novgorod).

I can say that my academic voyage started there. From 1961 to 1965, I studied mathematics and physics in the Chair of Oscillation Theory and Automatic Control, named after its founding director, Alexander A. Andronov. I was fortunate to take classes from distinguished professors like Dmitry A. Gudkov, Yuri I. Neimark, Nickolai A. Fufaev, and Nickolai A. Zheleztsov, the brilliant team of Andronov's disciples. In the meantime, in 1964, the University established a new large research unit, the Research Institute of Applied Mathematics & Cybernetics (NII PMK as abbreviated in Russian), directed by my future Ph.D. adviser, Liudmila N. Belyustina. As a master's student, I informally became a part

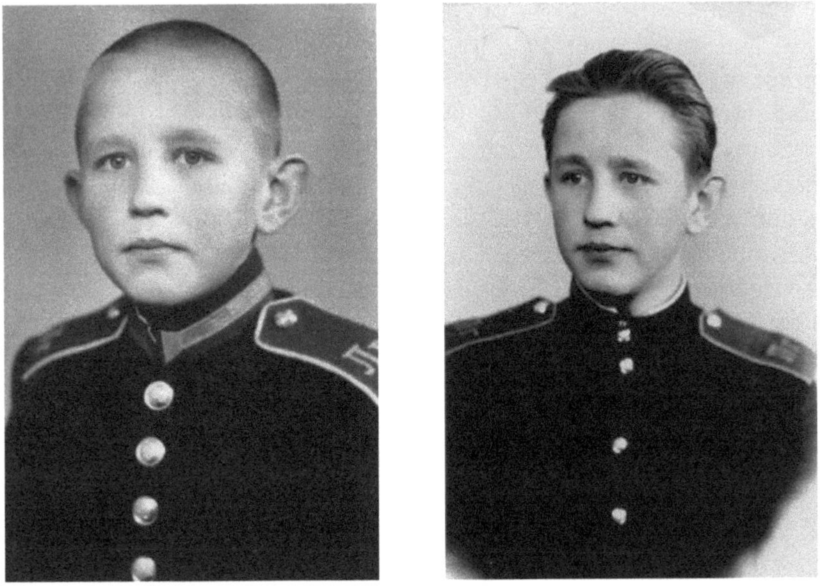

(a) (b)

Figure 11. (a) Vladimir Belykh at the military school, circa 1955; (b) circa 1960.

of this institute and started working on my diploma under the supervision of Vladimir D. Shalfeev (who was a Ph.D.-level researcher only two years older than me). My diploma research (an analog of an M.S. degree) was on the analysis of homoclinic orbits bifurcations in a two-dimensional pendulum-type dynamical system on a cylinder, modelling a phase-locked loop (PLL), a control system with a nonlinear filter. After graduating in 1966, I joined the NII PMK as a junior scientist. It was usual practice for fresh graduates to be placed in a research institute before they could begin working on their Ph.D. thesis. It was a stroke of luck and a turning point in my academic and personal life that I was accepted into the Division of Differential Equations headed by Evgeniya A. Andronova-Leontovich, Andronov's spouse and close collaborator and a sister of another celebrated Soviet physicist, Mikhail Leontovich of Moscow State University. At that time, I continued my work on periodic dynamics of two-dimensional planar systems [22] and had a poor understanding of the emerging concept of 'chaos,' instead viewed and called 'irregular oscillations.' Remember that only a few years have passed since the seminal paper's publication on a deterministic nonperiodic flow by Edward Lorenz [113].

I was fortunate to meet Leonid P. Shilnikov (L.P., as friends called him), who worked in the same division. Over time, we became friends, and 1 day L.P. explained to me his famous theorem on the bifurcation of a homoclinic orbit of a saddle focus, which became a widely used criterion for the existence of chaotic dynamics in saddle-focus dynamical systems [190]. I felt a lack of knowledge to understand the details of this complicated bifurcation structure fully. Similarly, the depth of my knowledge needed a boost, akin to my earlier encounter with Alexander N. Sharkovsky in 1965 when he presented the proof of his famous Sharkovsky's ordering at a research seminar run by Yuri I. Neimark in NII PMK. I got the main gist of the Sharkovsky's formal result but could not fully understand and appreciate its fundamental implications for observable chaotic dynamics due to my knowledge gap. These experiences were the primary motivations for my Ph.D. thesis, initiated in 1968 under the guidance of Liudmila N. Belyustina at NII PMK's Division of Dynamical Systems and Control Theory, headed by Neimark. My Ph.D. thesis focused on the existence and bifurcations of persistent homoclinic orbits, another criterion for chaos, in periodically driven systems of generalized pendulum equations, all without relying on small parameters or computer assistance. I was not sure how to reach the goal, but after lots of back and forth and head-scratching, I developed a novel method involving auxiliary two-dimensional systems to analyse homoclinic orbits. I first presented my findings at the 1971 conference on ordinary differential equations in Sverdlovsk (nowadays Ekaterinburg) and subsequently published the results [35–37]. A.N. Sharkovsky participated in this conference and attended my talk. He then agreed to be an external reviewer (called an 'opponent' in the Soviet system) for my Ph.D. thesis, which I defended in 1972.

1974 marked a significant year as I contributed to the theory of digital PLLs, yielding publications [29,30] and, along with colleagues Vladimir D. Shalfeev and Valery P. Ponomarenko, the prestigious Lenin Komsomol Prize (a Soviet science award for young researchers under 33, see Figure 12).

In 1975, with my Ph.D. student Vladimir I. Nekorkin, we rigorously proved the existence of Shilnikov's chaos in a three-dimensional autonomous phase system [28].

As part of the European-Soviet scientific exchange program, I spent the 1975–1976 academic year in the Electronics Laboratory at the Danish Technical University (DTU) in Lyngby, a quiet suburb of Copenhagen. During my stay at DTU, I primarily worked with

Figure 12. Belyustina's lab picture at the NII PMK circa 1978. Front row (from left to right): Vladimir Belykh wearing the Lenin Komsomol Prize medal, Liudmila Belyustina, Valery Ponomarenko, and Vladimir Shalveev.

George Bruun and Orla Christensen, two professors of electronics, studying discrete-time digital PLLs. In the meantime, two DTU experimental physicists, Niels F. Pedersen and Ole H. Soerensen, found out that I was an expert in the dynamics of pendulum equations and turned to me with a request to explain a random behaviour of the Josephson junctions' current density-voltage $J-V$ curves they obtained experimentally. Looking at the Josephson junction model equations, I quickly realized that they belong to the same class of pendulum questions studied in my Ph.D. thesis, and all my machinery can be applied to Niels's and Ole's experimental setup. Of the same age and mindset, we quickly became good friends with Niels and Ole and started working on the new project. In the end, we discovered that the mysterious phenomenon of random switching in $J-V$ curves is caused by an infinite, irregular set of bifurcations of homoclinic orbits. We published these results in two companion papers in Physical Review B in 1977 [32,33], with part I [32] and part II [33] devoted to the autonomous and non-autonomous cases, respectively. These papers were very well-received and cited within both experimental and theoretical physics communities. To the best of our knowledge, our results were the first theoretically validated observations of dynamical chaos in Josephson junctions.

In September 1976, following my return from Denmark, Valentin (Valya) Afraimovich, a Shilnikov's pupil, and L.P. showed me a paper on the original and structure of the Lorenz attractor, which they, together with Vadim Bykov, submitted to a reputable Russian journal [7]. Valya explained to me the action of the discontinuous Poincaré return map they constructed in a qualitative, implicit form. I soon realized that my model of a discrete-time PLL [23] in the case of a piecewise linear nonlinearity becomes a discontinuous 2D map that acts like theirs. I showed this map with a chaotic attractor to Valya and L.P., and Valya said: 'Here is the Belykh attractor.' L.P. confirmed. It was during this exchange that the Belykh attractor was coined. While there are several variations of the Belykh map [23,25,27], its standard form reads [31,34]

$$\begin{cases} \bar{x} = \lambda x \\ \bar{y} = \gamma y \end{cases} \qquad \text{if } L(x,y) \leq 0, \tag{1}$$

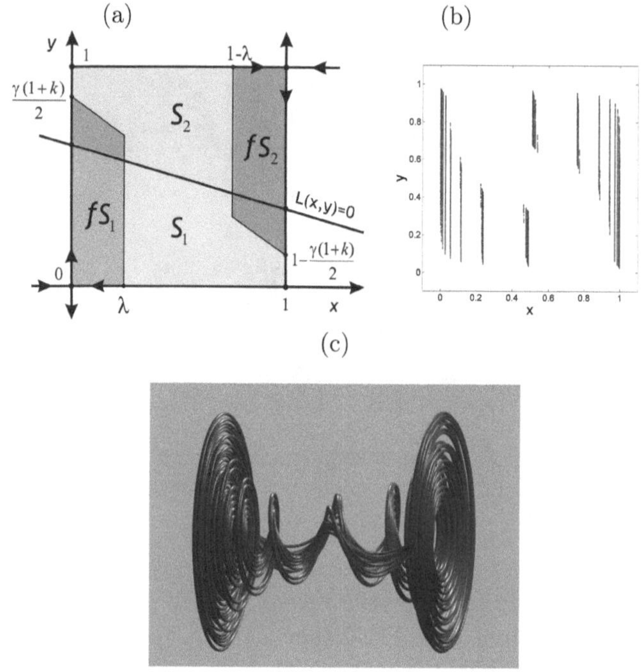

Figure 13. (a) First image of the unit square $S = S_1 \cup S_2$ under the map f (1)–(2), where S_1 and S_2 are separated by the discontinuity line $L(x, y) = 0$. One iterate of f transforms trapezoids S_1 (light brown) and S_2 (light blue) into fS_1 (dark brown) and fS_2 (dark blue), respectively. (b) Chaotic Belykh attractor generated by the map f with parameters $\lambda = 0.48$, $\gamma = 1.3$, and $k = 0.5$. (c) Embedding of the Belykh attractor into the 3-D phase space, see [31] for the corresponding 3D flow system.

$$\begin{cases} \bar{x} = \lambda(x-1)+1 \\ \bar{y} = \gamma(y-1)+1 \end{cases} \quad \text{if } L(x,y) > 0, \tag{2}$$

where $L(x, y) = k(2x - 1) + 2y - 1$. The map is defined on the square $S = \{(x, y) : 0 \leq x \leq 1, 0 \leq y \leq 1\}$. The square S is cut into two parts, S_1 and S_2, by the line $L(x, y) = k(2x - 1) + 2y - 1 = 0$ (see Figure 13). The dynamics of the map f is governed by (1) on the lower part S_1, where $L(x, y) \leq 0$, and by (2) on the upper part S_2. Please see the Scholarpedia paper [31] for further details on the map dynamics, and our paper [34] in this special issue for the map derivation from a continuous-time PLL system.

I initially dismissed the importance of this discontinuous, chaotic map due to its apparent simplicity, as its absorbing region contains an attractor composed of only saddle-type orbits. Later Valya Afraimovich informed Yakov G. Sinai about this new object of potential interest for the ergodic theory. After my talk at Sinai's seminar at Moscow State University, the map gained some prominence and became a subject of study in ergodic theory by young and well-established mathematicians, including Sinai's former student, Leonid Bunimovich. I also gave a talk at Anosov's seminar, and Dmitry V. Anosov's pupils (Yakov Pesin, Evgeny Sataev, and Nikolai Chernov) began their analysis of the ergodic properties such as the transitivity and invariant measures of multi-dimensional modifications of the Belykh map that laid the foundations for the 'generalized hyperbolicity' theory [146,165,166]. At that time, I was working on my second doctoral thesis, and

Figure 14. Vladimir Belykh (left) and Dmitry Anosov (right) at a conference in Moscow.

Anosov strongly recommended me to include the description of my map and its detailed analysis in the thesis. Later, in 1983, Anosov served as an opponent for my 'Doctor of Sciences' degree thesis (see Figure 14).

Giving back to my alma mater, in 1977, I started teaching a graduate course on qualitative methods in nonlinear dynamics in Andronov's Chair of Oscillation Theory and Automatic Control. In particular, this course covered homoclinic orbits, Shilnikov's saddle-focus theorem, and the dynamics of the Lorenz system [24]. This would be the standard content for an advanced nonlinear dynamics course by modern standards; in 1977, it was quite rare. I vividly remember the first group of excellent students I taught, including Arkady Pikovsky, who later became a leading expert in coupled oscillator theory.

In 1979, George Bruun, the Danish professor and my host during the 1975–1976 stay in Lyngby, invited me to spend a few months at DTU to continue our collaborative work. To remind the young reader, these were the times when most Soviet citizens were not allowed to travel abroad by the Soviet State, and an exit permit was required. Mine was delayed, and the entire trip to Denmark could quickly become elusive. To make it happen, George arranged for an invitation letter, signed by the Danish Queen, that opened all the bureaucratic doors and let me travel.

In 1980, I made a bet with Valya Afraimovich on a bottle of Cognac that I prove the existence of a homoclinic orbit in the genuine Lorenz system. Victor I. Yudovich, a professor from the University of Rostov-on-Don who 1 year before proved the existence of a homoclinic orbit in the Lorenz system, in a particular case of a small parameter, was our bet witness. It turned out that my original plan of attacking this problem was flawed. It took me about a month of trying and sleepless nights to find a solution and prove the statement. To fully convince Valya and other colleagues, I gave two long seminars to provide comprehensive, step-by-step proof of my theorem for the existence of a butterfly homoclinic linkage in the Lorenz system. The proof draft was published in [24] and the complete one in [26]. Valya lost the bet but was very happy with my success; we opened that bottle of Cognac together. Given the limited communication between the West and the Soviet Union, it took a while for this result to become known worldwide. Eventually, this theorem made its way

Figure 15. (From left to right): Vladimir Nekorkin, Valentin (Valya) Afraimovich, Lev Lerman, and Vladimir Belykh during Valya's short visit to NII PMK from Georgia Tech, one of the places he worked in the USA in the 1990s. Photo circa 1996.

into the state-of-the-art analysis of the Lorenz system and, in particular, was cited in the classical dynamical systems book by John Guckenheimer and Philip Holmes [79].

In 1983, I moved to the Institute of Water Transport Engineers (an engineering school in downtown Nizhny Novgorod, a five-minute walk from NII PMK). I took the position of its Mathematics Department's head, which became available after the retirement of Nickolai N. Bautin, Andronov's disciple, known in particular for the Bautin bifurcation and Bautin theorem [21]. Also, in 1983, I met Vadim Anischenko at a conference on the Oka River, with whom we quickly became life-long friends.

The Fall of the Iron Curtain intensified international collaborations and allowed easier exchange visits (see Figure 15). After meeting Leon Chua at a conference in Crimea, I started collaborating with him and spent a while at UC Berkeley working with Leon, Chai Wah Wu, and Ljupco Kocarev (see Figure 16). My research interests have broadened to the dynamics of networks and synchronization that later led to long-term collaborations with Erik Mosekilde, a physics professor at DTU, and Martin Hasler, an engineering professor at École Polytechnique Fédérale de Lausanne (EPFL). This important chapter of my scientific life deserves a separate description and will not be detailed in this essay.

I am 80 and still hold a professor position without immediate retirement plans. Reflecting upon my journey, I am profoundly grateful for the bonds I formed and the scientific camaraderie I shared with my late friends Leonid Shilnikov (see Figure 17), Valya Afraimovich, and Vadim Anishenko. Their enduring influence on my life and work remains a source of immense pride. Equally gratifying is the realization that my Ph.D. students have blossomed into accomplished and highly respected scientists who have left indelible marks in their respective scientific domains. Notably, Vladimir Nekorkin became a leader in nonlinear wave and coupled oscillator theory; Andrey Shilnikov, L.P.'s son, has

Figure 16. From left to right: Leonid Shilnikov, Diana and Leon Chua, Vladimir Belykh, Valery Ponomarenko, Lev Lerman, and Albert Morozov. Nizhny Novgorod, June 1992. The group picture was taken against the Research Institute for Applied Mathematics and Cybernetics (NII PMK) backdrop and a memorial plaque featuring A. Andronov's bas-relief.

Figure 17. Leonid Shilnikov (left) and Vladimir Belykh (right), 2005.

emerged as an authority in global bifurcation theory; and Nickolai Verichev was among the pioneers in the discovery of chaos synchronization back in 1986. Witnessing their success, as well as the accomplishments of the disciples of my dear old friends, I believe it is fair to say that the future of dynamical systems is in good hands.

I extend my heartfelt gratitude to René Lozi, whose enthusiasm, encouragement, and determination made this essay possible. As much as the Lozi [117] and Belykh attractors share in common, René and I have a profound mutual respect and friendship, bound by our shared admiration for the renowned Russian singer and humanitarian, Bulat Okudzhava whose song we once sang together at the L.P. Shilnikov's memorial conference in Nizhny Novgorod in 2013.

5. Thrilled by nonequilibrium phase transitions by Eckehard Schöll

As a school boy in a small German town, Nürtingen near Stuttgart in Southwest Germany, I was interested in a very broad spectrum of subjects, ranging from Latin, English, French, German via History, Music, Art to Mathematics and Physics. While I could not really decide whether my favourite subjects were languages, music, history, or mathematics, I was always fascinated by theoretical physics, since I did not really understand what it was about, but it sounded complicated and challenging. I loved repairing old clocks, building radios, and solving maths puzzles. But when I passed the final school exam (Abitur) with best grades in all subjects in 1970, it was clear that I wanted to study Physics at the University. During my undergraduate studies at the Universities of Stuttgart and Tübingen, I discovered that I enormously enjoyed pushing the borders of my knowledge to challenging new horizons, in particular in mathematics and theoretical physics.

In a Physics Colloquium in Tübingen in 1974, I heard a lecture by Werner Heisenberg on his unified field theory of which I did not understand anything, but I was fascinated by his personality. In Mathematics seminars by Rainer Nagel, together with a group of fellow students, Eberhard Müller and Burkhard Kümmerer, I studied ergodic theory, functional analysis, von Neumann algebras and C^* algebras which were cutting-edge topics at that time. At the same time, Professor Harald Stumpf (1927–2021) became my first mentor by offering me a Master Thesis (called Diploma Thesis at that time) in theoretical semiconductor physics (calculation of generation-recombination processes of electrons). For this, besides analytics, I had to do computer simulations of the quantum mechanical transition probabilities and the rate equations, i.e. differential equations. Already in the summer vacations 1973, I had taken a programming course in Fortran at Imperial College of Science and Technology in London, besides language courses at Exeter and Broadstairs/UK. At that time, the code had to be punched onto cards, and they were fed manually as a batch into a central computer. For my Master Thesis, I had to drive to the Computing Center of the University, punch my cards, and on the next day come back to collect the output on large sheets of paper, in most cases only to see that the code had not worked because of flaws. In Tübingen, a newly appointed professor, Werner Güttinger, introduced a course on nonlinear dynamical systems, where I made my first contact with this field. Probably the first conference which I attended when I was still an undergraduate student in Tübingen was the mathematically oriented 'Rencontre entre mathématiciens et physiciens théoriciens' in Strasbourg in 1976, where I was very impressed by a talk by David Ruelle. In summary, I studied a wide selection of basic and special courses in theoretical and experimental physics and pure and applied mathematics, and moreover also history of arts, musicology, and philosophy, and all through my life, I played the piano and sang in choirs, performing many concerts.

After finishing my Diploma degree in physics in 1976, I wanted to continue my studies in England, since I had always been fascinated by foreign cultures and languages. I obtained a grant by the prestigious German Academic Scholarship Foundation (Studienstiftung des deutschen Volkes). To explore different possibilities, since I was still undecided whether I wanted to focus on semiconductor physics or mathematical physics of von Neumann algebras, I travelled to Britain and visited two scientists who represented quite different fields. When I first met Professor Peter T. Landsberg in Southampton, he showed me his latest publication [106], and I was immediately thrilled by this, since it combined generation and recombination processes in semiconductor with nonequilibrium thermodynamics and a nonlinear dynamical systems approach. At this bifurcation point in my scientific life, I decided without hesitation to indulge into this project, and this determined my main field of research for the next 20 years, and influenced my scientific career up to the present day.

I started to work with Peter Landsberg (1922–2010) in Southampton, and he became a mentor and a friend to whom I owe a lot. I was accompanied by my girlfriend Viola, who later became my wife, and travelled with me to many scientific conferences and research visits throughout the world for 48 years until her much too early passing in 2023. Peter Landsberg, in his meticulous and ingenious way, introduced me to doing research and writing papers, and taught me attention to detail, which is equally essential for a researcher as having brilliant ideas. He has been a friend far beyond my Ph.D. times, and we continued to collaborate and visit each other until his death. Besides generation – recombination (g-r) processes of electrons and holes in semiconductors, a topic which he had pioneered [105] as well as the fundamentals of thermodynamics [104], he introduced me to semiconductor laser dynamics which subsequently and in parallel became my second topic [179]. Both cases were applications of nonlinear dynamics in terms of nonlinear ordinary differential equations. We also discussed difference equations (iterated maps), which represent a time-discrete version of the dynamics and exhibit rich nonlinear dynamic scenarios including a route to chaos, in spite of their surprisingly simple form, e.g. the logistic map

$$x(n+1) = ax(n)(1 - x(n)), \tag{3}$$

where $x(t)$ is the dynamical variable and a is the bifurcation parameter. This resulted in a paper on applications to semiconductors much later [107]. As a small note, the value of the famous Feigenbaum constant, which gives the universal asymptotic ratio of successive period-doubling bifurcations in the logistic map, was found by Siegfried Grossmann and Stefan Thomae [78] already before it was published by Mitchell Feigenbaum [67]. At Southampton, I attended also lectures by David Chillingworth who came from the Warwick dynamical systems group and put this research into the more mathematical framework of bifurcation theory and catastrophy theory, and by David J. Wallace on renormalization group theory and critical phenomena which was an exotic but interesting new physical topic for me. As it turned out, I was able to complete a Ph.D. Thesis in Applied Mathematics in Southampton on 'Nonequilibrium Phase Transitions in Semiconductors' in 1978 (see Figures 18 and 19).

Our first published papers dealt with models for first- and second-order g–r induced non-equilibrium phase transitions in semiconductors [177], which were built on the analogy of the nonlinear rate equations with chemical rate equations, in particular the famous Schlögl model [167]:

$$\dot{x} = -x(x - a)(x - 1). \tag{4}$$

Figure 18. Photo taken at my graduation ceremony in 1978 at Southampton University, UK.

Figure 19. Photo taken at my graduation ceremony in 1978 at Southampton University, UK. From left to right: Eckehard Schöll, Peter Landsberg, Academic Registrar of Southampton University.

I had previously made contact with Professor Friedrich Schlögl (1917–2011) at the RWTH Aachen (Aachen Institute of Technology, Germany), and he had offered to me to do a Ph.D. under his supervision. He became my third mentor, and from him I have learned a lot about teaching physics and about physical thinking. After finishing my British Ph.D. in Mathematics at Southampton, I moved to Aachen and prepared another Doctoral Thesis (Dr. rer. nat.) in Physics, which I completed in 1981. It dealt also with the field of dynamical systems, but developed the generation–recombination models in a different

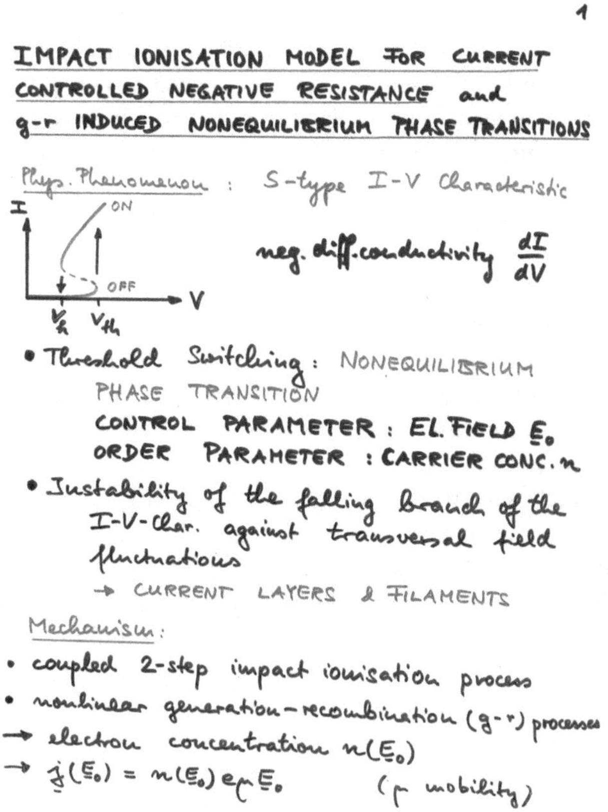

Figure 20. First slide of my talk at the International Conference on Hot Carriers in Semiconductors in 1981. At that time transparencies were written by hand and projected by an overhead projector.

direction. Studying a doped semiconductor, and considering impact ionization of electrons from ground state and excited state donors, I was able to develop a model which showed bistability in the S-shaped current–voltage characteristic (switching between the two stable branches is nowadays called tipping), with an unstable middle branch of negative differential conductivity (see Figure 20), and hence resulted in a plethora of self-organized spatial and spatio-temporal instabilities in the form of current filaments with a high conductivity core surrounded by low-conductivity [168,169]. The field of self-organization and spatio-temporal pattern formation based upon nonlinear partial differential equations had become a focus of research at that time, and examples from laser physics, chemistry (Brusselator developed by Nicolis and Prigogine and collaborators, Belousov–Zhabotinsky reaction), and fluid dynamics (Taylor and Bénard instabilities) were abundant, but semiconductors had not been noted in this context except for a few works by Russian authors. In the following years in Aachen I explored and developed this field in great detail and in many aspects. Hermann Haken had created the field of Synergetics which deals with these nonlinear systems far from thermodynamic equilibrium and their universal features. In his pioneering work, he interpreted the laser transition as a nonequilibrium phase transition [77] and discovered the analogy between the semiclassical laser equations and the Lorenz model in fluid dynamics, which gives rise to chaotic laser dynamics [82].

Concepts from thermodynamics and statistical physics have been applied to describe self-organization, spatio-temporal pattern formation, phase coexistence, critical phenomena, and first- and second-order nonequilibrium phase transitions [83]. Much more recently these ideas have been applied to networks of oscillators, where synchronization transitions may arise, giving birth to a plethora of partial synchronization patterns and complex collective behaviour, tipping transitions, explosive synchronization, nucleation, critical slowing down, etc., with applications to many natural, socio-economic, and technological systems [38,176,198]. A very recent application of that paradigm is heterogeneous nucleation of partially synchronized patterns in adaptive networks leading to two scenarios of first-order nonequilibrium phase transitions [68].

Hermann Haken became very important in my scientific career when he invited me to Stuttgart in 1983 to give a seminar on my work on nonequilibrium phase transitions and self-organization in semiconductors. He was very interested in this recent addition to the family of Synergetics and invited me to write a monograph for his Springer Series in Synergetics. I happily agreed, and the volume was published in English in 1987 [170], and translated into Russian in 1991. This book has been cited more than 600 times. This influenced my whole direction of research for the coming years up to now. I submitted the book as my habilitation thesis at the RWTH Aachen, and after this became a Lecturer at RWTH Aachen and started my own group of students. In those years, I personally met many pioneers of nonlinear dynamical systems, and I initiated several collaborations with colleagues who performed experiments on nonlinear dynamics in semiconductors, which led to very fruitful joint workshops, research visits, and many joint publications, in particular with the Regensburg group of Wilhelm Prettl [205], the Tübingen group of Joachim Peinke and Jürgen Parisi [145,180], the Japanese group of Kazunori Aoki [16], and later the Berlin group of Marion Asche [100].

In April 1989, a few months before the Berlin wall came down, I accepted an offer from the Technische Universität (TU) Berlin as Full Professor of Theoretical Physics, where I worked and taught for 30 years. During that time, I had the great chance to establish Nonlinear Dynamics, Complex Systems and their Control as a new internationally visible focus at TU Berlin. I organized many International Conferences on Control of Complex Systems and Networks (e.g. Palma de Mallorca 2012, Warnemünde 2014, Toronto 2015, Usedom 2016), and established collaborative research centres: I was Deputy Chairman and Principal Investigator of the Collaborative Research Centers on Semiconductor Nanostructures (SFB 296, 1994–2002), on Complex Nonlinear Processes (SFB 555, 1998–2010), and Founder and Chairman of SFB 910 on Control of Self-Organizing Nonlinear Systems (2011–2018). As a retired professor I am still active in research and in conferences worldwide, moreover I am Principal Investigator of the Bernstein Center for Computational Neuroscience Berlin (since 2010) and a guest scientist at the Potsdam Institute for Climate Impact Research (since 2020). Over the years I have been involved in many international collaborations; in 2000 I held a Fulbright Senior Scholar Award at Duke University, USA; in 2004 a Visiting Professorship of the London Mathematical Society; and in 2017 I received an Honorary Doctorate from Saratov State University, Russia, as a result of my very fruitful and active collaboration with the Saratov group of Vadim Anishchenko and Galina Strelkova. I am President of the International Physics and Control Society (IPACS, since 2019), a member of the German Physical Society (DPG), and a member of the Italian Society for Chaos and Complexity (SICC). Since 2021 I am Speciality Chief Editor of the new open access *Journal Frontiers in Network Physiology: Networks of Dynamical Systems.*

My first talks which I gave at International Conferences were in 1978 at the International Conference on the Physics of Semiconductors (ICPS) in Edinburgh, in 1980 at the German Physical Society (DPG) Annual Spring Meeting in Münster, and in 1981 at the International Conference on Hot Carriers in Semiconductors (HCIS) in Montpellier, France. All of these were parts of regular annual or biannual conference series, which I subsequently attended many more times. In particular, the DPG Spring Meetings have closely accompanied my life and my career for the last 45 years, since I first attended them as a student, later as a scientist giving talks and organizing Focus Sessions, and finally from 2008 to 2018 as the Local Organizer of the big Annual Spring Meetings in Berlin with over 6000 participants each time. I always enjoyed meeting personally all those people whose names I knew from their published works, but now we had stimulating discussions, and often new ideas arose from those. At the conference in Montpellier, I met Melvin P. Shaw from Wayne State University, Detroit, USA, whom I knew as author of famous papers and a book on semiconductor instabilities and in particular of a book on the Gunn effect. We arranged that I would visit him in the USA for a year at the Department of Electrical Engineering, and as a result of this year which I spent with him in 1983/84 as a Visiting Assistant Professor, I became a coauthor of the book *The Physics of Instabilities in Solid State Electron Devices* which was finally published in 1992 [188].

In 1986, I attended for the first time Dynamics Days in Enschede, Netherlands, a conference series which was founded by Robert Helleman in 1980, and which is organized annually up to the present day. The first seven meetings were held at Twente University, Enschede, as very small informal workshops, then they shifted to Düsseldorf under the chair of Gert Eilenberger for four years, and afterward moved all around Europe. Dynamics Days Europe is the oldest Conference in the Dynamics Days family, which now comprises Dynamics Days Europe, Dynamics Days US, Dynamics Days Asia Pacific, Dynamics Days Latin America and the Carribean, Dynamics Days Central Asia.

Over the years it became my main conference, and I have attended 23 editions of this series, organized one myself in Berlin in 2005 (see Figure 21), and have been a member of the International Advisory Committee since 2005, and its chair 2016–2019. As the focus of this conference over the years shifted from chaos, self-organization and nonequilibrium phase transitions to a wide spectrum of topics in mathematics, physics, life sciences, and engineering ranging from networks to machine learning, brain dynamics, power grids, climate modelling, also my focus of research has been expanding and developing over the decades. Often this shift has been fertilized by the interaction with other researchers on conferences and workshops. After extending nonlinear dynamics to semiconductor nanostructures [171,172] and quantum wires, quantum dots, quantum transport, growth kinetics of quantum dots and cell populations, electro-optical nonlinearities, nonlinear laser dynamics, I studied more general delayed complex systems and networks [71,92], chaos control [178] (the Handbook on Chaos Control has been cited more than 1000 times) and in particular time-delayed feedback control [181], synchronization on delay-coupled networks [173,174], partial synchronization patterns [176], chimera states [175,182,208] which are intriguing hybrid states where the network spontaneously splits into coexisting synchronized (coherent) and desynchronized (incoherent) domains, and applied these phenomena to brain dynamics and power grids. The methods involved both ordinary and partial differential equations and difference equations, and deterministic and stochastic effects. As an example, if the simple logistic map (Equation 4) is coupled nonlocally in a

Figure 21. Dynamics Days Europe in Berlin 2005. The photo was taken at the conference excursion on the river Spree. In front: Bernd Krauskopf (left), Eckehard Schöll (right).

ring network [143], intriguing chimera states arise:

$$z_i^{t+1} = f\left(z_i^t\right) + \frac{\sigma}{2P} \sum_{j=i-P}^{i+P} \left[f\left(z_j^t\right) - f\left(z_i^t\right)\right], \tag{5}$$

where z_i are real dynamic variables ($i = 1, \ldots, N$, $N \gg 1$ and the index i is periodic mod N), t denotes the discrete time, σ is the coupling strength, P specifies the number of neighbours in each direction coupled with the ith element, and $f(z)$ is a local one-dimensional map. We choose f as the logistic map $f(z) = az(1 - z)$ and fix the bifurcation parameter a at the value $a = 3.8$. This choice yields chaotic behaviour of the map f.

A typical scenario of the coherence–incoherence transition via chimera states is illustrated in Figure 22(a–f), where we fix the coupling radius $P/N = 0.32$ and decrease the coupling strength σ. First, in Figure 22(a), the solution profile z_i^t is clearly smooth for $\sigma = 0.43$. Thus the network dynamics is spatially coherent. For smaller σ, the profile z_i^t sharpens up and, at some value $\sigma \cong 0.40$, loses smoothness in two points x_1 and x_2 as shown in Figure 22(b). This is a bifurcation point for the coherence–incoherence transition: beyond this parameter value, the wave-like profile z_i^t splits up into upper and lower branches, and two narrow boundary layers of incoherence are born around the points x_1 and x_2 (shaded yellow stripes α_1 and α_2 in Figure 22c): a chimera state. The incoherence stripes become wider with further decrease of σ (Figure 22d) and, eventually, the dynamics becomes completely incoherent (Figure 22e,f).

Looking back, I have been very fortunate since I have been able to realize a lot of the goals which I set to myself during the past 50 years since I started studying physics. What are the essentials in life? In the centre of my professional life have always been the young people to whom I gave guidance, whom I helped to make their first steps in science, and develop their carrier. I have supervised over 150 Diploma, Master, and Bachelor theses, 36 Ph.D.s, and 4 Habilitations. This is my secret: since I have spent most time

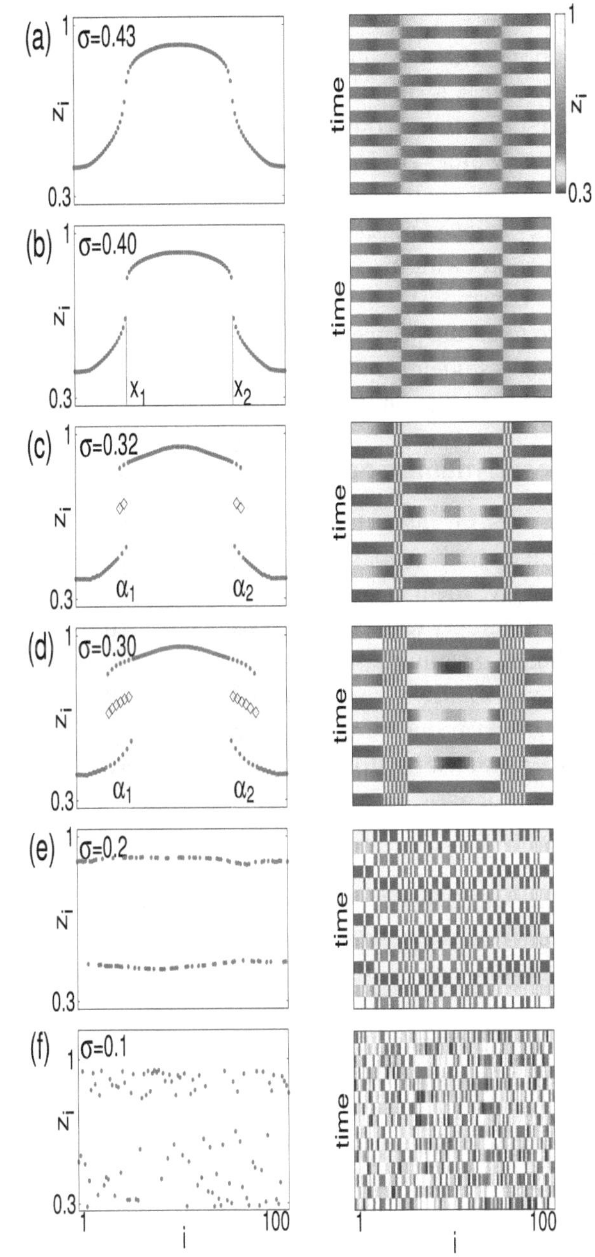

Figure 22. (Colour online) Coherence–incoherence bifurcation for coupled chaotic logistic maps for fixed coupling radius $P/N = 0.32$. For each value of the coupling parameter σ (decreasing from top to bottom, $\sigma = 0.43, 0.4, 0.32, 0.3, 0.2$, and 0.1, respectively) snapshots (left columns) and space–time plots (right columns) are shown.

with these young people, I have also stayed young! And there is an essential point: science, after all, is made by humans, and it is essential to stay human in this sometimes rough world.

Another essential thing in my life: Science is very international. I am happy to be part of this international family of physicists and mathematicians. This starts with student exchanges with foreign countries which I have always supported and encouraged. Then, international collaborations with colleagues all over the globe, and conference all over the world, where this family meets, form a network. Having lived in England and in the USA for several years, and in different regions in Germany, and having very active collaborations with Russia and many other countries, I feel at home in many places of this world, and in my heart I have taken something from every place. Life is more than science! I have always tremendously enjoyed music and poetry, playing the piano for over sixty years and singing in oratorio choirs with many concert performances during five decades, and writing poems. This has enormously enriched my life.

Finally, my thanks go to all the people who have stimulated and enriched my scientific life by their contributions as students, collaborators, and mentors. As Isaac Newton said: If I have seen further it is by standing on the shoulders of giants.

6. Impact of nonlinear dynamics on my scientific life by Galina Strelkova

During my high school years, my interests were quite broad. On the one hand, I really liked the natural sciences, such as mathematics and physics. These were my favourite subjects in school and I was ready to give them all my time. On the other hand, I was very attracted to medicine. I read books on surgery with great interest and sometimes I was even allowed to attend operations in clinics (my mother was an anesthesiologist and resuscitator). In addition, over all my school years I also seriously studied music (piano and vocals), composed music for poetry, took part in various competitions and became their laureate several times. It was expected that after graduating from school I would certainly continue my studies either at a medical institute or at the Conservatory. However, by the time I graduated from high school, I definitely decided for myself to become a school teacher in mathematics, physics, and computer science. The last subject was so curious for me (although there were no computers at school yet) that I just dreamed about it. I was incredibly fond of mathematics, but to a greater extent not as a beautiful and abstract science, but as an essential tool for describing natural phenomena and events and solving the most interesting physical problems.

I lived in a small city and after graduating from high school I went to Saratov (a big regional centre on the Volga river, 260 km from my native place) to enter the Pedagogical Institute. Since I graduated from high school with honours, I was advised to continue my studies at the Department of Radiophysics of the Faculty of Physics of Saratov State University. I entered there without exams and since then my entire conscious (professional and scientific) life has been inextricably linked with the Department of Radiophysics.

My first introduction in the field of dynamical systems came when I was a first-year physics student at the Department of Radiophysics (since 1995 it is the Radiophysics and Nonlinear Dynamics Department) of Saratov State University. This was in the early 90s. The Department was headed by Professor Vadim Anishchenko (1943–2020) who was also a scientific supervisor of a just organized Nonlinear Dynamics Laboratory. Stochastic dynamics or deterministic chaos was the key direction of research of this group. At the end of the first course, I had to prepare and present a course work that was dedicated to the transition to chaos in the Hénon map [85]:

$$\begin{cases} x_{n+1} = 1 - ax_n^2 + y_n, \\ y_{n+1} = bx_n. \end{cases} \tag{6}$$

I knew nothing about this system and chaos at all and that was my first investigation and attempt to figure out in this issue and to perform numerical analysis. My high school dream came true! I got the opportunity to learn how to use a computer and conduct numerical simulations. Of course, at that time computer technology was only developing in our country, but the Nonlinear Dynamics Laboratory had its personal computers. In the future, thanks to the cooperation and support of Professors Jürgen Kurths and Leon Chua, as well as to grants from International and Russian research foundations, our computer park was constantly expanded and strengthened.

After listening the lecture course on deterministic chaos given by Professor Anishchenko and reading a number of papers and reviews [85,152,160], some aspects in the topic of deterministic chaos became clear and I had decided to continue studying in this research direction. Moreover, I was impressed by the fact that a simple model system in the form of difference equations can demonstrate a nontrivial and, in some sense, exotic behaviour such as chaos, and the main properties and important regularities established for the maps can also be observed in more complicated systems described by ordinary differential equations. In particular, it was shown that the Hénon map is similar to a map generated in the Poincaré section of three-dimensional systems, such as the Rössler oscillator and the Anishchenko–Astakhov oscillator (the latter is a real radiophysical device) [14].

Going deeper into the study of the topic of dynamical chaos and its mathematical image – a strange attractor [160], the understanding came that chaotic self-sustained oscillations can be significantly different in their properties and thus, there are differences in the structures of the associated attractors. Strange attractors are robust hyperbolic objects [160,191] and satisfy a number of rigorous conditions [6]. These attractors are rather 'ideal' and only few mathematical models and physical devices can realize robust hyperbolic attractors [102,148,206]. However, real systems and devices typically demonstrate chaotic self-sustained oscillators which correspond to *quasihyperbolic* and *nonhyperbolic* attractors [191]. The first group of chaotic attractors includes quasihyperbolic Lozi, Belykh, and Lorenz-type attractors [25,113,116,146]. Nonhyperbolic attractors are realized in many dynamical systems, i.e. the Hénon map, the cubic map, the logistic map, the Rössler system, the Anishchenko–Astakhov oscillator, the Chua's oscillator, etc.

When I graduated from the University, Professor Vadim Anishchenko invited me to continue my postgraduate studies under his supervision and at the same time to work as a research engineer at the Department. Of course, I agreed with great pleasure! Since then, I also became a member of the Nonlinear Dynamics Laboratory. The Lab also included a few associate professors who defended later their Doctor of Sciences (Habilitation) theses (Vladimir Astakhov, Tatiana Vadivasova, Alexander Neiman, Dmitry Postnov, Alexey Shabunin) and became professors at the same Department or moved to another University. Several Ph.D. students were also involved in the research work in the Nonlinear Dynamics Laboratory and defended their Ph.D. theses (Olga Sosnovtseva, Natalia Janson, Alexey Pavlov, Alexander Balanov, Igor Khovanov, Alexander Silchenko, Alexander Nikitin). Many of them now are recognized experts in the nonlinear dynamics and statistical physics field and have permanent positions in various European and American universities and research centres.

To classify and understand the properties of chaotic attractors from a physical point of view, Professor Vadim Anishchenko suggested me to study this issue in the framework of my Ph.D. thesis. Exploring the dynamics of the aforementioned systems by using different

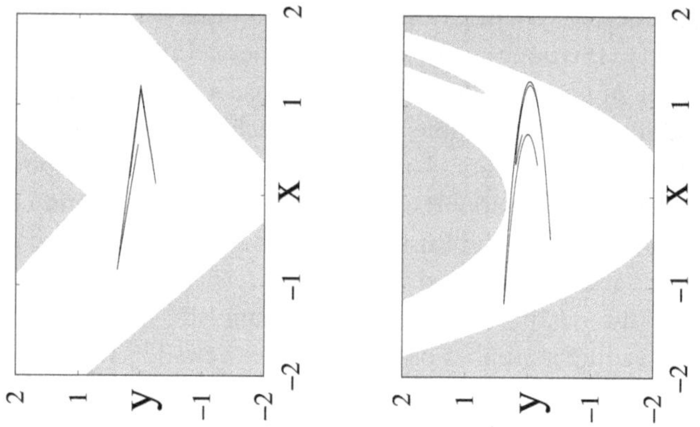

(a) (b)

Figure 23. Chaotic attractors (black curves) and basins of their attraction (white region) (a) in the Lozi map (7) for $\alpha = 1.5$ and $\beta = 0.3$ and (b) in the Hénon map (6) for $\alpha = 1.32$ and $\beta = 0.3$. Trajectories from the grey region go to infinity.

dynamical and statistical characteristics and measures, I was able to define the fundamental differences in the properties of quasihyperbolic and nonhyperbolic attractors [10]. The main model systems under study were the Lozi map (Figure 23a):

$$\begin{cases} x(n+1) = 1 - \alpha\,|x(n)| + y(n), \\ y(n+1) = \beta x(n). \end{cases} \tag{7}$$

and the Hénon map (6) (Figure 23b). I successfully defended my Ph.D. thesis in 1998 at Saratov State University. The results obtained were then highlighted in lectures on the theory of discrete-time systems and dynamical chaos and were also included into the book [14] published in Springer in 2014. As it turned out later, the results obtained on the classification of chaotic attractors were very useful in further research.

After defending my Ph.D. thesis, I was invited several times by Professor Jürgen Kurths for research visits in his Nonlinear Dynamics Working Group in Potsdam University. It was a great experience and we published several joint papers (e.g. [11–13]). The successful cooperation with Professor Kurths is still continuing and this is very important for us.

As I mentioned above, my professional career was inextricably linked with the Department of Radiophysics and Nonlinear Dynamics. Around 2002–2003, I moved to assistant professor, and in 2010, I was promoted to associate professor in the same department.

At the beginning of 2015, some very important and pleasant changes happened in the scientific life of our group. Thanks to the initiated collaboration and the activities of Professor Eckehard Schöll, Technical University of Berlin, our group was involved in Collaborative Research Center (CRC) SFB 910 (2011–2022): Control of Self-Organizing Nonlinear Systems, in the framework of the first Russian Project in a German CRC. Professors Vadim Anishchenko and Tatiana Vadivasova were Principal Investigators (PIs) of the Project and in 2019 I became a PI too.

The scientific cooperation with Prof. Eckehard Schöll and Dr. Anna Zakharova turned out to be very fruitful and successful. As a result, a new research direction appeared in our department, which was devoted to the studies of complex spatio-temporal structures, such

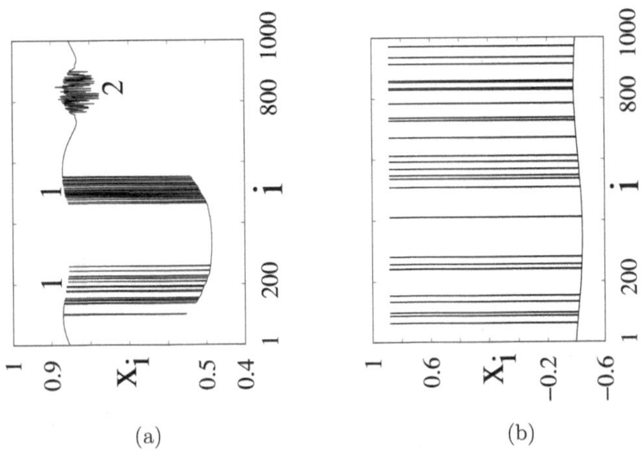

Figure 24. (a) Phase (1) and amplitude (2) chimeras in the network of nonlocally coupled logistic maps; (b) solitary states in the network nonlocally coupled Lozi maps.

as chimera states [101], in networks of coupled nonlinear oscillators. I remember very well with what enthusiasm and interest we searched and studied the relevant literature, how we organized and performed our own research on chimera states, how enthusiastically we discussed the results obtained and made joint research plans with our colleagues from the Technical University of Berlin during our numerous scientific visits. And here our knowledge on the peculiarities of dynamical systems played a key role. The first bright idea was expressed by Professor Vadim Anishchenko, namely whether chimera states always accompany a transition from complete synchronization (coherence) to spatio-temporal chaos (incoherence) independently on the type of individual oscillators in a network. Before, these states were found numerically in networks of nonlocally coupled chaotic logistic and Hénon maps (Figure 24a) [143,144]. However, our further studies showed that the ring network of nonlocally coupled Lozi maps demonstrates the transition from complete coherence to incoherence through the emergence of solitary states (Figure 24b) when the coupling strength between the elements decreases [184].

Working on this problem was very interesting and exciting. The members of the Nonlinear Dynamics Laboratory had an excellent opportunity to have research visits to the Technical University of Berlin, to give talks at scientific seminars of the groups supervised by Prof. Eckehard Schöll and Dr. Anna Zakharova and participate in various international conferences (see Figure 25). Together with our Berlin collaborators, we revealed the mechanisms of formation of amplitude and phase chimeras in networks of chaotic oscillators [40], explored the possibilities of external, mutual and relay synchronization of chimera and solitary states in multilayer networks of various coupled nonlinear dynamical systems [162,163], as well as the stability of chimera and solitary states towards noise excitations [185]. In particularly, it has been shown for the first time that repulsive interlayer coupling can induce anti-phase synchronization of spiral and target wave structures, including chimeras, in networks of self-sustained oscillatory and neural networks [189]. It has also been shown that solitary states and solitary state chimeras can be observed in neural networks [161,164].

At the beginning of 2019, I had a serious conversation with Professor Vadim Anishchenko, in which he suggested and, moreover, strongly recommended that I start

Figure 25. At the conference of the German Physical Society, Regensburg, April 2019. In the front row, Prof. Anna Zakharova is first on the right. In the second row, Prof. Vadim Anishchenko is first on the right, Prof. Eckehard Schöll is third on the right, Dr. Galina Strelkova is fourth on the right.

preparing a Doctor of Sciences (Habilitation) thesis on this topic based on joint scientific publications. By that time, indeed, a large number of new and important results had been obtained, which formed a coherent and logical picture of serious scientific work [195]. I immediately followed my mentor's advice. In December 2019, my thesis work 'Chimera structures in ensembles of nonlocally coupled chaotic oscillators' was accepted by the Dissertation council of Saratov State University for defence. After being rescheduled twice due to severe restrictions caused by the coronavirus pandemic, on 1 October 2020, my defence took place and was very successful. As Professor Vadim Anishchenko noted, my work was the first Doctor of Sciences thesis defended in Russia, on the topic of special spatio-temporal structures in ensembles of interacting oscillators.

After the death of Professor Vadim Anishchenko on 30 November 2020, I headed the Radiophysics and Nonlinear Dynamics Department of the Institute of Physics of Saratov State University, and our group is now successfully continuing to work in this direction in close and fruitful cooperation with our German colleagues.

In conclusion, I would note the *butterfly effect* which reflects a strong sensitivity of chaotic systems to initial conditions and which played a significant role in defining the direction in my scientific life.

7. My life to promoting difference equations by Saber Elaydi

Born in the sun-scorched land of Beer Sheba, Palestine, I emerged from humble beginnings that would shape my indomitable spirit. My family was ethnically cleansed in 1948 and escaped to a refugee camp in the Gaza strip, where I was raised amidst the trials and tribulations of a refugee camp. From a young age, I displayed an insatiable thirst for knowledge that transcended the confines of my circumstances. In a world plagued by uncertainty and limited resources, I sought solace and inspiration within the realm of mathematics. The numbers and theorems became my refuge, offering a sanctuary where possibilities knew no boundaries.

With limited access to formal education, my intellectual journey began as a self-taught endeavour. Armed with unwavering determination and an unquenchable thirst for learning, I delved into the world of mathematics, voraciously consuming every piece of knowledge I could find. Against all odds, my perseverance caught the attention of mentors and benefactors who recognized my talent. This unwavering dedication earned me a scholarship that paved the way for formal education, propelling me beyond the confines of the refugee camp. Embarking on a path of enlightenment, I pursued higher education with an unyielding resolve. And in 1964, I received a Bachelor's degree in mathematics and physics from Ain-Shams University in Cairo, Egypt.

Finally, in 1978, I received my Ph.D. from the University of Missouri in the USA, under the guidance of Ping-Fun Lam and David Carlson. My dissertation, titled 'preferred sets in topological dynamics', was focused on transformation groups. After my graduation, I was hired as an assistant professor at Kuwait University. There, I worked in three different projects with three different researchers. The first was with H. Farran on topics such as 'On weak isometrics and their embeddings in flows, Isometrics and certain dynamical systems'. The second was with F. Dannan on topics such as 'Lipschitz stability of nonlinear systems of differential equations I and II'. The third was with S. Kaul on topics such as 'Semiflows with global extensions I and II, Notions of negative stability, Stability of limit and prolongation sets in semiflows'. And in 1982, Otomar Hajek, from Case Western University, Cleveland, Ohio, visited Kuwait University to give a series of talks 'on discrete and continuous semi-dynamical systems'. I invited him with friends to celebrate the New Year's eve in my house. We all got drunk that evening drinking so much Scottish Whisky. He was very surprised, since alcohol is prohibited in Kuwait, and told me that he never had so much Whisky in his entire life. Two days later when we all got sober, I invited him to the faculty club for coffee and snack. Suddenly, he looked at me and said in a very serious tone 'Saber, you should get out of this rotten place'. I said 'I agree with you and I have been thinking about this for a year'. Two months later, I received an email from the department's chair in his University offering me a visiting appointment at Case Western Reserve University, and I accepted. Then in the summer of 1983, I moved with my wife and three children to Cleveland, OH. Hajek became my third mentor after my two doctoral advisors. My joint work with him was devoted to the study of dichotomy and trichotomy of nonlinear differential equations, two of the most important asymptotic properties of dynamical systems. In this direction, we published the papers: 'Exponential trichotomy of differential systems, Exponential dichotomy' and 'Trichotomy of nonlinear differential equations'. Several years later, these results were followed by my joint work with K. Janglajew, University of Bialystok, Poland, 'Dichotomy and trichotomy of difference equations'.

However, I was eager to return closer to my original interest. And in 1987, I found it in difference equations and discrete dynamical systems. Here we have an action of a group, the integers, on Euclidean, metric, or topological spaces. During a conference in Orlando, I participated in a special session on difference equations, where I felt that difference

Figure 26. ICDEA 2002, Changsha (China). Front row, fourth from the right: Bernd Aulbach, in the middle Saber Elaydi (with a white shirt and no tie).

Figure 27. ICDEA 2007, Lisbon. Front row, fifth from the right: Michał Misiurewicz, sixth from the left: Saber Elaydi. Second row third from the left Bernd Aulbach.

equations play second fiddle to differential equations. That was the moment when I decided to change that situation and bring respect to the area of difference equations. The first step I took, with the help of my friend Gerry Ladas, was to establish the *Journal of Difference Equations and Applications* (JDEA) in 1994. The second step was to hold the first international conference (ICDEA I) on difference equations in San Antonio, TX, in May 1994. ICDEA became a great success and has been an annual conference ever since, held on almost all continents (see Figures 26 and 27).

At ICDEA, I met two people that have had a great influence on my research and career, Bernd Aulbach of the University of Augsburg and Jim Cushing of the University of Arizona.

Figure 28. ICDEA 2000. From left to right Gerry Ladas, Bernd Aulbach with a beard, last on the right Saber Elaydi.

Bernd invited me to his University to give a colloquium talk and to do joint research. As a result we published a paper, with his colleague Zeigler, titled 'Asymptotic solutions of the Schrodinger equation arising from a Dirac equation with random mass'. Then he raised a question whether or not a converse of Sharkovsky's theorem holds true which led me to publish a proof of it later. Since then and until his death in 2003, we became very close friends. And in 2001, at the ICDEA meeting in Augsburg, Bernd and I discussed the formation of a new society which we called 'The international Society of Difference Equations (ISDE)'. At that meeting, Bernd was elected president and I was elected vice president (see Figures 28 and 29).

Jim Cushing has greatly influenced my decision to work on a new area, namely, mathematical biology. Through his work and personal discussion, I started slowly but surely switching my research interest to mathematical biology, particularly the area of ecology. I met Jim the first time in 1994, when he attended ICDEA I in San Antonio. During my visit to the University of Southern California, I met with Robert Sacker (the author of the famed Neimark–Sacker bifurcation). At that time, he quit doing research in mathematics as his main area was ordinary differential equations, which was replaced by partial differential equations. I presented to him the Cushing–Henson conjecture that deals with non-autonomous periodic Beverton–Holt model in ecology. This was a turning point for me as we started working jointly and laid the foundation of nonautonomous periodic difference equations using the technique of skew-product discrete dynamical systems. Then I continued working in the area of mathematical biology focusing on developing the mathematical foundation of various open problems in ecology. And at the joint math meeting on 13th January 2006, I had a dinner with Jim Cushing at the River Walk in San Antonio, TX. After a few beers, I suggested to Jim to start a new journal titled '*Journal of Biological Dynamics*' with emphasis on ecology and epidemiology. After some hesitation, Jim agreed to jointly submit a proposal to a publisher that I been working with through *Journal of Difference Equations and Applications*, Taylor and Francis. Our proposal was approved and

Figure 29. ICDEA 2001, Augsburg, Germany, 30 July- 3 August 2001. Front row, fifth from the right: Bernd Aulbach, to his right Alexander Sharkovsky, to his right Saber Elaydi (with a cap). On the top line in the middle, Gerry Ladas (to the left of the woman).

we started the journal. This led to the initiation of a new conference titled 'International Conference on Mathematical Modeling and Analysis of Populations in Biological Systems'.

Two more people have had great influence on my research. The first was Lord Robert May. In the spring of 1995, I suggested to my (former) dean, John Dickey to invite him to give a talk as part of the 'Distinguished Science Lectures Series' at Trinity University. He gave a lecture titled 'How Many Species Are There?'. It was a huge success with over 1500 in attendance. The day after the conference, I had the pleasure of asking him several questions about chaos and the evolution of species. And at dinner, I discussed with him my new book project on discrete dynamical systems. He liked the title 'Discrete Chaos' but he asked me to add 'with applications in science and engineering' to distinguish it from Devaney's book. Then he, graciously, suggested some valuable criticism and suggestions and promised to write the preface of the book, which he did. The second person is Jim Yorke. I met Jim Yorke at the fifth ICDEA, which was held in Temuco, Chili, in January 2000. I worked hard with Gerry Ladas to convince Yorke to travel to Chili since he was worried about travelling overseas at that time. At that meeting, Jim gave a very convincing argument that difference equations are more general than differential equations, contrary to the prevailing conception. Since that time we have developed a lasting friendship exchanging ideas and engaging in discussions relevant to our common interests.

Throughout my career, I mentored numerous Ph.D. students and authored various publications, including *Discrete Chaos: Applications in Science and Engineering*, *An Introduction to Difference Equations*, *Upside-Down: The Interplay between Life and Chaos*, and co-authored with Jim Cushing *Discrete Mathematical Models in Population Biology: Ecological, Epidemic, and Evolutionary Dynamics*, among others.

While my mathematical achievements have been recognized, I never forgot my roots. With the help of my friend, Ulrich Eckern, University of Augsburg, we helped organize biennial conferences in mathematics and physics at universities in the occupied West Bank.

8. My road to dynamics by Jim Michael Cushing

I was raised in Cheyenne, Wyoming, in the western part of the United States. Despite being a small town in a large and sparely populated state, Cheyenne had an outstanding K-12 education system. Its broad curriculum offerings included courses in all the basic sciences and mathematics, up through calculus, all of which I took while a student there in the late 1950s. I was so enamoured with mathematics that I would check out the text book from the local Carnegie library during the summer recess before I took a math course to teach myself, as best I could, the topic prior to the upcoming school year. In this way, I was introduced to not only the standard courses in algebra, but plane geometry (using Euclid's Elements), solid geometry, trigonometry, and single variable calculus. During summer school recesses I would also occupy myself with applications of mathematics (measuring heights of trees using trigonometry, etc.), which was the beginning of a lifelong love of applied mathematics. I would add that, given the lack of activities available to young people in such a small town, I also spent a great deal of time studying piano performance, practicing sports (I played school basketball, football, and baseball teams), and reading philosophy (a lot of Aristotle).

Upon graduation from high school in 1960, I had various opportunities that included a scholarship offer in piano performance and basketball (of all things!). However, I felt I needed to ground myself more before making any heavy career decisions, so I left for a freshman year of college at the University of Oklahoma, where I had a good friend. There I skipped over introductory calculus and focused on courses in physics, chemistry, and geology. I transferred to the University of Colorado in Boulder for the last 3 years of my undergraduate education where I continued with courses in these sciences as well as astronomy and psychology. I needed to declare a major at this point and so I reflected back on my school subjects to see what it was I most enjoyed doing academically, with the result that I declared myself a mathematics major. I finished the major requirements for my bachelor's degree before finishing my senior year, during which I enrolled in postgraduate mathematics classes. During that year I also participated, as part of the University's team, in the preeminent national William Lowell Putnam Mathematical Competition for undergraduate students (our team placed in the top 10). While at the University of Colorado I also continued my piano performance studies, which included coaching from the celebrated concert pianist Maria Clodes who was visiting the University at that time.

After graduation from the University of Colorado (1964), I was accepted for graduate school in mathematics at the University of California at Berkeley. However, at the last minute I received a phone call from the assistant director of the Institute of Fluid Dynamics and Applied Mathematics at the University of Maryland who offered me a National Defense Act Fellowship that would fund my graduate studies. At the time UC Berkeley didn't offer financial support to first-year students and I had little financial resources available to me. To get through my undergraduate studies, I had worked several jobs ⋯ from a handy man in a women's dormitory to a upper-class student advisor in a men's dormitory. So I turned my beat-up car from west to east and drove to College Park, MD, where I became a graduate student in applied mathematics.

My studies at Maryland, for the first year, were solidly concentrated in pure mathematics, as was the fad those days, both in analysis and algebra. (In my oral Ph.D. candidacy examination, I remember being asked to outline the proof of the unsolvability of quintics by using Galois field extensions.) It was after passing the written and oral candidacy examination that I got heavily into dynamics. My studies ranged from ordinary and partial differential equations to abstract dynamical systems and from basic theoretical matters (e.g. existence and uniqueness of weak solutions and distributions) to practical numerical analysis and coding (computers were just coming into use).

My Ph.D. dissertation advisor was Monroe H. Martin, who at the time was the director of the Institute of Fluid Dynamics (and former chair of the Mathematics Department at Maryland). He took on dissertation students only one at a time (one of whom, before me, was Simon A. Levin, who is currently the director of the Center for BioComplexity at Princeton University). Several former students of Professor Martin's, as well as I, worked on a conjecture of T. Levi–Civita concerning deep water waves and the uniqueness of waves corresponding to a certain physical parameter involving the wave length, the square of the wave profile's speed, and the gravitational constant. None of us settled the conjecture in our dissertations, but each made various contributions towards its resolution. My years at Maryland (1964–1968) were mathematically quite exciting and fruitful, in large part because of numerous high level faculty members and student classmates (one of whom was the celebrated James Yorke, who went on to coin the word 'chaos' in dynamical systems).

Upon completion of my Ph.D., I celebrated by taking a summer off to roam around Europe (with friends in the counter culture remnants of the beatnik generation), from Scotland to Italy and Spain, before taking up a position as assistant professor of mathematics at the University of Arizona in Tucson in the Fall of 1968. On the faculty there at the time was the celebrated fluid dynamicist L. M. Milne-Thomson and some of his students. Milne-Thomson had written several books out of which I had studied (including *Theoretical Hydrodynamics and Theoretical Aerodynamics*) and I was thrilled to meet him. Although the teaching load was quite heavy, by today's standards, research publication was still expected and so I remained engrossed in Levi–Civita's conjecture until I settled it in the affirmative, for sufficiently small amplitude waves, in my first publication after that of my dissertation [48].

In 1970, Milne–Thomson moved on from Arizona and that same year I took a post-doctoral position at the IBM Thomas J. Watson Research Center, just north of New York City in Yorktown Heights. There I continued research on existence and uniqueness theorems for nonlinear elliptic partial differential equations, as they depend on model parameters, which naturally led me into a study of bifurcation theoretic methods. I also attained a renewed interest in dynamic stability while at IBM, mainly through stimulating discussions with Charles Conley (of Conley index fame), who was my office mate for a while.

Upon return to the University of Arizona, I began a collaboration with John Bownds, a colleague in the mathematics department, on stability theory for integral and integro-differential equations. We found interesting applications of our work in theoretical ecology, which for me began a career long adventure in mathematical biology and theoretical ecology. Integro-differential versions of the famous Lotka–Volterra models for population dynamics and species interactions appear in Volterra's original work [183]. As models, these types of equations (which were the topic of my first book [49]) often result from time delays related to demographic structure in a population. There are two broad types of models used for the dynamics of demographic structure, one utilizing continuous time and structure and the other utilizing discrete time and structure. Models of the first type are described by a first order hyperbolic partial differential equation (the von Forester equation) under nonlinear, nonlocal boundary conditions, on which I focussed my research activity in the 1970s. Such equations typically involve highly technical mathematical issues with regard to even the basic questions of existence of solutions and, because of this, can provide difficulties for theoretical biologists whose interests lie in other directions (ecological questions involving asymptotic dynamics, stability, bifurcations, etc.).

The second type of structured population models are the discrete time matrix models originally popularized by Lewis and Leslie in the 1940s. These models are described by systems of difference equations and do not entail mathematical difficulties surrounding the well-posedness of initial value problems. An introduction to both types of structured population dynamic models can be found in my second book [50], which resulted from a series of lectures I gave at North Carolina University in 1997, sponsored by the Conference Board of the Mathematical Sciences.

The period of my professional life, from the late 1970s to the early 1990s, was an active one that involved many studies and publications in theoretical ecology in conjunction with several postdoctoral students (Steve Simmes, Mohamed Saleem, Jia Li, and Yicang Zhou) and several graduate students (Zakaria Alawneh, Kathleen Crowe, Guillermo Uribe, Mona Alameddine-Roddier, BingXu, Utith Inprasit and Kebenesh Blayneh). The overall goal of my work during this period was to investigate the extent to which basic principles of classical theoretical ecological dynamics holdup (or do not hold up) when demographic structure within a population is included in the models – principles such as logistic equilibration, predator–prey oscillations, competitive exclusion, and so on. My growing involvement with population and ecological dynamics was greatly enhanced by the foundation of two programs in 1976 at the University of Arizona: the Interdisciplinary Program in Applied Mathematics (of which I have been a member since its inception) and the Department of Ecology and Evolutionary Biology (EEB), one of the first departments of its kind in the world. EEB had faculty members who were prominent advocates of mathematical modelling, including Michael Rosenzweig (of predator–prey 'paradox of enrichment' fame) and William Schaffer, both students of Robert MacArthur at Princeton University who is considered a founding father of ecology and evolutionary biology. Bill Schaffer made scientific headlines with his research in epidemics and especially from his claim that certain measle outbreak data displayed chaotic dynamics. Bill and I never published together, but we team taught graduate and undergraduate level modelling courses in the EBB and mathematics departments and, as a result, I learned a great deal from him. That interaction sparked my first interest in chaos theory (along with the famous seminal papers of Robert May).

In 1989 and 1990, I attended and spoke at workshops and seminars at the Department of Environmental Studies, University of California at Davis, where I met biologist Robert Costantino, who was visiting there from the University of Rhode Island (to collaborate with Alan Hastings). Bob is an expert on flour beetles (*Tribolium sp.*). Besides

Figure 30. The Beatle Team at a meeting in Tucson, AZ, USA, in 1998. From left to right: Robert Desharnais, Robert Costantino, Brian Dennis, me, and Shandelle Henson.

being an agricultural pest, these beetles have been a laboratory experimental animal for over a half century. Bob and his former post-doc Robert Desharnais had done some modelling on (indeed wrote a book on) modelling the dynamics of *Tribolium* and approached me about a possible collaboration. The result was the creation of a team of four interdisciplinary researchers (the Beatle Team) consisting of Bob Costantino, Bob Desharnais and me together Brian Dennis, an ecologist and statistician. We later joined by Shandelle Henson and Aaron King (see Figure 30).

A basic issue for the Beetle Team (yes, we were cognizant of the fab four ⋯ the famous Beatles) was an uneasiness with the state of mathematical population ecology in that it was long on theoretical concepts and principles (stability, equilibrium, limit cycles, chaos, etc.) but short in convincing and predictive explanations of real biological systems. Ecologists rightly asked whether a low-dimensional, nonlinear (or even linear) population model had ever been a convincing representation of a real ecological system. We were hard pressed to find any examples in which population models constituted reliable explanations and predictions of population dynamic phenomena. This view was held by many, if not most, ecologists [3].

In 1991, we approached the National Science Foundation (NSF) with a grant proposal to establish a rigorous and descriptive, but more importantly, predictive model for the dynamics of a biological population (namely, the flour beetle), which would be validated by carefully designed, controlled, and replicated laboratory cultures. The main goal, in our mind, was to show that mathematical models, even low dimensional models, could be predictively accurate when confronted with data. NSF responded with a year's support to re-write the proposal so as to highlight the use of the model (if we could establish one) to demonstrate model predicted complex (nonequilibrium) dynamics and, in particular, a bifurcation route-to-chaos (a hot topic at the time). This we did and gained the support of NSF for over a decade, during which we developed and parameterized a discrete time (Leslie) model using historical data, studied the predictions of the model using analysis and numerics, determined bifurcation sequences that included chaotic dynamics and experimentally feasible manipulations that place replicated cultures along that route-to-chaos, and statistically validated (without re-parameterization) the model predictions against the observation data obtained. That project (which involved 8 years of uninterrupted counting of hundreds of individuals every 2 weeks!) is considered the first unequivocal demonstration that a biological population can exhibit chaotic dynamics, as predicted by a mathematical model.

What type of mathematical model did the Beetle Team use? We began simple and made the model more complicated only when necessary to obtain an adequate statistical validation of model predictions (using already available historical data). The well-known Ricker model (which has the famous period doubling route-to-chaos) failed miserably to parameterize adequately (using maximum likelihood and conditioned least squares methods) for the dynamics of the flour beetle. We extended the model to include juvenile stages, as a time delay, but this two dimensional Leslie model also failed to parameterize adequately. A successful parameterization was obtained by including two juvenile stages, larval and pupal, and an adult stage (which our data luckily included). The result was a three dimensional Leslie model, now known as the LPA model,

$$\begin{cases} L_{t+1} = bA_t \exp\left(-c_{el}L_t - c_{ea}A_t\right) \\ P_{t+1} = \left(1 - u_l\right) L_t \\ A_{t+1} = P_t \exp\left(-c_{pa}A_t\right) + \left(1 - \mu_a\right) A_t \end{cases}$$

which served as a basis for the team's research for two decades. A report on the route-to-chaos project is given in the book [53]; also see [51].

The beetle chaos project accomplished much more than just a tight model-data fit and a demonstration of biological chaos. Numerous other dynamic phenomena were explained by the model, often with jaw-dropping accuracy (see Figure 31). Some of these model predicted phenomena were unknown to us prior to the experiments and some even seemed impossible or highly unlikely to the biologists, and yet were validated by experiments. The model explanation of many observations utilized variations of the LPA model: stochastic, periodically forced, coupled competing species, and evolutionary (Darwinian) versions (see Figure 32).

For example, repetitious patterns observed by Bob Costantino in the long data time series for the chaotic attractor corresponded beautifully to the unstable periodic orbits lying on the attractor [98]. The data followed orbits that (repeatedly, due to stochastic perturbations) fly-by the unstable (saddle) cycles on the chaotic attractor, a remarkable observation attesting to the accuracy of the *LPA model.* There was, however, a notable exception, namely, a distinct period 6-cycle pattern that is nowhere to be found on the chaotic attractor. An explanation for this cycle was found by realizing that all life cycle stages in real beetles come in whole numbers, but that the LPA state space is continuous. When the LPA model was placed on an integer lattice, we found that the attractor was then 6-cycle, in exactly the three dimensional phase space configuration observed by Bob. However, chaos cannot be present in a dynamical system on a finite state space lattice, since all bounded (deterministic) orbits are necessarily periodic. So, in what sense did the flour

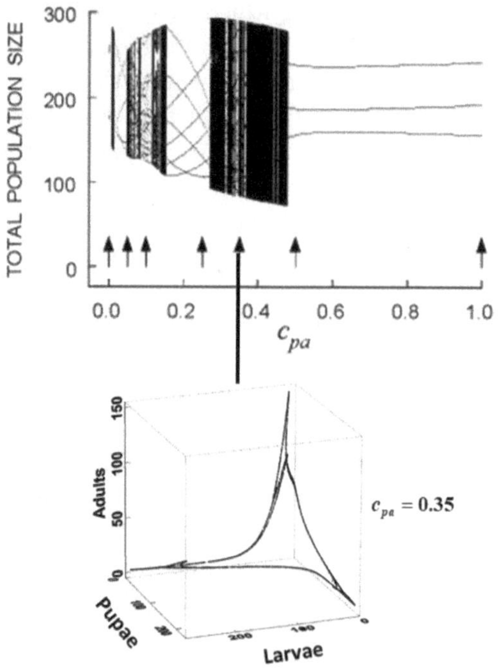

Figure 31. The bifurcation diagram from the LPA model that was used in the route-to-chaos experiments, conducted over a period of 8 years, shows numerous types of attractors, including a chaotic attractor at $c_{pa} = 0.35$.

Figure 32. Lecturing at the 25th ICDEA conference held at University College London in 2019 on a Darwinian version of the famous Ricker model.

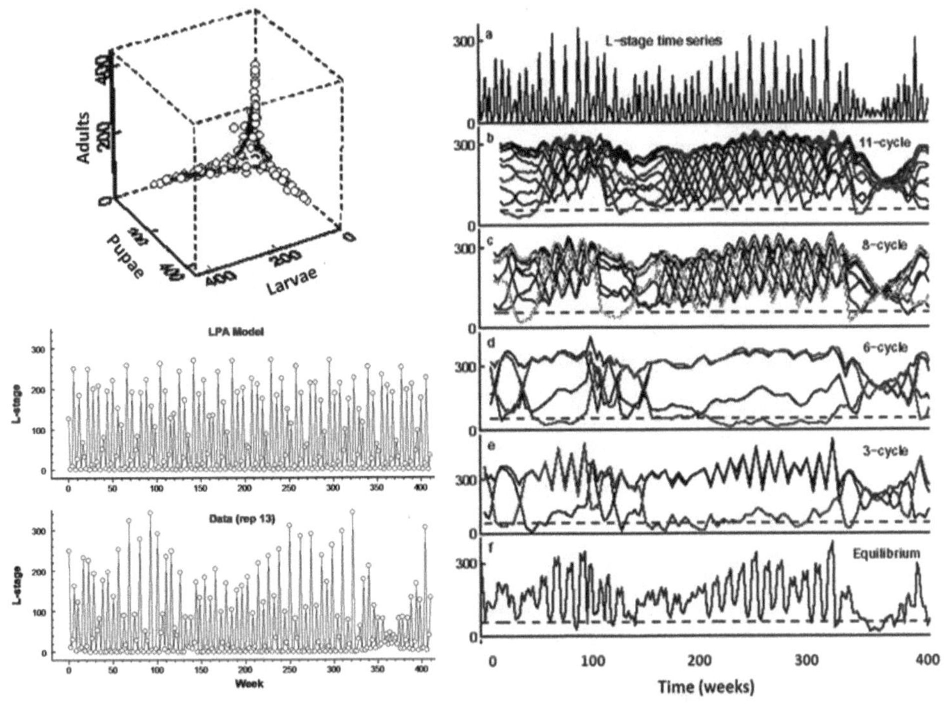

Figure 33. Beetle data, from one of the replicates set at the model predicted chaotic attractor, cluster around the attractor in state space. The time series of the larval classes from the LPA and the replicate data are compared, the latter of which shows a fly-by of the unstable equilibrium during weeks 350–400. The graphs on the right are lag metric measures demonstrating the beetle population flying by various unstable periodic orbits on the chaotic attractor of both the LPA model and the lattice LPA model.

beetle experiment demonstrate the existence of chaotic dynamics? As shown in [87] it is a *stochastic version of the lattice LPA model* that explains the patterns observed in the experiment: although model predicted orbits tend to a 6-cycle, continual stochastic perturbations cause transient dynamics which, even though on a discrete lattice, have properties of the chaotic attractor of the underlying continuous state space LPA model (including fly-bys of the cycles on the chaotic attractor).

The stochastic lattice version of the LPA model is but one example of extensions of the model that provide remarkable explanations of patterns observed in the Beetle Team's experiments (see Figure 33).

Other examples include saddle fly-by's in non-chaotic dynamics and stochastic phase shifts in oscillatory dynamics for stochastic versions (e.g. see [52]); resonance (as observed in experiments with flour beetles) and attenuance for periodically forced versions (e.g. see [46]); the evolution of genetic polymorphisms for Darwinian (evolutionary game theoretic) versions (see [154]); and challenges to the classical competitive exclusion principle based on coupled LPA models [56]. Much of this research is covered in the book [53]. A book that more comprehensively covers the work of the Beetle Team is currently in preparation (with Shandelle Henson as lead author), also see [55]. The LPA model and its numerous versions that we developed provide, to this day, many open and challenging problems for mathematicians interested in discrete time dynamics described by difference equations.

Figure 34. Members of the Seabird Ecology Team, hiking on the north shore of Protection Island, from left to right: me, Shandelle Henson, and James Hayward.

The research of the Beetle Team involved laboratory cultures of flour beetles. A rigorous connection between mathematical models and ecological data is obviously more difficult for populations in nature, where biological and environmental interactions are more abundant, complex, and not so controllable (if at all). My most recent interdisciplinary collaboration has been with Shandelle Henson and her husband, James Hayward, a field ecologist who has worked and gathered an immense amount of data on marine animals and birds on Protection Island, a United States Natural Wildlife Refuge managed by the US fish & Wildlife Service (see Figure 34). I was privileged to joint Shandelle and Jim, and their many colleagues in the Seabird Ecology Team, for over a decade of collaborations that modelled marine animal behaviour and population dynamics. While we use differential equation models for some projects, we also make use of difference equation models, particularly matrix models for structured populations and periodically forced and evolutionary game theoretic versions. Some projects are in the spirit of the Beetle Team research in that, after obtaining from data statistical approximation of parameters and correlating them with environmental factors, we made predictions for what behaviour activities will be observed the following next season, predictions that were subsequently observed in the field. This work was done with nesting colonies of the glaucous-winged gull (*Larus glaucescens*). Other projects involve using models, in a proof-of-concept manner, to test hypotheses formulated from field observations. Of particular interest are phenomena related to climate change (specifically, rising mean sea temperatures in the Strait Juan de Fuca where Protection Island is located). For example, egg cannibalism by adult glaucous-winged gulls significantly increases during el Niño years (a kind of natural experiment for future climate warming). Such behaviour, it would seem, should be eliminated by natural selection over time; however, our models show that cannibalism can be (under certain circumstances) a long-term, evolutionarily stable strategy. This result utilizes Darwinian versions of a discrete time matrix model for juvenile–adult (victim–cannibal) structured population. We have recently published a book covering the nearly two decades of interdisciplinary research by the Seabird Ecology Team [54].

My collaboration with the Beetle and Seabird Ecology Teams not only greatly affected my research career but also my mentoring activities, especially with regard to discrete time matrix model analysis and applications in population, ecological, disease, and evolutionary dynamics. Postdoctoral students with whom I have worked, until my recent retirement, include Shandelle Henson, Olav Skarpaas, Aaron, King, Kehind Salau and Alex Farrell. Graduate students whose dissertations I directed include Martin Gildardo Garcia-Alvardo, Maref Alzoubi, Jeffrey Edmunds, Nakul Chitnis, Sheree Arpin (LeVarge), Rosalyn Rael, Suzanne Robertson, Amy Veprauskas, and Emily Meissen.

My road to dynamics has been a long one. And one that was quite varied, starting from early high school days of Euclidean dynamics and calculus, through pure mathematics training as an undergraduate, to applied mathematics and fluid dynamics in graduate school, and on to population, ecological and evolutionary dynamics with interdisciplinary teams of colleagues. After serving 53 years as a professor of mathematics at the University of Arizona, I am now retired with emeritus status. As a final word, I might add that during my travels on this long career path, I never stopped playing the piano. (I even worked with string students and their professors in the University's music department for several years.) And that I plan to keep doing so, as well as remaining active in research projects for as many years as I have left.

9. My scientific youth with one-dimensional dynamics by Lyudmila Efremova

I am originally from Nizhny Novgorod (between 1932 and 1990, the town was named Gorky, in honour of the famous Russian writer Maxim Gorky, who was born there in 1868). In the mathematical (and (radio)physical) world, Nizhny Novgorod is well known for Andronov's school of nonlinear oscillations and dynamical systems. In 2023, for the 90th anniversary(!) of the school, a book devoted to the history of Andronov's school [140] was published (see Figure 35).

Andronov's great merit is the use and development of fruitful, but forgotten (in the 20s and the 30s of the twentieth century) Poincaré's ideas in the theory of differential equations for the needs of radiophysics and the theory of radio transmitters (see, e.g. [18]). One of Andronov's ideas is based on the use the Poincaré secant (when it exists) for the description of trajectories dynamical behaviour of ordinary differential equations. Further development of this idea led to the creation of the 'point mapping' method, which was integrated later into the theory of discrete dynamical systems. Here I will refer only to the articles on one-dimensional dynamics by Maier (he worked at Gorky State University) on rough

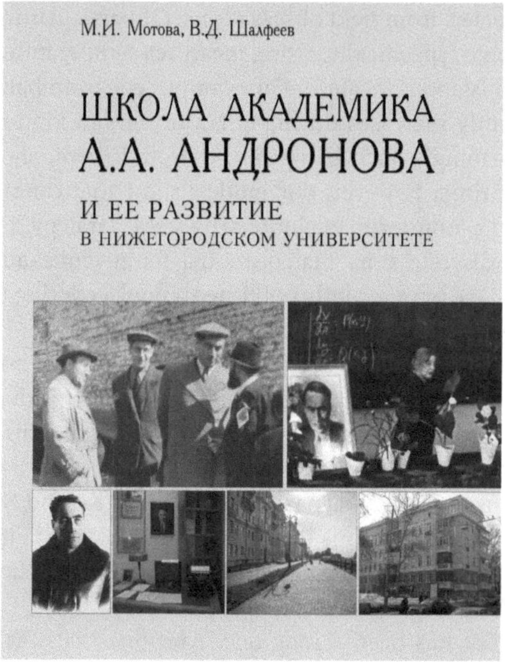

Figure 35. In the upper photo on the left, Academician of the USSR Academy of Sciences Andronov (second from right) and Professor Maier (first from right) with Ph.D. students (Ya. Nikolaev and S. Bellustin). In the upper photo on the right, Professor Leontovich–Andronova presents a report on Andronov's mathematical works.

(structurally stable) diffeomorphisms of a circle [127] and by N. N. Leonov (he worked at the Research Institute of Physics and Technology at Gorky University) on discontinuous maps of a straight line [108,109].

It must be said that throughout its more than a century of history, Nizhny Novgorod (Gorky) University remains one of the largest educational and scientific centres of the country. On 20 March 1956, by a decree of the Presidium of the Supreme Soviet of the Russian Federation, Gorky University was named after the great Russian mathematician N.I. Lobaschevskii, native of Nizhny Novgorod.

When I was in school, I not only loved solving difficult mathematical problems but also made attempts to write music (I also studied at a music school). One of my compositions for the symphony orchestra was performed at a major concert at my music school (to the delight of my parents). As a result, choosing which path to follow was very difficult for me. Nevertheless, in the struggle between my two serious hobbies, mathematics and music, mathematics won, and I entered the Faculty of Mechanics and Mathematics of the University.

When I was a second-year student, Dr. Rakhmankulov invited me to read the Sharkovsky's paper [187]. My impression was huge. I thought: 'It is necessary to only know the classical Bolzano–Cauchy intermediate value theorem for continuous functions and nothing else! But how did he come up with his own 'dynamical' order on the set of natural numbers? It is very beautiful!' Now after many years I can say that my naive first impression of Sharkovsky's theorem determined my scientific biography. Throughout my work

Figure 36. The Department of Differential Equations and Mathematical Analysis of Lobaschevsky University (1975–76). In the front row, Rakhmankulov is first on right, Otrokov is fourth on right. In the second row, Efremova is fifth on left.

in mathematics, from time to time I return to various generalizations of this wonderful theorem [60,61]. In the third year, I received from Dr. Rakhmankulov the topic of my student research work. The exact setting of the problem was as follows. *Let f be a continuous circle map with a periodic point with period $q \geq 1$ and suppose that the map f has periodic points with period p for $p \neq q$. It is necessary to describe the set of all the (least) periods of f-periodic points.* Rakhmankulov himself solved this problem for $q = 1$ [155,156], and I had to consider the cases of $q > 1$. At the time, attitudes to this problem oscillated between two extreme points of view: some believed that there was nothing to do here, since Sharkovsky did everything, while others said that it was impossible to solve this problem because it was very complex. I was young, carefree, and instead of these conversations, I listened with a feeling of happiness to lectures by major mathematicians (S.P. Novikov, Ya. G. Sinai) and physicists (V.L. Ginzburg, A.V. Gaponov-Grekov) who worked in Gorky or came here to give talks, as well as lectures by famous musicologists at the Gorky Conservatory on music theory, V.G. Blinova and M.M. Valentinov, the main director of the Gorky Opera and Ballet Theater.

My years as a student were filled with a constant sense of joy at learning new things. At the end of my studies, I obtained some partial results on the coexistence of periods of periodic points of continuous circle maps and graduated from Gorky University with honours. In graduate (Ph.D.) school, I continued to deal with the problem of the coexistence of periods of periodic points of continuous circle maps. Prof. Otrokov and his former graduate student, Dr. Rakhmankulov, were my scientific supervisors (see Figure 36). Otrokov is a major specialist in the theory of limit cycles of differential equations. He was never a graduate student of Professor Andronov, although he always considered him to be his teacher. In 1940, Otrokov defended his Ph.D. thesis under Andronov's supervision. I assume that the influence of another famous mathematician, Professor Braitsev, specialist in the theory of functions, defined the fact that Otrokov used a functional approach for studying limit cycles. Otrokov had a vivid imaginative mindset, and communication with him always gave the prospect of further research.

When I was a Ph.D. student in the second year, Otrokov decided that it would be useful to show the results obtained to Professor Sharkovsky. So, after a phone conversation with Otrokov, Sharkovsky invited me to present the report at his seminar at the Institute of Mathematics of the Academy of Sciences of Ukraine. That was in 1977. Of course, I do not remember all the details of this report, but I remember my surprise that Sharkovsky was sitting at the back of the room where the seminar was held, not at the front, as seminars leaders usually do. I now know for sure that the back of the room is the best place to listen to reports! There was a lively discussion with Sharkovsky after that seminar, and I have had the opportunity to sometimes go to Kiev to present and discuss new results. It was always been very interesting!

It was at this time that the idea arose to apply the concept of the degree of mapping to the description of the periods of periodic points of continuous circle mappings [57,58] (see also [39]). It was a fruitful idea which transformed the original setting of the problem, preserving it only for circle maps of degree 1. It turned out, for example, that the periods of periodic points of continuous maps of a circle of degree -1 satisfy the Sharkovsky's order (see the earlier works [39,58]). As for continuous maps of a circle of degree 1, although the rotation of the trajectories of such maps was apparent in my proofs and examples at the time, I did not have enough experience to introduce the concept of a set of rotation for a circle map of degree 1. This was done later in the article [141]. Note that in [137] and [8] rotation numbers of individual periodic orbits are used.

Analysis of the examples I have built has led me to the idea of using a horseshoe when considering circle maps of degree 1 with periodic orbits of periods $q \geq 2$ and $p > q$ for $p/q \notin \{2^i\}, i > 0$. It was the second fruitful idea, which gave a form of representation of periodic orbits periods of circle maps of degree 1. This form looks like the decomposition of a vector by basis vectors [58,64]; in addition, it turned out that the periods of periodic points of maps of degree 1 for $q = 1$ and $p > 1$ admit such a representation [62]. The use of the horseshoe also made it possible to prove the criteria for the existence of homoclinic points of circle maps of all degrees and analyse the role of transversality-type conditions for the existence of homoclinic trajectories of circle maps of degree 1 [58,64]. Sharkovsky recommended this work to the Programme Committee of the Ninth International Conference on Nonlinear Oscillations (ICNO – IX, Kiev, 1981), and as a young scientist, I delivered a report on the results described above at this conference. It was the first major conference in my life, and I was very worried. Professor Plykin was the chairman of the section where I presented my report. Thanks to his goodwill, the report was successful. After that there were many interesting discussions, first of all with Professors Plykin and Szlenk. So, Szlenk informed me that the idea of representing the periods of periodic points of circle maps of degree 1 having periodic points with mutually simple periods $p > q > 1$, similar to the one described above, was also expressed by Professor Nitecki (unpublished).

As for Plykin, then later, in the collection of poems by poets of the science city of Obninsk, I saw his poems. So I found out that the author of the Plykin's attractor was also a poet! Moreover, in his young years Plykin was engaged in mountaineering and some mathematicians (e.g. Professor Sinai) climbed mountains with him. A few months after ICNO – IX, I submitted and defended my Ph.D. thesis (see Figure 37) of which Sharkovsky and

Figure 37. This photo was taken after the presentation of the results of my Ph.D. thesis at a seminar on differential equations led by Professors G.A. Leonov and Matveev at the Leningrad (St. Petersburg) Pedagogical Institute (1981).

Belykh were the official opponents (referees). With great warmth, I also recall the very lively discussions of my Ph.D. thesis with Professor Belykh. Many years later, Belykh came out as one of the referees of my Doctor of Sciences thesis (physics and mathematics). Later, on the recommendation of Professors Plykin, Szlenk and Nitecki, I made a report on the above results at the Sinai seminar. The room in the main building of the Moscow State University, where the seminar was held, was crowded with people. Among the listeners were not only mathematicians and physicists, but also, as Plykin told me, meteorologists from the Hydrometeorological Center. As a member of the editorial board of the journal *Russian Mathematical Surveys*, Sinai submitted my short article to this journal.

While still in graduate school, I started working at the Gorky Pedagogical Institute. This work continued after defending my Ph.D. thesis in 1981 until 1987, when I received an invitation to work at Lobachevsky University at the department of Otrokov. So I went back to my home university and I am working there now.

After defending my Ph.D. thesis, I thought about continuing my work in the theory of dynamical systems. It should be said here that the last 25 years of the twentieth century were a golden period of one-dimensional dynamics. All the major achievements in this field relate to this period of time. But another period was coming, in which the most interesting areas of activity seemed to me, first, the applications of one-dimensional dynamics to the study of discrete dynamical systems on manifolds of dimension at least two; and, second, the creation of dynamical systems theory on complicated one-dimensional ramified continua, which do not allow order topology. I started studying skew products of one-dimensional maps on finite-dimensional cells, cylinders and tori, choosing the first line of research. The second problem was formulated for my first graduate student, E. Makhrova who is currently docent at Nizhny Novgorod University. She found very interesting and successful results in this direction (Figure 38).

Figure 38. At the conference 'Mathematical Physics, Dynamical systems and Infinite-Dimensional Analysis – 2023' (Dolgoprudny, Moscow Region, 2023). Efremova is fourth from the left, Makhrova is third from the left.

At about the same time (in the early 90s of the twentieth century), representatives of Ukrainian, Czech, Spanish, Italian, and other mathematical schools began to study the topological dynamics of skew products of interval maps. All these studies were aimed at solving the following problem. *Which properties of one-dimensional maps do skew products preserve, and which ones do not?* This approach allowed to construct examples of skew products of maps of an interval in dimension two with properties other than those of continuous interval maps. And although I also built examples, I was more interested in the question of why this is happening.

One of the main principles of A.A. Andronov was the principle of 'mobilization of information' [140]. Therefore, when I started working on skew products of maps of an interval in dimension 2, I read, in particular, all the works on the study of cylindrical cascades, starting with a small text in Poincaré's memoir [149]. And again, among the ideas of Poincaré (he is truly inexhaustible!) there was one, the modification of which, in relation to the skew products of the maps of the interval in dimension two, made it possible to conduct a systematic study of such maps [59]. Professor Anosov, Academician of the Russian Academy of Sciences, was the first mathematician who drew attention to the relationship between the study of cylindrical cascades and the solution of Hilbert's 5th problem. In the process of studying skew products of interval maps, it also turned out that the study of such maps adjoins the range of issues related to Hilbert's 13th problem (Figure 39).

In 2003, the book 'Mathematical Events of the XX-th Century' was published in Russian, under the editorship of Academician of the Russian Academy of Sciences, Professor Arnold. This book opens with an article by Anosov, 'Dynamical Systems in the 60s: The Hyperbolic Revolution' [15]. The article made a powerful impression on me. This is a mathematical article, in which the author managed to convey the experiences of a person creating a completely new knowledge!

Figure 39. (This is a photo from the family archive of D.V. Anosov). D.V. Anosov gives a lecture at the Workshop 'Contemporary Mathematics' in Dubna.

It was at that time, when the idea arose of my internship at the Steklov Mathematical Institute of the Russian Academy of Sciences under the supervision of Anosov. The opportunity for a 2-week internship was presented only in 2009. Anosov agreed to be the scientific supervisor of my internship. In that time the feeling of happiness of learning new things returned to me, which I experienced during my student and graduate years. There was another 2-week scientific internship under the guidance of Anosov in 2011. And all this time since 2009 I interacted with Anosov (until the beginning of 2014, and in August when he died). I have repeatedly made presentations at the seminar 'Dynamical Systems and Ergodic Theory' under the guidance of Anosov and Stepin, where the results were thoroughly discussed (Figure 40).

I want to tell only one story here, which vividly characterizes the high human qualities of both Professors Anosov and Stepin. One day I called Dmitry Victorovich Anosov in the autumn of 2013. He was already feeling unwell, and during the conversation he suddenly asked me: 'Which of the Moscow mathematicians would you like to cooperate with?' After some silence, I said that I would like to cooperate with Stepin. Then the conversation turned to another topic, and Anosov did not return to this issue anymore. And it was only from Stepin's speech at the defence of my Doctor of Sciences Thesis in 2018 that I learned that Anosov spoke to him and asked him to treat me carefully and help me. And this was done by a man who understood that he was leaving··· Even now, when I write about it, it seems incredible to me! Stepin fulfilled everything that Anosov had talked to him about shortly before his death. In particular, he submitted my survey on skew products of maps of an interval in dimension 2 to the editorial board of the journal 'Russian Mathematical Surveys', and after a careful review the article was published (Figure 41). He then recommended this work to Ya. G. Sinai (Abel Prize, 2014). After that, I exposed my Doctor of Sciences thesis at the seminar of the Dobrushin Laboratory at the Institute of Information Transmission Problems of the Russian Academy of Sciences. The work was accepted, and I defended my Doctor of Sciences thesis in 2018 at this Institute in the Council, which was

Figure 40. The photo was taken by Professor Ryzhikov before the meeting of the seminar on ergodic theory and dynamical systems. In the front row, Professor Oseledets is first on right, Professor Gurevich is first on left. In the second row, Professor Anosov is the first on right, Professor Stepin is first on left.

Figure 41. The photo was taken by Professor Ryzhikov. Professor Stepin is listening to a talk at the seminar 'Ergodic Theory and Dynamical Systems'.

chaired by Sinai. The official referees of my Doctor of Sciences thesis were Belykh (one of the most prominent representatives of the school of Academician Andronov), Professor A. Zhirov (a former Ph.D. student of Professor Plykin) and the famous Russian topologist Professor S. Bogatyi.

In 2018, I received an invitation to work at the Moscow Institute of Physics and Technology. Currently, I am also full professor at this Institute. This work, communication with specialists in mathematical physics attracted my attention to the study of quantum mechanics, and my husband Professor Sakbaev and I applied the technique of studying the Ω-blow

Figure 42. This photo is taken during the Conference on Dynamical Systems (Atlanta, USA, 2011). From right to left: Professors Fournier–Prunaret, Efremova, Sakbaev, my Ph.D. student Andrey Filchenkov.

up in skew products of interval maps to the classification of blow ups in the set of solutions of differential equations [63].

My active work in low-dimensional dynamics continues (Figure 42). Along the way, I met many outstanding people who played a big role in my scientific biography. I remember this with great warmth and gratitude. I am writing this article thinking about a new generation of mathematicians. They are different, but I get a great pleasure from working with the last generation of my students. When I see the sincere interest in their burning eyes, I believe that the development of the theory of dynamical systems and its applications will continue.

The author thanks Dr. E. Andronova, Dr. A. Klimenko, Dr. E. Makhrova and Professor V. Ryzhikov for their help in the selection of photographic materials.

10. How I got caught up in chaos studies by René Lozi

Is everyone's life predestined? this question has troubled humanity for millennia. Personally, I don't think so, but I wonder about the circumstances that can lead someone to choose their path in life. The term 'math bump' originates in the nineteenth century, in a pseudoscience called 'phrenology'. Founded by the German neurologist Franz Joseph Gall, it argues that brain capacities are distributed in well-defined areas of the brain, and that the shape of the skull then reflects these capacities. According to this false theory, which is still more or less popular, people gifted in mathematics would have a visible 'bump'; or at least this expression is used in a metaphorical sense. A friend of my mother's touched my baby's head, and she decided that this was the case for me. This preposterous statement was often repeated to me during my early childhood, and I wonder if hearing it convinced me that I had a predisposition to study mathematics!

What is remarkable is that, like Michał Misiurewicz did and does today (Section 2), I counted many things around me during my childhood and until now (how many friends I had at school?, how many seconds is a year?, a century?, how many stairs I have to climb to reach my office?, etc.). My whole family laughs about it! Besides, 70 years later, I also have bad memories of my preschool year: my teacher didn't let me play during recess. She made me do math exercises that the other students weren't able to do. When I was 13, one of my cousins, an engineer, introduced me to the base 10 logarithms. I was fascinated. I constructed a cardboard slide rule, grading it with my older sister's logarithms chart, and used it at school, much to the astonishment of my classmates and the teacher. Forty years later, many of them still remembered it!

I feel sometimes uneasy to recall what exactly happened nearly 50 years ago in my academic life. In what follows, I hope to reconstitute fairly accurately the beginning of my research in bifurcation and chaos theories at University of Nice (now University Côte d'Azur) where I spent all my life, even under different administrations of research (C.N.R.S.) or education. In fact, my office was always inside or in the same small neighbourhood of the Faculté des sciences, where stands the Institute for mathematics Jean-Alexandre Dieudonné, in honour to the famous 'Bourbakist'.

Being a mathematician is very special, not only do people think you are different from normal people, not keeping your feet on the ground; but you can develop a decades-long friendship with researchers from other countries, speaking the same common mathematical language. This is why, beyond mathematical considerations, I will take the time to talk about friends of mine and outstanding mathematicians that I have met over the years.

I started my studies at the University of Nice – which is a distance of 20 km from Antibes, my home town – in October 1967, in mathematics and physics. It was classical studies in a world that had not known many upheavals since the second world war, except the conquest of space started in 1957 with the launch of the Russian satellite 'Sputnik' and the journey of the first astronauts, the Russian Yuri Gagarin in 1961 and the American John Glenn in 1962. I was fascinated by particle physics and I dreamed to make research in theoretical physics, without knowing which courses to attend to reach this goal. At the university I took my first programming course about FORTRAN IV in 1968, using punched cards. The decades that followed and of which I will try to describe the evolution that I personally perceived in the limited field of the study of bifurcations and chaos, has seen great changes, such as the conquest of space with Neil Armstrong, the first man who walked on the moon on 20 July 1969, the wide spread of Personal Computers, wireless phones, the birth of Internet, Artificial Intelligence, the easy plane travels, etc.

During my studies at the University of Nice, I took at the same time courses in theoretical physics and applied mathematics. Graduated with a certificate in fields and particles, I no longer liked this discipline but I discovered with fascination the methods of numerical integration of ordinary differential equations (ODEs). It was during the academic year 1970–71, I was following my bachelor's degree under the supervision of Professor Martin Zerner (1932–2017). During my previous studies, I had been taught that there was a list of ODEs written by Bernoulli, Lagrange, Clairaut, Riccati, etc. and a list of solving methods. A clever student in mathematics should be able to know which method and which trick was efficient to solve a given equation among a huge list of them available in handbooks. No physical sense, in fact no meaning at all, was attached to these academic exercises. No numerical method was taught. Moreover, between professors, there was a strict separation between 'pure mathematicians' and the few 'applied mathematicians' who were able to use a computer. It must be said that only one computer 'IBM-1130' was available at the university of Nice (another computer IBM-7040 was installed at the Observatory[5]). Martin was considered as weird by the other professors because he was using a computer (although his Ph.D. supervisor was Laurent Schwarz and he was chosen, when working as 'attaché de recherches' at C.N.R.S., to help Jacques Hadamard to finalize the last books of his complete works in 1958–59). I knew him fairly well, because during the famous 'student revolution' of May 1968, students and professors of the small faculty of science of Nice had exchanged a lot, breaking down the usual hierarchical barriers. With Martin we even guarded together, against hypothetical attacks, the 'Château de Valrose', building of the Institute of mathematics, during many nights of this month of May, perched on its roof, like medieval warriors in their dungeon.

During his lectures, my mind knew a breakthrough that changed the paradigm in which I conceived the theory of differential equations: the set of all the equations I was taught, were of zero measure in the set of all ones existing. No closed formula of solution can be found for most of them. Only numerical methods like Runge–Kutta, Adams–Bashforth, etc. were able to provide approximate solution. Of course, in this scope, computers were essential. Moreover, ODEs were useful to model physical, chemical or even biological situations (at the end of the year, to be graduated, I presented a memoir on the Yukawa potential [202]). This new paradigm has guided my research career throughout my whole life. Looking for a supervisor for my Ph.D. thesis in 1972, I had the chance to meet Professor Giuseppe 'Pippo' Geymonat, professor at the Politecnico in Turin (Italy) in sabbatical in Nice, who first supervised my 'Diplôme d'Études Approfondies' and proposed me to study the particular nonlinear ODE

$$\begin{cases} u''(x) - \left[u'(x)\right]^2 + cu'(x) = 0, \\ u(0) = u(1) = 0, \end{cases} \tag{8}$$

which belongs to the class of two-point boundary value problems [96].

The study of (8) led me to defend in 1975, my Ph.D. thesis, formally supervised by Jean Céa, entitled 'Analyse numérique de certains problèmes de bifurcation' [115]. Just after the defence I published a short article on, perhaps the first example of continuation method, at least in the context of the bifurcation theory [116]. While preparing my Ph.D., the name 'bifurcation' was largely unknown in the communities of mathematical and numerical analysis in France. When people in Nice and other universities were asking me on my research topics, they were astonished by this name. Of course, as one can read in Abraham's [110] that the term bifurcation was introduced 90 years before, by Henri Poincaré [150] who used the term 'form of bifurcation' to describe a form of equilibrium belonging to two different linear series of differential equation, which does not correspond exactly to what it is referred nowadays as the solution to nonlinear differential equation. Furthermore, we must consider that the decade 1960–70 was in France the golden age of the Bourbaki group, whose philosophy was drastically opposed to Poincaré's way of thinking. The bedside books of every young student in maths at university were 'Éléments de mathématiques' [41]. Moreover, Jean Alexandre Dieudonné, one of the founders of the Bourbaki group, arrived at Nice in 1964, just after the University of Nice was created. He was the most prominent professor from the department of mathematics. During the academic year

[107] LIAPUNOV A.M. Sur les figures d'équilibre peu différentes des ellipsoïdes d'une masse liquide homogène douée d'un mouvement de rotation. Zap.Akad. Nauk. Vol 1 (1906) pp 1 - 225;puis (1908) pp 1 -175 (1912) pp 1 - 228 ; (1919) pp 1 - 112

[130] POINCARE H. Sur l'équilibre d'une masse fluide animée d'un mouvement de rotation. C.R.A.S. de Paris p346 ,9 fev. 1885 ; p 1068 , 20 avr. 1885 ; p 307 , 27 juil. 1885; p 857 ,12 avr. 1886 ; (note de MATTHIESSEN); p970,27 avr. 1886; p 622,7 mars 1887 (relative à Liapunoff); p 1571,4 juin 1888 ; p 655,16 mars 1903.

[131] ________ Idem. Acta Math. 1 (1885) pp 259 - 380

Figure 43. The references to Lyapunov and Poincaré in the bibliography of my Ph.D. thesis.

1969-70, I attended his lectures about the integration theory, with only a dozen of students. Three years after, upon Martin Zerner's request, he spent an entire afternoon in helping me on some problem related to my doctoral work. Poincaré's works were therefore not at all in my mind, although I eventually quoted him [150] in the analytical bibliography of my Ph.D. thesis (Refs. [130, 131] in [115]), together with the papers of A. M. Lyapunov (Ref. [107] in [115]) (see Figure 43). Abraham [110] referring to [197] mentions also that 'The word bifurcation is indexed by Thom on 18 pages', but René Thom, Fields medalist in 1958, was doing, in my opinion, theoretic mathematics at a level I never could reach. Moreover, his book in English was published during the middle of my thesis, and at this epoch, it took months, or rather years, for a new published book in English will be delivered to our library. Oddly, few years later beginning in 1982, Thom assisted me many times for my 'Thèse d'État', later for my Habilitation and to apply for higher position in the university.[6]

Hence, the first paper in French on this subject (however in the fields of mechanics) was published in 1971 by Gérard Iooss [88] who obtained later, in 1974, a permanent position in the University of Nice. After my Ph.D., I asked him for a new subject for my 'Thèse d'État'. We hence worked together on the famous dynamo problem explaining the origin of the magnetic earth field [90]. During this period, we both attended the 'International Conference on Mathematical Problems in Theoretical Physics', organized by the university of Roma, Italy (6–15 June 1977), with a team of researchers from the university of Nice (Claude Bardos, Alain Chenciner, Pierre Coulet, etc.). The opening talk of this conference was given by David Ruelle [158], famous for coining the name 'strange attractor' with his colleague Floris Takens [160]. In his talk, Ruelle conjectured that, for the Hénon attractor, the theoretical entropy should be equal to the characteristic exponent. It is how I learned the first example of chaotic and strange attractor (6) whose fractal properties were highlighted by Hénon and astonished the research community. Although Michel Hénon was astronomer at the Nice Observatory, few kilometres abroad from my office, I did not know him at this time.

Beyond bifurcation problem, my main interest was focused to discretization problem and finite element method (f.e.m.) in which nonlinear functions are approximated by piecewise linear ones. During the Roma conference, I tried unsuccessfully to apply the spirit of the f.e.m. to the Hénon attractor. Back to Nice on June 15 in the morning, during the defence of a Ph.D. by a colleague of mine, I eventually decided, using paper and pencil, to change the square function of the Hénon attractor, which is U shaped, into the absolute value function, which has a V shape, implying folding property. Few minutes after, in my office, I tested this modification, on my small desktop computer HP 9820 linked to the HP 9862 plotter. I shifted the parameter value a from 1.4 to 1.7 and b from 0.3 to 0.5 (why? I do not remember!) and plotted what is known today as the 'Lozi map'. Iooss and Chenciner encouraged me later to publish the formula (7) which appeared in the proceedings of a 'Meeting between physicists and mathematicians about nonlinear problems and their applications' held in Nice on 26–30 September 1977 [117]. This was for me the very beginning of my career in chaotic dynamical systems. I was convinced that few weeks would be enough to explain and give a proof of the structure of a so simple attractor, but I failed.

In the next 2 years, I attended two workshops on iteration theory: the first one on 21–23 May 1979 at La Garde-Freinet (a small town in the south of France), where Alain Arnéodo, Igor Gumowski, Michel Hénon, Christian Mira and Yves Pomeau were also present, and where Michał Misiurewicz, after some questions at the end of my talk, jumped on the stage, and on the blackboard gave some clues of his forthcoming results presented at the famous New York conference, 7 months later [135]. The second meeting is a summer school in physics on July 1979 in Cargèse (Corsica) where I used the genuine non differentiable Lozi map to prove the existence of one homoclinic point for a smooth version of it

$$T_{z_\epsilon}(x,y) = (y + 1 - az_\epsilon(x), bx), \tag{9}$$

$$\begin{cases} |z_\epsilon(x) - |x|| \leq \epsilon & \text{for } |x| < \epsilon, \\ z_\epsilon(x) = |x| & \text{for } |x| \geq \epsilon, \\ z_\epsilon(\cdot) \in C^k, & k \geq 2, \end{cases}$$

with ϵ small enough.

Then, applying a theorem proposed by Stephen Smale [192], I proved the existence of an invariant Cantor set for (9) [118]. Before these two workshops, I met Smale in Orsay, and at the end of his talk, I discussed with him about the validity of my proof. During the workshop in Cargèse, I sympathized with Joe Ford from Georgia Institute of Technology who, in the 70s, conceived the idea of disseminating *Nonlinear Science Abstracts*. Eventually his efforts resulted in the foundation in 1980 of the first journal devoted to nonlinear dynamics, *Physica D* [201]. I met him again in New York (see below) and during the Dynamics Days, Twente University, 25–26 May 1982. After Cargèse, took place the International Conference on Nonlinear Dynamics, patronized by the New-York Academy of Sciences, on 17–21 December 1979, in the Barbizon–Plaza Hotel theatre, near the south bound of Central Park. During this conference I am proud to have shaken the hand of Edward Lorenz,[7] the father of chaotic attractors, and I listen with a mix of anxiety and curiosity to the first proof by Misiurewicz for the existence of a chaotic attractor for the map I discovered 2 years and six months before [135]. The speakers and the session chairmen were the worldwide foremost figures of nonlinear analysis or experimenters in physical sciences or chemistry: M. V. Berry, M. J. Feigenbaum, J. Ford, J. P. Gollub, J. Guckenheimer, D. D. Joseph, A. Katok, J. B. Keller, N. Kopell, O. Lanford III, P. D. Lax, E. N. Lorenz, B. B. Mandelbrot, J. E. Marsden, M. Misiurewicz, J. Moser, S. Newhouse, Y. Pomeau, M. I. Rabinovich, D. Ruelle, M. Shub, Ya. G. Sinaï, S. Smale, E. A. Spiegel, H. L. Swinney, Y. Ueda, C. Vidal, S. M. Ulam, J. Yorke, and many others. From Nice, Uriel Frisch, Claude Froeschlé, Michel Hénon (these three from the Observatory of Nice), Gérard Iooss, Charles Tresser, as I remember, attended the conference. I knew personally some speakers: Ford, Frisch, Joseph, Marsden,

Misiurewicz, Moser, Newhouse, Pomeau, Ruelle, Smale, Spiegel, and Vidal. The topics of the conference was split into six sections: Turbulence, Ergodic and integrable behaviour, Physics and chemistry, Chaotic maps and flows, Chemical and fully developed turbulence, Strange attractors.

I was interested in the session devoted to turbulence due to the concept of strange attractor developed by Ruelle and Takens [159] to overcome the classical theory of Landau promoting an infinite sequence of bifurcations in tori of increasing dimension corresponding to quasiperiodic motion with an increasing number of frequencies [103]. The talk by Vidal [203] on the Belousov–Zhabotinsky reaction was of a so great interest for me, because I had visited in the early 1979 his Paul Pascal Institute in Bordeaux, during which Claude Lobry who was teaching there, at the mathematics department, introduced me to the team of chemists working on this oscillating reaction.

Of course, the talk by Misiurewicz [135] was a kind of ecstasy for the young researcher that I was. Surprisingly the talk was entitled 'The Lozi mapping has a hyperbolic structure', differently from the corresponding paper in the Annals, and nested in the session 'chaotic maps and flows', not in the last session 'strange attractors'. Even if it was possible to listen every lecture, I did not attend all, nor did Michel Hénon that I met many times visiting computer shops in the streets of New York city. One has to say that in 1979, there was a huge technological gap between France and the USA. USA was a kind of *magic kingdom* for me. I saw for the first time microwave oven and cloth dryer, 1 year before in California. Therefore, Hénon who was a geek before the word 'geek' was invented was looking for to buy a new pocket calculator, the HP 25 which was then unavailable in France.

Beside the interest of the scientific program, the most memorable memory of my stay in New York is to have crossed the city by night, after dinner, walking during 3 hours from Battery Park to Central Park, via Greenwich Village, with my colleague Claude Froeschlé! I remember also the first snowfall of the winter with 3.5" of snow fell, on Wednesday 19, on New York City which forced me to buy warm clothes and gloves! [210].

Let me go back to the early 70s and also to fill the gap between July 1977 and May 1979. As I said above, in France, the name bifurcation was largely unknown in the math community of analysts at the beginning of this decade. Referring to [110], the situation was greatly different in the USA. However, the situation was slowly changing in my country. At the end of March 1973, on the incentive of Geymonat, and with the help of Zerner, I invited to give a seminar in Nice, Paul H. Rabinowitz, a former student of Jürgen Moser (the M of the KAM theorem), who travelled from the 'Scuola Normale di Pisa' to Paris. He spoke about his last results 'Bifurcation from simple eigenvalues' [47]. However, the name bifurcation was not yet very popular, it was in competition with 'nonlinear eigenvalue problems', 'branching' and 'buckling' (and 'diramazione' in Italian).

On 16–25 June 1974, I attended a C.I.M.E. (Centro Internazionale Matematico Estivo, sponsorized by NATO) seminar in Varenna, Italy, entitled 'Eigenvalues of nonlinear problems.' The title of the talk given by Rabinowitz was 'Variational methods for nonlinear eigenvalue problems' [153]. The lecture by R. E. L. Turner was entitled 'Positive solutions of nonlinear eigenvalue problems' [199], Paul Fife, in his talk about 'Branching phenomena in fluid dynamics and chemical reaction-diffusion theory' [70], indicated that 'The study of the bifurcation and branching behaviour of problems in partial differential equations is a relatively new endeavour and is rapidly gaining momentum'.

The arrival of Iooss in our department of Mathematics in September 1974 was the turning point of the development of the bifurcation theory in Nice [89]. He was in touch with many researchers working on the topic including Klaus Kirchgässner, Daniel D. Joseph, Hans True, and many others. Alain Chenciner arrived the next year and stayed 3 years in Nice. Some researchers from the USA were visiting our department like J. F. K. Auchmuty [20] and H. B. Keller [95]. Nevertheless, the study of bifurcation phenomena was often marginally nested in workshop, seminar or conference of nonlinear analysis and application or physics. From 18 to 24 January 1976, I was invited by Kirchgässner with Iooss and Claude Bardos (visiting professor at Nice in 74–76), at Mathematisches Forschungsinstitut Oberwolfach for a seminar on 'Nichtlineare Funktionalanalysis und irhe Anwendungen',[8] organized by him, H. Amann (Bochum) and N. Bazley (Köln) [9]. I spoke about the numerical computation of branches of bifurcation [115] and Iooss about the 'Conjectures on the dynamo problem for the geomagnetic field' [90]. Some bifurcation problems were exposed by N. Chafee (Hopf Bifurcation and Arbitrary Perturbations of a Differential Equation), E. N. Dancer (Bifurcation from infinity), R. J. Magnus (The odd multiplicity criterion of bifurcation theory), and R. Nussbaum (A Hopf Global Bifurcation Theorem for Retarded Functional Differential Equations). There were only six talks on bifurcation among a total of 27.

All the walls of the conference room at Oberwolfach support gigantic blackboards. An amazing time during this seminar was when, at the end of the talk by Bardos (Hölder estimates and time regularity for the Euler equation), all these blackboards were covered with formulas and theorems and that Ernst Hölder (Mainz), the son of Otto Hölder, the mathematician whose name was given to the Hölder spaces, stood up, went to the blackboards and before each word 'Hölder' wrote either E. or O., with reference to his father's work, or to his own! Unfortunately, I did not understand the explanations he gave in German.

When as young student, I studied analysis, the name of the mathematicians given to the functional spaces, like Frechet, Banach, Hilbert, Hölder, Sobolev, Besov, had a scent of history, except for Sobolev (1908–1989) who, Céa reminded me from time to time, attended to the first colloquium on numerical analysis he had organized in Paimpol (France) in 1967.[9] However, concerning Hölder, I simply realized that he was not so far in the past as I imagined.

In September 1976, after 2 years as an assistant (the first grade of permanent position in the French university) I obtained one of the very few positions in France at C.N.R.S. as 'Attaché de Recherche' with Céa for superviser. In September 1977, the group around Iooss, Chenciner, Coste, Frisch, Coulet, Tresser, was sufficiently strong to organize a 'Meeting between physicists and mathematicians about non-linear problems and their applications', gathering around 30 speakers, including Hénon [86]. In 19–27 June 1978, this group attended the C.I.M.E seminar on Dynamical systems in Bressanone (Italia). Chenciner gave one of the five lectures during this event [42], along with the reputed researchers Guckenheimer, Moser, Misiurewicz and Newhouse. I sympathized with Newhouse and I drove him in my car to the nearby mountains of Trentino. On 10–11 July of the same year, Marsden [131] organized a 'Symposium on nonlinear analysis and mathematical physics' at the university of California, Berkeley, in the Evans building recently achieved. I attended this workshop together with Chenciner, Bardos, Iooss and Jean-Michel Lemaire (geometrician) from Nice. I remember that David Chillingworth who studied also the dynamo reversal of the magnetic Earth field, was present [43]. It was my first visit to this university and my first stay in the USA. During a picnic lunch, Marsden tried to initiate us to play frisbee; however, the French team was never able to play as a true team, like our American colleagues.

In October 1978, Céa, incited me, from an idea of J. L. Lions, to organize a conference on bifurcation in Nice. The Institute of mathematics provided all the funds. I managed a 4-day conference entitled 'Didactique et recherche en bifurcation', on December 4–7. Among the participants was Jean-Christophe Yoccoz (Introduction to the work of Pesin on differentiable manifolds). It was the first ever invitation of Yoccoz, future Fields medalist (1994), to give a talk in a conference. In fact, I tried to invite his well-known supervisor Michael Herman that I visited in Paris; however, Herman was not comfortable in travelling by train (at this time, travelling from Paris to Nice by air was very expensive). He said me 'I know a very gifted student in third year of the École [Normale Supérieure (Paris)], he can speak as well as me'.[10]

During the year 1979, I worked on the research of a realistic model for chaos, coming either from physics or chemistry. At this time, I encountered Claude Lobry, at the university of Bordeaux (he moved the next year to Nice), who was working with the group directed by Adolphe Pacault on the Belousov–Zhabotinsky reaction at the Paul Pascal Institute including Christian Vidal and Jean-Claude Roux. Then I collaborated for several years with him. We were interested in 'slow–fast' mathematical models for the Belousov–Zhabotinsky reaction, like the Brusselator model made of two ODEs developed by Nicolis and Prigogine in 1977 [142], the Oregonator devised by Field and Noyes working at the University of Oregon which is composed of five coupled elementary chemical stoichiometries [69], and the model with even more ODEs developed at Bordeaux [203,204]. To model the complex bifurcation diagram observed in very precise chemical experiments in Bordeaux, I introduced the idea of bifurcation of 'motifs' (patterns) in a note to the Académie des Sciences de Paris, presented by Thom [119]. This was my first indirect exchange with Thom. I developed this model with Lobry [112], and on 29 June 1983, I defended my 'Thèse d'État' that Thom accepted to review [121]. He was very interested by the ontology of the slow-fast metaphoric model I introduced.

I had also a long scientific relationship with José Argémi (1933–1985) [17] who worked at C.N.R.S. in Marseilles and his doctoral student, Bruno Rossetto (now professor emeritus) from the University of Toulon. Argémi was a specialist in slow–fast ODEs systems. He coined *co-folded line* and *pseudo-singular point*, very useful mathematical tools for such systems. During 7–17 September 1982, the three of us attended the colloquium 'Logos et théorie des Catastrophes' organized by Jean Petitot to celebrate René Thom in Cerisy–la–Salle [147]. Numerous renowned mathematicians and physicists like Michael Berry, Chenciner, Chillingworth, Adrien Douady, Georges Lochak, Ruelle, Timothy Poston, Bernard Teissier, and Sir Cristopher Zeeman gave a talk. All those, and scientists of other disciplines, philosophers, psychoanalysts, linguists, were very impressed by Thom. Me too much more. It was my second in-person meeting with him. The first one took place in May of the same year, during a meeting in Toulouse [120]. It was precisely during the colloquium in Cerisy-la-Salle, where I gave a talk entitled 'Bifurcation de motifs dans la réaction de Belousov–Zhabotinsky', that Thom accepted to review my 'Thèse d'État'. During the speech of conclusion talk Thom said [147] (p. 513)

'[⋯] So that even where there is apparently chaos, there are always morphological elements to study. The presentations that we have heard subsequently, for example that of Lozi very recently, are in exactly the same direction: there is always interest in trying to understand how things evolve morphologically and, if we cannot, of course, one can always dive into statistics.[11]'

Thom was very kind to suggest to Ruelle to invite me to give a talk on the same subject during a meeting on week turbulence held at I.H.E.S. on 18–19 November 1983. In May 1989, I was one of the height mathematicians (with M. Chaperon, A. Chenciner, J. Martinet, J.P. Ramis, P. Moussa, R. Moussu, F. Pham,) invited to pay tribute to Thom on his retirement during the 'Rencontre entre mathématiciens et physiciens théoriciens' in Strasbourg [122,157].

During spring 1981, I received the surprise visit of Hiroshi Fujii [72] from Kyoto university, who was spending 1 year in INRIA near Paris. He was interested by my Ph.D. dissertation. He gave me the names and addresses of Shigehiro Ushiki, and his supervisor Masaya Yamaguti [207], both also from Kyoto university. I sent to Ushiki some papers on July 1981 who, in turn, sent me back his own ones on August 1981. I met him 1 year later, during the meeting in Toulouse. He was fluent in French. Then we worked very closely for 12 years, first on slow–fast systems, and hereafter on the Chua circuit.

In 1983, thanks to Professor Yamaguti, I got 3 weeks grant from the Japanese Society for the Promotion of Science for an internship at the Faculty of Science of the imperial University of Kyoto. I met also Hiroshi Kawakami who I knew since the conference in Toulouse in 1982 and with whom I will discover the Alpazur oscillator in 1991 [93,94]. On 31 August to 1st September, I travelled to Hiroshima to visit Masayasu Mimura. Discussing with him, I suddenly realized that my vocation in research will be building mathematical models, rather than proving theorems, what I did until today. I turned back to Kyoto on the precise day the Korean Air Lines Flight 007 which was a scheduled flight from New York City to Seoul via Anchorage Alaska, was shot down by a Soviet Su-15 interceptor. All 269 passengers and crew aboard were killed, arousing strong indignation in Japan. The last week of my stay in Japan was devoted to the IUTAM symposium on 'turbulence and chaotic phenomena in fluids' (September 5–10) in Kyoto (see Figure 44). During the symposium, Yamaguti introduced me to Benoît Mandelbrot who remembered meeting me at Cerisy-la-Salle.

After my first stay in Japan, I could find some grants from C.N.R.S. allowing Ushiki to spend several months every year in Nice between 1984 and 1992. We worked on slow–fast models mimicking the Lorenz attractor. The Lorenz model does not belong to this class of ODE. To link this model to slow–fast equations, I proposed an ideal geometrical representation of the Lorenz attractor on a cusp (see Figure 45) [121]. Hiroe Oka and Hiroshi Kokubu, two Ph.D. students of Yamaguti and Ushiki, investigating systematically chaotic attractors modelled by constrained systems, using implicit differential equation (also called generalized vector field), found the set of equations

$$\begin{cases} \epsilon\dot{x} = y - \alpha xz - x^3, \\ \dot{y} = Ax + By + \beta xz, \\ \dot{z} = Cz + \gamma x^2, \end{cases} \tag{10}$$

that produces a similar attractor (see Figure 46) to the model I draw by hand. With Ushiki we studied thoroughly equations linked to (10) in this paradigm, using some tools introduced by Argémi: pseudo singular points, fold and co-fold curves, together with others

Figure 44. Masaya Yamaguti, Shigehiro Ushiki and René Lozi during the IUTAM symposium on 'Turbulence and chaotic phenomena in fluids' (5–10 September 1983) in Kyoto.

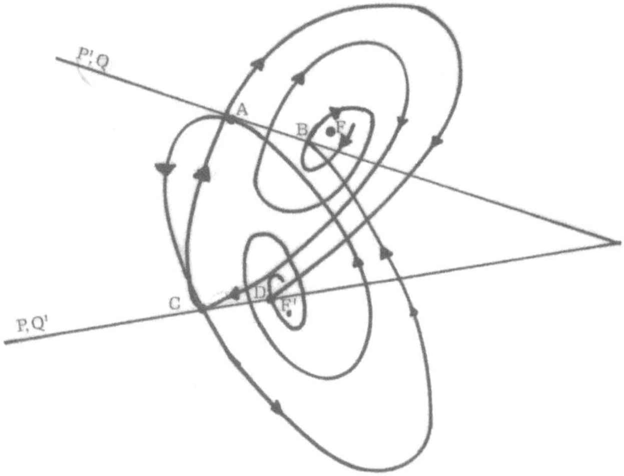

Figure 45. Sketch of a slow–fast construction for the Lorenz attractor.

defined by Takens: induced and reduced vector field [196]. Especially, we studied the family of differential equations constrained on a cusp surface (11), similar to (10) [200].

$$\begin{cases} \epsilon\dot{X} = Y + \dfrac{3}{4}XZ - X^3, \\ \dot{Y} = \tau^2 AX + \tau BY + \dfrac{3\tau}{4\alpha}\beta XZ, \\ \dot{Z} = \tau CZ - \dfrac{4\alpha\tau}{3a}\gamma X^2, \end{cases} \tag{11}$$

My personal search for a good realistic model for chaos comes to an end with the knowledge of the chaotic electric system discovered by Leon Chua in August 1983 when he was visiting professor in the laboratory of Takashi Matsumoto, at Waseda University, Tokyo [132]. I first presented the results on constrained Lorenz model (11) during the conference on chaotic phenomena at Issy-les-Moulineaux near Paris, on 3–5 December

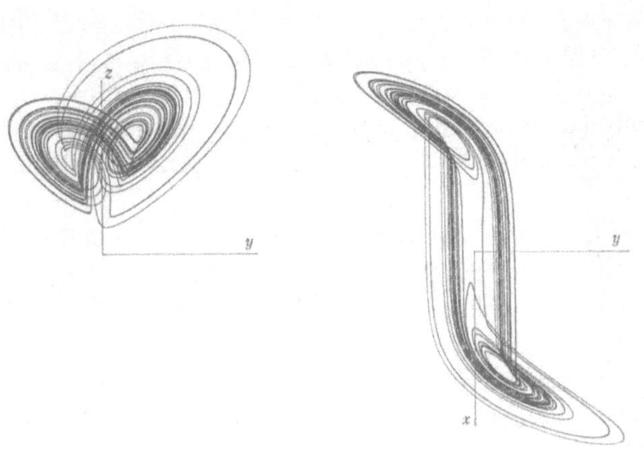

Figure 46. Chaotic attractor produced by the slow–fast system (10). Parameter values: $\epsilon = 0.03$, $A = 0.7, B = 0.7, C = -1.0, a = 0.7, \beta = -1.0$ and $\gamma = 1.0$.

Figure 47. Leon Chua and René Lozi on West Lake in Hangzhou, China, in 2014.

1986 [124]. During this conference, Leon Chua gave a talk on his electric circuit [44]. It was the first time I met him. One year later, on 22 December, travelling in the south of France, he visited me briefly in Nice and invited me to visit him. This was the beginning of a very long cooperation which lasts more than thirty height years with Chua (see Figure 47). With Ushiki, we quickly switched our attention to the Chua circuit, because, from our knowledge, it was the first realistic model for chaos, with only three variables, the minimum number required to produce chaotic solutions.

To conclude this recollection, I would like to mention my relationships with four professors (one Ukranian and three Russians) who made major contributions to chaotic dynamical systems during the decades 1960–80: Vadim Anishenko, Vladimir Belykh (Belykh's attractor), Alexander N. Sharkovsky (Sharkovsky's order), Leonid Shilnikov (homoclinic bifurcations) and my brief encounter with Romen Plykin and Evgenii Sataev.

Before the fall of the Berlin Wall on 9 November 1989, it was uneasy to meet researchers from USSR. Even if it was possible for them to travel to West (M. I. Rabinovich, Ya. G.

Sinaï and few other Russians from the Landau Institute, Moscow, attended the New York conference 10 years before). Very often they did not get their visa on time or they had a lack of money to travel, and they missed the conference. Moreover, most of them did not speak English. They published in Russian. It took years for few of their results to reach the West community through translation.[12] A prototypal example of the difficulty to make known their results is the differential treatment of the Li and Yorke theorem 'period three implies chaos' published in 1975 [111] and the 'Sharkovsky's order' published in an Ukrainian journal in 1964 [186,187]. In the 1970s, the studies of chaotic phenomena were soaring. Many amazing results were published like the Feigenbaum's constants for one-dimensional maps [66], and the Hénon map in dimension 2 or the Lorenz and the Rössler attractors for ODEs. In many seminars, the result of Li and Yorke was hailed as an important mathematical discovery. It is only 2 years after, in 1977, that the Sharkovsky's order was published in English by Stefan [194]. After that, this order overshadowed the Li–Yorke theorem which was perceived as a peculiar case of this order, published 9 years before. However, it is not exactly the case as Yorke himself reminded me during the banquet of the International Conference on Complex Systems and Applications, in Le Havre on 28 June 2009. I was sited in front of him and near Leon Chua. Yorke said that this order was not giving any result of chaotic properties of continuous function, contrarily to the theorem he proved with Li 34 years before, their paper analysing a situation in which the sequence of iterates of a map is non-periodic and might be called 'chaotic'.

It is during my third 1-month visit at Berkeley from November to December 1993, that I have had the opportunity to meet both Shilnikov and Sharkovsky. At this time, Chua was planning the special workshop 'Chua's circuit: chaotic phenomena and applications' during the International Symposium on Nonlinear Theory and its Application (Nolta'93) in Honolulu, December 5–10. As usual, he was a terrific organizer. To get a large audience, he invited in Berkeley many speakers to stay 1 month in his laboratory and chartered a flight for all of them to Honolulu. Then, early in the morning, Sunday 5 December 1993, with Sharkovsky, Shilnikov, Anishenko (Saratov, Russia), Martin Hasler (Lausanne, Switzerland), Maciej Orgozalek (Krakow, Poland), and many others, we started our trip until to be lodged at the Sheraton Waikiki, Honolulu for 1 week (see Figure 48)! I did not published with Vadim Anishenko, however, I met him several times after Hawaii, in Warsaw, Nizhny Novgorod, etc., and discussed with him with great pleasure.

Diana Chua, Leon's wife organized our dispatching in the rooms. I was not smoking, nor Sharkovsky, contrarily to Anishenko and Shilnikov who shared the same room. Therefore, she planned that I must share the room with Sharkovsky. Consequently, we both discussed a lot during the week, even if he was not very fluent in English. This was the beginning of a many long years of cooperation! During the following years, with Bruno Rossetto and Ahmed Aziz-Aloui (my first Ph.D. student, now full professor in Normandy university), we found several grants between 1994 and 2002 to invite Sharkovsky in Toulon, Le Havre and Nice. I published three papers, two with him, Aziz-Alaoui and A. D. Fedorenko from Kiev, and one with him and J. Sousa Ramos from Lisbon. Sharkovsky invited me to the Summer school on dynamical system, Katsiveli, Crimea, in August 2000, and to the European Conference on Iteration Theory (ECIT) held in Yalta, Crimea, in September 2008. I worked with him and Sousa Ramos in Lisbon in May 2004. I last met him during the ECIT 2012 in Ponta del Gada, Azores in September 2012, and received his last personal mail few months before his death, on 5 July 2022.

Figure 48. Alexander Nikolaevich Sharkovskii, Leonid Shilnikov and René Lozi upon their arrival in Honolulu Airport on December 5, 1993.

Figure 49. From left to right, René Lozi, Romen Vasilievich Plykin and Evgenii Sataaev on August 2000, during the Summer school on dynamical system, Katsiveli, Crimea.

I had the pleasure, during the 10 days spent in Katsiveli, to meet two other important figures of the dynamical systems in USSR: Evgenii Sataaev and Romen Vasilievich Plykin (1935–2010) [99], both from Obninsk State Technical University (see Figure 49). The venue of the Conference, the building of the Ukrainian academy of sciences, was standing near the beach, on the Black Sea. The attendants were spending the free long hours in the afternoon, swimming during this hot summer, before the scheduled evening sessions. The banquet took place on a pontoon during the night of August 28th. I strongly disappointed Plykin by not accepting his insistent invitation to dive down with him from the pontoon into the Black Sea, around midnight, after the traditional speeches washed down with vodka!

The year after the Honolulu workshop, Shilnikov visited Christian Mira in Toulouse in 1994. From there, he phoned me to visit my Institute. I arranged a stay for him and his

Figure 50. Vladimir Belykh and René Lozi, singing popular songs of Bulat Okudzhava during the banquet of the Conference NOMA–2017 on June 2017, in Nizhny Novgorod.

assistant Dimitri Turaev from Nizhny Novgorod, on 18–23 February 1994. Shilnikov gave a talk on 18 Friday. The next day, I drove them to visit Monte-Carlo. Shilnikov absolutely wanted to have a picture of him and me in front of the famous Casino where Anton Tchekov and his friend Ignaty Potapenko lost all their money in 1898. Next, visiting the Oceanographic Museum of Monaco, he told me about his fishing trips in Siberia north of the polar circle. To work together we applied twice unsuccessfully for INTAS programs. I never met him again. I attended the conference Dynamics, Bifurcations and Strange Attractors dedicated to his memory, held at the Lobachevsky State University of Nizhny Novgorod, Russia, on 1–5 July 2013, where I met for the first time Vladimir Belykh. I was very impressed to finally meet him after knowing his attractor for decades. Since then, we have become friends and shared our admiration for the songs of Bulat Okudzhava we sang together (see Figure 50).

Fifty years after the beginning of my academic life, I realize that I was incredibly lucky to begin my career as the same time as the emergence of chaos studies in dynamical systems. I had a great feeling of freedom, being able to discover a new world of extreme richness, as undoubtedly explorers of America had it hundreds of years before me. I had also the privilege to discuss with outstanding researchers, among them five Fields medalists: Thom, Yoccoz, Smale, Alain Connes (within the national committee of C.N.R.S. to which I belonged for 9 years) and Pierre-Louis Lions (during seminars in Paris), one Abel prize laureate: Yves Meyer (in the same committee), and many renowned scientists: Belykh, Chua, Hénon, Lorenz, Rössler, Sharkovsky, Shilnikov, Yorke, etc. In this recollection, I have described roughly half of my research life, between 1972 and the beginning of the third millennium. Since the 2000s, I focused my research on concrete applications of chaos theory like pseudorandom number generators [125], cryptography [91], global optimization [84], etc. I explored the new ideas of fractional mappings [97] and fractional derivative [1,2], and recently built model of Tuberculosis with Saber Elaydi. I also had a 30-year long cooperation with Abul Hassan Siddiqi, president of the Indian Society of Applied and Industrial Mathematics (ISIAM) which I helped him establish in 1990, and Pammy Manchanda Secretary of this society belonging to The International Council for Industrial and Applied Mathematics (ICIAM) [129,130]. Since 2013 I have published many paper with Guanrong

Figure 51. Ron Chen and René Lozi, in the south of France in 2013.

(Ron) Chen, Editor in Chief of the *International Journal of Bifurcation and Chaos*, founded by Chua (see Figure 51).

Nowadays, the Belykh, Hénon, Lorenz, Lozi maps, the Chua, Lorenz, Rössler systems and many others are routinely used for many applied purposes. They are living their own life, without the help of their 'fathers', like emancipated children.[13] The most amazing fact for me is that, in 1975 and during many years, I tried unsuccessfully to find an efficient algorithm for global optimization. Now, the map I introduced in an entirely different aim is often used within evolutionary algorithms to find global solution of optimization problems involving hundreds of variables [123]. This emphasizes the intimate unity of mathematics.

I am indebted to the following friends and colleagues who helped to refresh my memory: Ahmed Aziz-Alaoui, Lenore Blum (Carnegie Mellon University, Pittsburgh), Jean Céa, David Chillingworth, Brian Quigley (Head, Engineering & Physical Sciences Division/Mathematics, Statistics & Computer Science Librarian, University of California, Berkeley), Jean Petitot, Paul H. Rabinowitz, Shigehiro Ushiki, Monique and Jeanne Zerner, wife and daughter of Martin Zerner, to Stéphane Junca for shrewd remarks on continuation methods, to Leon Chua for stimulating discussions on the history of nonlinear circuit theory.

11. Conclusion

We often hear that the language used by mathematicians is universal. This means that they write their equations with the same symbols, regardless of the alphabets, ideograms, or other symbols they use to write the text in their theorems. Moreover, they employ the same objects to describe their reasoning: group, functions, topologies, spaces, etc., and the same logic for their proofs. We could infer from these similarities that they think the same way in their lives. A question then arises: are there similarities in the paths to their university careers and the way in which they chose it?

We asked nine experienced, long-career mathematicians specialized in discrete dynamical systems, from five countries (seven if we include their country of origin), to describe how they started their careers long time ago (8 of them are over 70 years old but remain very active). We can point out some similarities and diversities in their backgrounds.

Mathematics was always present from their youth, to varying degrees. It was the main focus of Michał Misiurewicz since childhood. His first encounter with this discipline which he remembered (without having the memory of how old he was then) was counting the cars of a moving train. The urge to count various things remained with him for his whole life until now. By the age of 5, when he learned to read, his favourite book was the math textbook for the first grade. Later he was fascinated by the book *Lîlâvatî* by Szczepan Jeleński. In the high school, he found a lot of interesting problems in the Hungarian journal *Matematikai Lapok*, that a friend of him was subscribing. He participated in two International Mathematics Olympiads during his high school years (calling them a 'great adventure').

Jim Michael Cushing remembered that he was so enamoured with mathematics when he was a student in the late 1950s, that he would check out the textbook from the local Carnegie library during the summer recess before he took a maths course in order to teach himself, as best he could, the topic prior to the upcoming school year. In this way, he was introduced to not only the standard courses in algebra, but plane geometry (using Euclid's Elements), solid geometry, trigonometry, and single variable calculus. During summer school recesses, he would also occupy himself with applications of mathematics (measuring heights of trees using trigonometry, etc.), which was the beginning of a lifelong love of applied mathematics.

René Lozi indicated that like Misiurewicz, he was counting many things around him in his early childhood. During his preschool year, his teacher did not let him play during recess. She made him do math exercises that the other students were not able to do. When he was 13, one of his cousins, an engineer, introduced him to the base 10 logarithms. He was fascinated and constructed a cardboard slide rule, grading it with his older sister's logarithms chart, and used it at school, much to the astonishment of his classmates and the teacher.

Saber Elaydi sought solace and inspiration within the realm of mathematics in a world plagued by uncertainty and limited resources. The numbers and theorems became his refuge, offering a sanctuary where possibilities knew no boundaries. Armed with unwavering determination and an unquenchable thirst for learning, he delved into the world of mathematics, voraciously consuming every piece of knowledge he could find.

The decision to study mathematics was taken by Laura Gardini when she was 12 years old. There was not a specific event that lead to her decision, she simply loved mathematics and realized that she was happy studying it, and that it was easy for her. At the lyceum she was always the first in mathematics, and at the University in Bologna she was always among the best students in her class.

Vladimir Belykh was a gifted child in all areas and after completing his primary education he was accepted into an elite military boarding school in St. Petersburg. However, a massive reorganization of the Soviet army in 1960 led to the closure of this school. So he graduated from high school in his hometown and was after accepted to Gorky State University where he studied mathematics and physics in the chair of theory of oscillations and automatic control, named of its founding director, Alexander A. Andronov.

However, mathematics was for some in competition with other topics. Eckehard Schöll recalled that as a schoolboy he was interested in a very broad spectrum of subjects, ranging from Latin, English, French, German via History, Music, Art to Mathematics and Physics. He also loved repairing old clocks, building radios, and solving maths puzzles. But when he passed the final school exam (Abitur) with best grades in all subjects in 1970, it was clear that he wanted to study physics at the University.

Lozi was fascinated by particle physics and dreamed to make researches in theoretical physics, without knowing which courses to attend to reach this goal. However, when graduated with a certificate in fields and particles, he no longer liked this discipline but he discovered numerical analysis.

Galina Strelkova said that during her high school years, her interests were quite broad. She liked mathematics and physics but was very attracted to medicine, she also seriously studied music. She expected that after graduating from school she would certainly continue her studies either at a medical institute or at the Conservatory. However, by the time she graduated from high school, she definitely decided to become a school teacher in mathematics, physics and computer science. She was incredibly fond of mathematics, but to a greater extent not as a beautiful and abstract science, but as an essential tool for describing natural phenomena and events and solving the most interesting physical problems.

Lyudmila Efremova when she was in school not only loved solving difficult mathematical problems but also made attempts to write music (she also studied at a music school). Therefore, the choice of the further way was very difficult for her. Nevertheless, in the struggle between her two serious hobbies, mathematics and music, mathematics won, and she entered the Faculty of Mechanics and Mathematics of Nizhny Novgorod University.

A second similarity is music that was of great importance to Jim Michael Cushing, Lyudmila Efremova, Eckehard Schöll, and Galina Strelkova. Cushing spent a great deal of time studying piano performance. Upon graduation from high school in 1960 he had various opportunities that included a scholarship offer in piano performance. During his travels on this long career path, he never stopped playing the piano. He even worked with string students and their professors in the University's music department for several years. At school, one of Efremova's compositions for the symphony orchestra was performed at a major concert at her music school. Later she listened lectures by famous musicologists at the Gorky Conservatory on music theory, V.G. Blinova and M.M. Valentinov the main director of the Gorky Opera and Ballet Theater. During his undergraduate studies at the Universities of Stuttgart and Tübingen, Schöll studied special courses in history of arts and musicology and all through his life, he played the piano and sang in choirs, performing many concerts. Strelkova during all her school years seriously studied music (piano and vocals), composed music for poetry, took part in various competitions and became the winner of them several times.

Another key common point appears when reading this memory: for these nine mathematicians, many outstanding researchers had a great influence at the beginning and during their careers. They met them, either in person or by reading their books and several times interacted with them. These mathematicians also interacted with each other, forming an intriguing social network worldwide, across all borders of nationality and languages.

We can find the names of Abraham, Andronov (Andronov–Hopf bifurcation), Anosov (Anosov diffeomorphisms, Anosov flow), Arnold (nominated for the 1974 Fields Medal), Bautin (Bautin bifurcation and Bautin theorem), Dieudonné (one of the founders of the

Bourbaki group), Ginzburg (Nobel Prize of physics 2003), Hölder (Hölders' space), Kolmogorov (Kolmogorov or Kolmogorov–Sinai entropy), Leontovich–Andronova, Mazur (Stefan Banach Prize, 1949), Mira, Neimark (Neimark–Saker bifurcation), Novikov (Fields medalist, 1970), Pesin (Pesin theory or theory of non-uniformly hyperbolic dynamical systems), Plykin (Plykin attractor), Sousa Ramos, Saker, Sataev, Shilnikov (Shilnikov bifurcation), Sinai (Abel Prize, 2014), Sharkovsky (Sharkovsky' order), Sobolev (Sobolev's space), Strelcyn, Szlenk (of Szlenk indices fame), Thom (Fields medalist, 1958) and many other scholars in the recollections of Vladimir Belykh, Lyudmila Efremova, Laura Gardini, René Lozi, and Michał Misiurewicz.

For example, René Lozi said that in 1969–70, he was a student of Jean Alexandre Dieudonné at the university of Nice. His Ph.D. advisor Jean Céa recalled him recently the visit of Sobolev who was fluent in French and knew a lot about the history of France, at the first colloquium on numerical analysis he had organized in Paimpol (France) in 1967. In 1976, at Oberwolfach he listened a conference of Ersnt Hölder, the son of Otto Hölder, the mathematician whose name was given to the Hölder spaces.

Vladimir Belykh followed the teachings of Yuri I. Neimark, Dmitry A. Gudkov, Nickolai A. Fufaev, and Nickolai A. Zheleztsov, the brilliant team of Andronov's disciples. He was accepted into the Division of Differential Equations headed by Evgeniya Alexandrovna Andronova–Leontovich, Andronov's spouse and close collaborator and a sister of another celebrated Soviet physicist, Mikhail A. Leontovich of Moscow State University (Lenin Prize 1958). He became friend with Leonid P. Shilnikov who worked in the same division, and who one day explained to him his famous theorem on the bifurcation of a homoclinic orbit of a saddle focus. In 1974, Belykh shared the prestigious Lenin Komsomol Prize (a Soviet science award for young researchers under 33) with his colleagues Vladimir D. Shalfeev and Valery P. Ponomarenko. In 1976, Valentin (Valya) Afraimovich, a Shilnikov's pupil, explained to him the action of the discontinuous Poincaré return map of the Lorenz attractor he constructed in a qualitative, implicit form with Shilnikov and Vadim Bykov. Belykh realized soon that his model of a discrete-time phase-locked loop (PLL) in the case of a piecewise linear nonlinearity was comparable to their map and showed this model with a chaotic attractor to them. Afraimovich said: 'Here is the Belykh attractor' and later informed Sinai about this new object of potential interest for the ergodic theory. Belykh also gave a talk at Anosov's seminar, and Dmitry V. Anosov's pupils (Yakov Pesin, Evgeny Sataev, and Nikolai Chernov) began their analysis of the ergodic properties of such map. Anosov strongly recommended him to include the description of this map and its detailed analysis in his second doctoral thesis. Later, in 1983, Anosov served as an opponent for his 'Doctor of Sciences' degree thesis. The same year he moved to the Institute of Water Transport Engineers in Nizhny Novgorod, where he took the position of its Mathematics Department's head, which became available after the retirement of Nikolaï N. Bautin.

Lyudmila Efremova listened to the lectures of Sergueï Petrovitch Novikov, Yakov Grigorievitch Sinai, and physicists Vitaly Lazarevitch Ginzburg, Andrei Viktorovich Gaponov-Grekhov (Ph. D. student of Andronov) in Nizhny Novgorod. Her survey on skew products of maps of an interval in dimension two was published in *Russian Mathematical Surveys* and recommended to Sinai by Professor Stepin. In 2009 and again in 2011, D.V. Anosov was the scientific supervisor of her internship at the Steklov Mathematical Institute of the Russian Academy of Sciences. She interacted with him until the beginning of 2014, and in August 2014 when he died. In the autumn of 2013, although he was feeling unwell, he recommended her to professor Stepin who helped her until the defence of her Doctor of Sciences Thesis in 2018 that she defended at the Institute of Information Transmission Problems of the Russian Academy of Sciences the council of which was chaired by Professor Sinai. The official referees were Belykh (one of the most prominent representatives of the school of Andronov), A. Zhirov (a former Ph.D. student of Plykin) and the famous Russian topologist S. Bogatyi.

Michał Misiurewicz described the origin of the dynamical systems in Warsaw, when Stanisław Mazur sent two young Warsaw mathematicians, Wiesław Szlenk and Karol Krzyżewski to Andrey Kolmogorov who addressed them to Sinai. They returned to Warsaw after learning the basics of the dynamical systems. Sinai followed them and gave a cycle of lectures in ergodic theory. Another member of this Warsaw group was Jan Maria Strelcyn, who was taking notes from Sinai's lectures, that were later available as a booklet. Around 1969, Strelcyn left Poland for France, but Szlenk, and later also Krzyżewski, started to build a group in this new area of mathematics. The first two students from the new generation were Michał Krych and Misiurewicz. Next students that joined were Feliks Przytycki and Maciej Wojtkowski, and after them many others. In 1977, Misiurewicz wrote with Szlenk a paper, where they proved some basic results in the theory of interval maps.

In 1994, during a visit to the University of Southern California, Saber Elaydi met Robert Sacker (the author of the famed Neimark–Sacker bifurcation) and started working jointly. He met Jim Yorke at the fifth ICDEA, which was held in Temuco, Chili, in January 2000.

In 1975, Belykh with his Ph.D. student Vladimir I. Nekorkin, rigorously proved the existence of Shilnikov's chaos in a three-dimensional autonomous phase system.

In 1994, at Katsiveli, in Crimea, at the conference 'Differential Equations, Bifurcations and Chaos' organized by Yuri and Vladimir Maistrenko, and Iryna Sushko (Ph.D. student of Y. Maistrenko and Alexander Sharkovsky) Laura Gardini met Leonid Shilnikov and his son Andrey in 1994, and soon she became friends with them. They visited her in 1995 in Urbino and asked her to explain her result on homoclinic bifurcations of expanding cycles in $\mathbb{R}^n$. In September 1996, the famous *Blue sky catastrophe model* was presented (for the first time outside Nizhny Novgorod) by Leonid Shilnikov at the *European Conference on Iteration Theory* that she organized in Urbino.

Lozi met Leonid Shilnikov and Alexander Sharkovsky in the Leon Chua' laboratory in San Francisco in 1993. Shilnikov after having worked with Christian Mira in Toulouse in 1994 visited him with Dimitri Turaev in the university of Nice in 1994 and gave a talk in the mathematics department.

The study of some books by Thom, Zeeman, Postom and Stewart, Collet and Eckmann, Guckenheimer and Holmes decided Laura Gardini to spend her academic life in the research of dynamical systems. Thom accepted to review the 'Thèse d'État' of Lozi in 1983 being interested by the ontology of the slow–fast metaphoric model of the Belousov–Zhabotinsky reaction proposed in it. He suggested to Ruelle to invite him to give a talk at I.H.E.S the same year. In 1986, Thom reviewed the Lozi's memoir for the recently created Habilitation diploma and assisted him many times in his applications to higher position in the university. In 1990, Lozi was one of the height mathematicians (with M. Chaperon, A. Chenciner, J. Martinet, J.P. Ramis, P. Moussa, R. Moussu, F. Pham,) invited to pay tribute to Thom on his retirement during the 'Rencontre entre mathématiciens et physiciens théoriciens' in Strasbourg.

Belykh met first Alexander N. Sharkovsky in 1965 when he presented the proof of his famous numbers ordering at a research seminar run by Neimark in Gorki. In 1972, Sharkovsky agreed to be external reviewer of his Ph.D. thesis initiated in 1968 under the guidance of Lyudmila N. Belyustina at NII PMK's Division of Dynamical Systems and Control Theory, headed by Neimark.

N. F. Otrokov and his former graduate student Dr. Rakhmankulov were the scientific supervisors of Lyudmila Efremova. N. F. Otrokov was a major specialist in the theory of limit cycles of differential equations and he defended his Ph.D. thesis under Andronov's supervision in 1940. When she was a second-year student, Rakhmankulov invited her to read the Sharkovsky's paper. Now after many years she can say that her first impression of this theorem determined her scientific biography. Throughout her work in mathematics, from time to time she returns to various generalizations of it. In 1977, when she was a Ph.D. student of the second year, Sharkovsky under the request of Otrokov invited her to present her research at his seminar at the Institute of Mathematics of the Academy of Sciences of Ukraine. After that, she has had the opportunity to visit several times Sharkovsky in Kiev to present and discuss new results. She had also many interesting discussions with Professors Plykin and Szlenk. She recalls, with great warmth, the very lively discussions of her research with Belykh who was with Sharkovsky, the official referees of her Ph. D. thesis.

After meeting Alexander Sharkovsky in 1993, Lozi began a long collaboration with him. Between 1994 and 2002, he found several grants to invite Sharkovsky in France (in Nice, Le Havre and Toulon). He published three papers, two with him and A. Aziz–Alaoui from Le Havre and A. D. Fedorenko from Kiev, and one with him and J. Sousa Ramos from Lisbon. Lozi was invited by Sharkovsky to the Summer school on dynamical system, Katsiveli, Crimea, in August 2000, where he met Romen Plykin and Evgenii Sataev (both from Obninsk State Technical University), and to the European Conference on Iteration Theory (ECIT) held in Yalta, Crimea, in September 2008. He worked with Sharkovsky and Sousa Ramos in Lisbon in May 2004. He last met him during the ECIT 2012 in Ponta del Gada, Azores in September 2012, and received his last personal mail few months before is death, on 5 July 2022.

In the spring 1991, Laura Gardini was invited by Christian Mira to visit him in Toulouse. That started her long-time friendship and collaboration with him. The same year in September, he invited her to participate in the *European Conference on Iteration Theory* in Lisbon, organized by Sousa Ramos where she met Jaroslav Smital, Sergei Kolyada, Ludomir Snoha, Francisco Balibrea and Jaume Llibre. In 1992, Laura Gardini presented some results on particular homoclinic bifurcations at two more conferences, one in Germany and one in Italy where, besides Christian Mira, also Ralph Abraham was invited, and with him she also started a long friendship and collaboration. He encouraged her research work and was fascinated by the dynamics of noninvertible maps. He was visiting professor in Urbino for some periods (six times from 1992 to 2000). She also worked closely with Christos Frouzakis and Ioannis Kevrekidis whom she met in Minneapolis in March 1995.

Not only others outstanding mathematicians like Anishchenko, Bownds, Chillingworth, Conley, Martin, Yorke, Milne–Thomson had a great influence on the authors of this article, there are also the astronomer Hénon, biologists (Hájek, Levin), ecologist (MacArthur, May, Rosenzweig), physicists (Chua, Haken, Helleman, Heisenberg, Kocarev, Martin, Ruelle, Shaw) and meteorologist (Lorenz) and many others who inspired the authors of this article.

In 1976, Michel Hénon published a paper about his attractor. After spending about half a year on trying to prove that it is really an attractor, Misiurewicz decided that temporarily he will investigate interval maps and only when he gain substantial knowledge of them, he will return to the Hénon's attractor. In fact, he remained stuck with one-dimensional dynamics. In 1977, spurred by a presentation by David Ruelle of the Hénon attractor, Lozi proposed a piecewise linear model of this attractor.

In 1983, under the guidance of Otomar Hajek (von Humboldt award at TU Darmstadt), Saber Elaydi wrote several papers on the dichotomy and trichotomy of nonautonomous differential equations. With the help of his friend Gerry Ladas, he established the *Journal of Difference Equations and Applications* (JDEA) in 1994. They organized the first international conference (ICDEA I) on difference equations in San Antonio, TX, in May 1994, which has since been an annual conference. Elaydi met Jim Yorke at the fifth ICDEA, which was held in Temuco, Chili, in January 2000. In 2001, with Bernd Aulbach they created the International Society of Difference Equations (ISDE). In the spring of 1995, he suggested to the Trinity University to invite Lord Robert May (Copley medalist, 2007) to give a talk as part of the 'Distinguished Science Lectures Series'. The day after the conference, he had the pleasure of asking him several questions about chaos and the evolution of species, and discussed with him his new book project on discrete dynamical systems.

Jim Michael Cushing recalls that his Ph.D. dissertation advisor was Monroe H. Martin (Guggenheim Fellowship 1959), who at the time was the director of the Institute of Fluid Dynamics and former chair of the Mathematics Department at Maryland. Martin took on dissertation students only one at a time (one of whom, before Cushing, was Simon A. Levin (National Medal of Science – 2016), who is currently the director of the Center for BioComplexity at Princeton University). The years of Cushing at Maryland (1964–1968) were mathematically quite exciting and fruitful, in large part because of numerous high-level faculty members and student classmates (one of whom was the celebrated James Yorke (Japan Prize 2003), who went on to coin the word 'chaos' in dynamical systems). In the Fall of 1968, he took up a position of assistant professor of mathematics at the University of Arizona in Tucson where was the celebrated fluid dynamicist Louis Melville Milne–Thomson (Milne–Thomson circle theorem) and some of his students. In 1970 having a post-doctoral position at the IBM Thomas J. Watson Research Center, in Yorktown Heights he attained a renewed interest in dynamic stability, mainly through stimulating discussions with Charles Conley (of Conley index fame), who was his office mate for a while. Back in Arizona in 1976, he began a collaboration with John Bownds (who introduced the concept of ROW (Rest of the World) to the search and rescue community and received the National Award from the National Association For Search And Rescue in 1991), on stability theory for integral and integro-differential equations, however, his growing involvement with population and ecological dynamics was greatly enhanced by the foundation of two programs in 1976 at the University of Arizona: the Interdisciplinary Program in Applied Mathematics and the Department of Ecology and Evolutionary Biology (EEB) where he met Michael Rosenzweig (of predator–prey 'paradox of enrichment' fame) and William Schaffer, both students of Robert MacArthur at Princeton University who is considered a founding father of ecology and evolutionary biology. In 1989, he met biologist Robert Costantino, of the University of Rhode Island. The result was the creation of a team of four interdisciplinary researchers (the Beatle Team) consisting of Bob Costantino, Bob Desharnais and him together Brian Dennis, an ecologist and statistician, later joined by Shandelle

Henson and Aaron King. His most recent interdisciplinary collaboration have been with Shandelle Henson and her husband, James Hayward, a field ecologist who has worked and gathered an immense amount of data on marine animals and birds on Protection Island, a United States Natural Wildlife Refuge managed by the US fish & Wildlife Service.

Eckehard Schöll in Tübingen in 1974 in Mathematics seminars by Rainer Nagel, studied ergodic theory, functional analysis, von Neumann algebras, and C^* algebras. Werner Güttinger introduced a course on nonlinear dynamical systems, where he made his first contact with this field. Harald Stumpf became his first mentor by offering him a Master Thesis (called Diploma Thesis at that time) in theoretical semiconductor physics. In the same year, he heard in a Physics Colloquium a lecture by Werner Heisenberg (Nobel Prize 1932) on his unified field theory. He was also very impressed by a talk by David Ruelle (Holweck Prize 1993, who coined the term 'strange attractor' with Floris Takens in 1970) at the mathematically oriented 'Rencontre entre mathématiciens et physiciens théoriciens' in Strasbourg in 1976. The same year he travelled to Britain and started to work with Peter T. Landsberg who became a mentor and a friend to whom he owes a lot. At Southampton, he attended also lectures by David Chillingworth who came from the Warwick dynamical systems group. Hermann Haken (Max Born Medal and Prize 1976) who had created the field of Synergetics which deals with these nonlinear systems far from thermodynamic equilibrium and their universal features became very important in his scientific career when he invited him to Stuttgart in 1983 to give a seminar on his work on nonequilibrium phase transitions and self-organization in semiconductors. In 1986, he attended for the first-time Dynamics Days in Enschede, Netherlands, a conference series which was founded by Robert Helleman in 1980, and which is organized annually up to the present day. He has attended 23 editions of this series. In 2017, he received an Honorary Doctorate from Saratov State University, Russia, as a result of his very fruitful and active collaboration with the Saratov group of Vadim Anishchenko and Galina Strelkova.

In 1983, Belykh met Vadim Anishchenko (1943–2020, Alexander von Humboldt Research Award 1999) from Saratov State University at a conference on the Oka River, and they quickly became life-long friends. Lozi met for the first time Anishchenko at the International Symposium on Nonlinear Theory and its Application (Nolta'93) in Honolulu. He never published with him, however, he met him again several times after Hawaii, in Krakow, etc. and discussed with him with great pleasure.

Galina Strelkova was a first-year physics student at the Department of Radiophysics (since 1995 it is the Radiophysics and Nonlinear Dynamics Department) of Saratov State University in the early 90s. The Department was headed by Anishchenko who was also a scientific supervisor of a just organized Nonlinear Dynamics Laboratory. When she graduated he invited her to continue her postgraduate studies under his supervision, asking her to understand the properties of chaotic attractors from a physical point of view. She was able to define the fundamental differences in the properties of quasihyperbolic (like Lozi, Belykh, and Lorenz) and nonhyperbolic attractors (like Hénon map, the cubic map, the logistic map, the Rössler system, the Anishchenko–Astakhov oscillator, the Chua's oscillator). The main model systems under study were the Lozi and the Hénon maps. After defending her Ph.D. thesis, she was invited several times by Professor Jürgen Kurths for research visits in his Nonlinear Dynamics Working Group in Potsdam University. Her professional career was inextricably linked with the Department of Radiophysics and Nonlinear Dynamics of Saratov. Thanks to the initiated collaboration and the activities of

Schöll, Professor of the Technical University of Berlin, this department was involved in Collaborative Research Center (CRC) SFB 910 (2011–2022): Control of Self-Organizing Nonlinear Systems, in the framework of the first Russian Project in a German CRC. Anishchenko and Tatiana Vadivasova were Principal Investigators (PIs) of the Project and since 2019 she became a PI too. After the death of Professor Vadim Anishchenko on 30 November 2020, she headed this department.

Electronics research inspired several authors of this recollection. As part of the European-Soviet scientific exchange program, Belykh visited the Electronics Laboratory at the Danish Technical University (DTU) in Lyngby, during the 1975–1976 academic year where he worked with George Bruun and Orla Christensen, two professors of electronics, studying discrete-time digital PLLs. In the meantime, two DTU experimental physicists, Niels F. Pedersen and Ole H. Soerensen, found out that he was an expert in the dynamics of pendulum equations and asked him to explain a random behavior of the Josephson junctions' current density-voltage $J-V$ curves they obtained experimentally. The research interests of Belykh having later broadened to the dynamics of networks and synchronization led to long-term collaborations with Erik Mosekilde, a physics professor at DTU, and Martin Hasler, an engineering professor at École Polytechnique Fédérale de Lausanne (EPFL).

The first talks which Schöll gave at International Conferences were in 1978 at the International Conference on the Physics of Semiconductors (ICPS) in Edinburgh, in 1980 at the German Physical Society (DPG) Annual Spring Meeting in Münster, and in 1981 at the International Conference on Hot Carriers in Semiconductors (HCIS) in Montpellier, France. At this last conference, he met Melvin P. Shaw from Wayne State University, Detroit, author of famous papers and a book on semiconductor instabilities and in particular of a book on the Gunn effect. He visited him for a year in 1983/84 at the Department of Electrical Engineering, and became a coauthor of the book *The Physics of Instabilities in Solid State Electron Devices* finally published in 1992.

After meeting Leon O. Chua at a conference in Crimea, Belykh spent a while at UC Berkeley working with him, Chai Wah Wu, and Ljupco Kocarev. Lozi met Chua near Paris in 1986 3 years after the discovery by this professor of electronics of his famous double scroll circuit. One year later, on 22 December, travelling in the south of France, Chua visited him in Nice. This was the beginning of a very long cooperation between them which lasts since more than 38 years. Lozi had also the privilege to discuss with outstanding researchers, among them five Fields medalists: Thom, Yoccoz, Smale, Alain Connes (within the national committee of C.N.R.S. to which he belonged for 9 years) and Pierre-Louis Lions (during seminars in Paris), one Abel prize laureate: Yves Meyer (in the same committee), and many renowned scientists: Belykh, Chua, Hénon, Lorenz, Rössler, Sharkovsky, Shilnikov, Yorke, etc. Recently he built model of tuberculosis with Saber Elaydi. he also had a 30-year long cooperation with late Abul Hassan Siddiqi, president of the Indian Society of Applied and Industrial Mathematics (ISIAM). Since 2013, he has published many papers with Guanrong (Ron) Chen, Editor in Chief of the *International Journal of Bifurcation and Chaos*, founded by Chua.

From this nonexhaustive summary of the interactions of the authors of this article with dozens of researcher, we can highlight that Science is very international and conclude like Eckehard Schöll 'I am happy to be part of this international family of physicists and mathematicians. This starts with student exchanges with foreign countries, then, international collaborations with colleagues all over the globe, and conference all over the world, where this family meets, form a network'. No mathematician, even if he is gifted in this discipline, can flourish alone. To assert his talent, he must be guided and then cooperate with many other researchers.

Notes

1. See https://polona.pl/item/lilavati-rozrywki-matematyczne,OTgwNzY0NjQ; it seems that it has not been translated to English; a pity.
2. Define a broken line as a union of finitely many closed straight line segments, and consider only those broken line that are not homeomorphic to a circle. Does there exist a broken line B and n points in the plane such that every straight line through any of those points intersects B at exactly two points? Continuing traditions of the Polish school of mathematics, Włodek set a prize. For a solution for $n = 5$, a half of a chocolate bar. For $n = 6$, a quarter, etc. Then for a solution for every n the prize would be a whole bar.
3. with honours
4. Recall that the existence of a homoclinic orbit, in invertible and noninvertible maps, is a proof of the existence of invariant chaotic sets of the map.
5. I used it for my Ph.D. facing the same problems encountered by Eckehard Schöll (Section 5.)
6. I'm impressed by what Smale wrote about Thom's influence at the beginning of his career, in 1956 [193].
7. 28 years after the New York conference, Lorenz mentioned the Lozi map in his last article [114], publicated after his death.
8. Nonlinear functional analysis and its applications.
9. Céa, who was very close to Jacques-Louis Lions, his former supervisor, was at that time professor at the university of Rennes. He said me recently that Lions phoned to him to invite Sergueï Lvovitch Sobolev, visiting professor at Collège de France, at this colloquium. Sobolev was fluent in French and knew a lot about the history of France. He was willing to meet young students. At first, in Paimpol, all the researchers wanted to stroll with him in the city, but soon they changed their mind because Sobolev was reading the names of the streets and told a lot about the incumbents, while the French mathematicians did not know so much. So, not to show their ignorance, there were fewer candidates to walk with this great man.
10. In his thesis, Yoccoz improved theorems of his supervisor Herman by giving simpler proofs but also obtaining the same results under weaker hypotheses. After that, he made major progress on an important conjecture about Benoît Mandelbrot's fractal ensemble, leading to a conjecture by A. Douady and J. Hubbard, which earned him the Fields Medal in 1994. In this aim, he introduced the famous 'Yoccoz puzzles' [126].
11. Translated from French
12. For instance *Doklady Mathematics* was a journal consisting of English translations of articles published in Russian.
13. A strange thing is when I attend a conference in which one lecture uses explicitly this map, and that I speak with young researchers who read my name which is quite uncommon on my badge, they never make the connection between the map and me. I did the same when I was in Oberwolfach with Ernst Hölder in the room!

Acknowledgement

The authors thank C. Guillope for polishing the English.

Disclosure statement

No potential conflict of interest was reported by the author(s).

Funding

The essay of Validimir Belykh on the history behind his early results was prepared in conjunction with the research paper [34] on the dynamics Lozi-Hénon and Belykh maps published in the special issue of the *Journal of Difference Equations and Applications* and supported by the Ministry of Science and Higher Education of the Russian Federation under Grant No. 0729-2020-0036, by the Russian Science Foundation under Grant Nos. 22-21-00553.

References

[1] M.-S. Abdelouahab, R. Lozi, and L.O. Chua, Memfractance, A mathematical paradigm for circuit elements with memory, Int. J. Bifurc. Chaos 24 (2014), p. 1430023.

[2] M.-S. Abdelouahab, R. Lozi, and G. Chen, Complex canard explosion in a fractional-order Fitzhugh–Nagumo model, Int. J. Bifurc. Chaos 29 (2019), p. 1950111.

[3] J.D. Aber, Why don't we believe the models?, Bull. Ecol. Soc. Amer. 78 (1997), pp. 232–233.

[4] R. Abraham and C.D. Shaw, *Dynamics the geometry of behavior, part 1; periodic behavior, 1982; part 2, chaotic behavior, 1983; part 3, global behavior, 1985; part 4, bifurcation behavior, 1988*, Aerial Press, Santa Cruz, CA, pp. 1981–1988.

[5] R. Abraham, L. Gardini, and C. Mira, Chaos in Discrete Dynamical Systems (A Visual Introduction in 2 Dimensions), TELOS, Springer-Verlag, New York, 1997.

[6] V.S. Afraimovich, *Attractors*, in *Nonlinear Waves*, A.V. Gaponov, M.I. Rabinovich and J. Engelbrechet, eds., Springer, Berlin, Heidelberg, 1989, p. 6.

[7] V.S. Afraimovich, V.V. Bykov, and L.P. Shilnikov, On the origin and structure of the Lorenz attractor, Akad. Nauk SSSR Dokl. 234 (1977), pp. 336–339.

[8] L. Alsedà and J. Llibre, A note on the set of periods for continuous maps of the circle which have degree one, Proc. Amer. Math. Soc. 93(1) (1985), pp. 133–138.

[9] H. Amann, N.W. Bazley, and K. Kirchgässner (eds.), *Nichtlineare funktionalanalysis und ihre anwendungen*, Workshop Report, Vol. 5, Mathematisches Forschunginstitut Oberwolfach, 1976. Available at https://oda.mfo.de/handle/mfo/1405.

[10] V.S. Anishchenko and G.I. Strelkova, Irregular attractors, Discrete Dyn. Nat. Soc. 2 (1998), pp. 53–72.

[11] V.S. Anishchenko, A.S. Kopeikin, J. Kurths, T.E. Vadivasova, and G.I. Strelkova, Studying hyperbolicity of chaotic systems, Phys. Lett. A 270 (2000), pp. 301–307.

[12] V.S. Anishchenko, T.E. Vadivasova, A.S. Kopeikin, J. Kurths, and G.I. Strelkova, Effect of noise on the relaxation to an invariant probability measure of nonhyperbolic chaotic attractors, Phys. Rev. Lett.87(5) (2001), p. 054101.

[13] V.S. Anishchenko, T.E. Vadivasova, A.S. Kopeikin, G.I. Strelkova, and J. Kurths, Peculiarities of the relaxation to an invariant probability measure of nonhyperbolic chaotic attractors in the presence of noise, Phys. Rev. E 65(3) (2002), p. 036206.

[14] V.S. Anishchenko, T.E. Vadivasova, and G.I. Strelkova, Deterministic Nonlinear Systems. A Short Course, Springer, Berlin, 2014.

[15] D.V. Anosov, *Dynamical systems in the1960s: The hyperbolic revolution*, in *Mathematical Events of the XX-th Century*, Springer-Verlag, Berlin, 2006, p. 117.

[16] K. Aoki, K. Yamamoto, N. Mugibayashi, and E. Schöll, Complex dynamical behavior and chaos in the Hess oscillator, Solid State Electron. 32 (12) (1989), pp. 1149–1153.

[17] J. Argémi, Points singuliers non élémentaires de quelques classes de systèmes dynamiques plans, Ann. Mat. Pura Appl. 89 (1971), pp. 321–351.

[18] V.I. Arnold, Forgotten and neglected theories of Poincaré, Uspekhi Mat. Nauk 61(1) (2006), pp. 1–18. [*English translation*], Russian Math. Surv., original work published 2006.

[19] D. Aubin and A.D. Dalmedico, Writing the history of dynamical systems and chaos: Longue durée and revolution, disciplines and cultures, Historia Math. 29 (2002), pp. 273–339.

[20] J.F.G. Auchmuty and G. Nicolis, Bifurcation analysis of nonlinear reaction-diffusion equations – I. Evolution equations and the steady state solutions, Bull. Math. Biol. 37 (1975), pp. 323–365.

[21] N.N. Bautin, *On the number of limit cycles appearing from an equilibrium point of the focus or center type under varying coefficients*, in *Transl. Amer. Math. Soc.*, American Mathematical Society, Providence, RI, USA, Vol. 100, 1954.

[22] V.N. Belykh, Periodic solutions of a second-order dynamic system with continuous piecewise-smooth right-hand sides, Radiophys. Quantum Electron. 11(10–11) (1968), pp. 980–981.

[23] V.N. Belykh, On models of phase synchronization systems and their study, Dinamika Syst. 11 (1976), pp. 23–32.

[24] V.N. Belykh, Qualitative Methods of Nonlinear Oscillation Theory in Point Systems, Gorky State University Press, Gorky, U.S.S.R., 1980.

[25] V.N. Belykh, *Models of discrete systems of phase synchronization (in Russian)*, in *Systems of Phase Synchronization*, V.V. Shakhildyan and L.N. Belyustina, eds., Radio I Svyaz, Moscow, 1982, pp. 161–216.

[26] V.N. Belykh, Bifurcation of separatrices of a saddle-point of the Lorenz system, Differ. Equ. 20(10) (1984), pp. 1184–1191.

[27] V.N. Belykh, Chaotic and strange attractors of a two-dimensional map, Math. USSR Sbornik 186 (1995), pp. 311–326. https://doi.org/10.1070/sm1995v186n03abeh000018.

[28] V.N. Belykh and V.I. Nekorkin, A qualitative investigation of a system of three differential equations in the theory of phase synchronization, Prikl. Mat. I Mekh. 39 (1975), pp. 642–649.

[29] V.N. Belykh and V.P. Maksakov, Dynamics of the simplest digital system of phase synchronization, J. Commun. Technol. Electron. 21(10) (1976), pp. 2155–2163.

[30] V.N. Belykh and V.P. Maksakov, Dynamics of discrete-time systems of digital phase synchronization (Russian), Dinamika Syst. 11 (1976), pp. 130–143.

[31] V.N. Belykh and I.V. Belykh, Belykh map, Scholarpedia 6(10) (2011), p. 5545. https://doi.org/10.4249/scholarpedia.5545.

[32] V.N. Belykh, N.F. Pedersen, and O.H. Soerensen, Shunted-Josephson-junction model. I. The autonomous case, Phys. Rev. B 16(11) (1977), pp. 4853–4859.

[33] V.N. Belykh, N.F. Pedersen, and O.H. Soerensen, Shunted-Josephson-junction model. II. The nonautonomous case, Phys. Rev. B 16(11) (1977), pp. 4860–4871.

[34] V.N. Belykh, N.V. Barabash, and D.A. Grechko, Existence proofs for strange attractors in piecewise-smooth nonlinear Lozi–Hénon and Belykh maps, J. Differ. Equ. Appl. 29(9–12) (2023), pp. 994–1014.

[35] L.N. Belyustina and V.N. Belykh, On a nonautonomous phase system of equations with a small parameter containing invariant tori and crude homoclinic curves, Radiophys. Quantum Electron. 15(7) (1972), pp. 793–800.

[36] L.N. Belyustina and V.N. Belykh, Qualitative investigation of a dynamical system on a cylinder, Differentsial'nye Uravneniya 9(3) (1973), pp. 403–415.

[37] L.N. Belyustina and V.N. Belykh, The global structure of the decomposition of the cylindrical phase space of a certain nonautonomous system, Differentsial'nye Uravneniya 9(4) (1973), pp. 595–608.

[38] R. Berner, S. Yanchuk, and E. Schöll, What adaptive neuronal networks teach us about power grids, Phys. Rev. E 103 (2021), p. 042315.

[39] L. Block, J. Guckenheimer, M. Misiurewicz, and L.S. Young, *Periodic points and topological entropy of one dimensional maps*, in *Global Theory of Dynamical Systems (Northwestern Univ., Evanston, IL 1979)*, Lecture Notes in Mathematics Vol. 819, Springer, Berlin, 1980, pp. 18–34.

[40] A.S. Bogomolov, A.V. Slepnev, G.I. Strelkova, E. Schöll, and V.S. Anishchenko, Mechanisms of appearance of amplitude and phase chimera states in ensembles of nonlocally coupled chaotic systems, Commun. Nonlinear Sci. Numer. Simul. 43 (2017), pp. 25–36.

[41] N. Bourbaki, *Éléments de mathématiques*, Masson, 1959–2019, collective work.

[42] A. Chenciner, *Hopf bifurcation for invariant tori*, in *Dynamical Systems: Lectures Given at a Summer School of the Centro Internazionale Matematico Estivo*, J. Guckenheimer, ed., Springer–Verlag, Berlin, Heidelberg, Germany, 1980, pp. 197–207.

[43] D.R.J. Chillingworth and P.J. Holmes, Dynamical systems and models for reversals of the earth's magnetic field, J. Intern. Ass. Math. Geol. 12 (1980), pp. 41–59.

[44] L. Chua, A rigorous proof of the existence of chaos in double scroll, Ann. Télécom. 42 (1987), pp. 239–246.
[45] P. Collet and J.P. Eckmann, Iterated Maps on the Interval As Dynamical Systems, Birkhauser, Boston, 1980.
[46] R.F. Costantino, J.M. Cushing, B. Dennis, R.A. Desharnais, and S.M. Henson, Resonant population cycles in temporally fluctuating habitats, Bull. Math. Biol. 60(2) (1998), pp. 247–273.
[47] M.G. Crandall and P.H. Rabinowitz, Bifurcation from simple eigenvalues, J. Funct. Anal. 8 (1971), pp. 321–340.
[48] J.M. Cushing, On a conjecture of Levi–Civita, J. Math. Mech. 19(12) (1970), pp. 1047–1050.
[49] J.M. Cushing, Integrodifferential Equations and Delay Models in Population Dynamics, Lecture Notes in Biomathematics Vol. 20, Springer, Berlin Heidelberg New York, 1977. (reprinted in 2013, ISBN 978-3-642-93073-7).
[50] J.M. Cushing, An Introduction to Structured Population Dynamics, Conferences Series Applied Mathematics Vol. 71, SIAM, Philadelphia, 1998. ISBN 0-89871-417-6.
[51] J.M. Cushing, *Matrix models and population dynamics*, in *A Chapter in Mathematical Biology*, IAS/Park City Mathematics Series Vol 14, Mark Lewis, A.J. Chaplain, James P. Keener, Philip K. Maini eds., American Mathematical Society, Providence, RI, 2009, pp. 47–150.
[52] J.M. Cushing, B. Dennis, R.A. Desharnais, and R.F. Costantino, Moving toward an unstable equilibrium: Saddle nodes in population systems, J. Animal Ecol. 67(1) (1998), pp. 298–306.
[53] J.M. Cushing, R.F. Costantino, B. Dennis, R.A. Desharnais, and S.M. Henson, Chaos in Ecology: Experimental Nonlinear Dynamics, Theoretical Ecology Series Vol. 1, Academic Press (Elsevier Science), New York, 2003. ISBN: 0-12-1988767.
[54] J.M. Cushing, S.M. Henson, and J.L. Hayward, Modeling Behavior and Population Dynamics: Seabirds, Seals, and Marine Iguanas, Interdisciplinary Applied Mathematics Vol. 57, Springer Nature, Switzerland, 2023.
[55] R.A. Desharnais, S.M. Henson, R.F. Costantino, and B. Dennis, Capturing chaos: A multidisciplinary approach to nonlinear population dynamics, J. Differ. Equ. Appl. (2023), pp. 1–37. https://doi.org/10.1080/10236198.2023.2260013.
[56] J. Edmunds, J.M. Cushing, R.F. Costantino, S.M. Henson, B. Dennis, and R.A. Desharnais, Park's Tribolium competition experiments: A non-equilibrium species coexistence hypothesis, J. Animal Ecol.72 (2003), pp. 703–712.
[57] L.S. Efremova, Periodic orbits and the degree of a continuous map of the circle (Russian), Differ. Integral. Uravn. 2 (1978), pp. 109–115.
[58] L.S. Efremova, *Periodic motions of discrete semidynamical systems*, Ph.D. diss., Gorkii State University, Gorkii, 1981, p. 117 (Russian).
[59] L.S. Efremova, Dynamics of skew products of interval maps, Russian Math. Surv. 72(1) (2017), pp. 101–178.
[60] L.S. Efremova, Small perturbations of smooth skew products and Sharkovsky theorem, J. Differ. Equ. Appl. 26(8) (2020), pp. 1192–1211. Special issue on the occasion of the 82nd birthday of O.M. Sharkovsky.
[61] L.S. Efremova, Geometrically integrable maps in the plane and their periodic orbits, Lobachevskii J. Math. 42(10) (2021), pp. 2315–2324.
[62] L.S. Efremova and R.G. Rakhmankulov, Theorems on the coexistence of periodic orbits of endomorphisms of the circle (Russian), Differ. Integral. Uravn. 4 (1980), pp. 116–118.
[63] L.S. Efremova and V.Z. Sakbaev, Notion of blowup of the solution set of differential equations and averaging of random semigroups, Theor. Math. Phys. 185(2) (2015), pp. 1582–1598.
[64] L.S. Efremova and E.N. Makhrova, One-dimensional dynamical systems, Russ. Math. Surv. 76(5) (2021), pp. 821–881.
[65] P. Fatou, Séries trigonométriques et séries de Taylor, Acta Math. 30 (1906), pp. 335–400.
[66] M.J. Feigenbaum, *Universality in complex discrete dynamics*, Los Alamos Theoretical Division Annual Report, 1976, pp. 98–102.
[67] M.J. Feigenbaum, Quantitative universality for a class of nonlinear transformations, J. Stat. Phys. 19 (1978), pp. 25–52.

[68] J. Fialkowski, S. Yanchuk, I.M. Sokolov, E. Schöll, G.A. Gottwald, and R. Berner, Heterogeneous nucleation in finite size adaptive dynamical networks, Phys. Rev. Lett. 130 (2023), p. 067402.
[69] R.J. Field and R.M. Noyes, Oscillations in chemical systems. IV. Limit cycle behavior in a model of a real chemical reaction, J. Chem. Phys. 60 (1974), pp. 1877–1884.
[70] P.C. Fife, *Branching phenomena in fluid dynamics and chemical reaction-diffusion theory*, in *Eigenvalues of Non-Linear Problems*, Vol. 67, G. Prodi, ed., Springer, Berlin, Heidelberg, 1974, pp. 23–83.
[71] V. Flunkert, I. Fischer, and E. Schi£¡ï£¡ï£¡ï£¡ll, Dynamics, control and information in delay-coupled systems, Phil. Trans. R. Soc. A 371 (2013), p. 20120465.
[72] H. Fujii, M. Mimura, and Y. Nishiura, A picture of the global bifurcation diagram in ecological interacting and diffusing systems, Physica D 5 (1982), pp. 1–42.
[73] C.F. Frouzakis, L. Gardini, Y.G. Kevrekidis, G. Millerioux, and C. Mira, On some properties of invariant sets of two-dimensional noninvertible maps, Int. J. Bifurc. Chaos 7(6) (1997), pp. 1167–1194.
[74] L. Gardini, Some global biforcations of two-dimensional endomorphisms by use of critical lines, Nonlinear Anal. Theory Methods Appl. 18 (1992), pp. 361–399.
[75] L. Gardini, *Global analysis and bifurcations in two-dimensional endomorphisms by use of critical curves*, in *European Conference on Iteration Theory*, Vol. 91, J.P. Lampreira, J. Llibre, C. Mira, J. Sousa Ramos, and G. Targonski, eds., World Scientific, Singapore, 1992, pp. 114–125.
[76] L. Gardini, Homoclinic bifurcations in n-dimensional endomorphisms, due to expanding periodic points, Nonlinear Anal. Theory Methods Appl. 23(8) (1994), pp. 1039–1089.
[77] R. Graham and H. Haken, Laserlight – first example of a second-order phase transition far away from thermal equilibrium, Z. Physik 237 (1970), pp. 31–46.
[78] S. Grossmann and S. Thomae, Invariant distributions and stationary correlation functions of one-dimensional discrete processes, Z. Naturforsch. A 32 (1977), pp. 1353–1363. https://doi.org/10.1515/zna-1977-1204.
[79] J. Guckenheimer and P.J. Holmes, Nonlinear Oscillations, Dynamical Systems, and Bifurcations of Vector Fields, Applied Mathematical Sciences Vol. 42, Springer-Verlag, New York, 1983.
[80] I. Gumowski and C. Mira, Dynamique Chaotique, Cepadues Editions, Toulouse, 1980.
[81] I. Gumowski and C. Mira, Recurrences and Discrete Dynamic Systems: An Introduction, Lecture Notes in Math Vol. 809, Springer, Berlin, Heidelberg, Germany, 1980.
[82] H. Haken, Analogy between higher instabilities in fluids and lasers, Phys. Lett. 53A (1975), pp. 77–78.
[83] H. Haken, Synergetics, An Introduction, 3rd ed., Springer, Berlin, 1983.
[84] T. Hamaizia, R. Lozi, and N.-E. Hamri, Fast chaotic optimization algorithm based on locally averaged strategy and multifold chaotic attractor, Appl. Math. Comput. 219 (2012), pp. 188–196.
[85] M. Hénon, A two-dimensional mapping with a strange attractor, Comm. Math. Phys. 50 (1976), pp. 69–77.
[86] M. Hénon, Une transformation á deux dimensions avec un 'attracteur étrange', J. Phys. Colloq. 39(Suppl. 8) (1978), pp. C5-5–C5-5.
[87] S.M. Henson, A.A. King, R.F. Costantino, J.M. Cushing, B. Dennis, and R.A. Desharnais, Explaining and predicting patterns in stochastic population systems, Proc. R. Soc. Lond. B 270 (2003), pp. 1549–1553.
[88] G. Iooss, Bifurcation des solutions périodiques de certains problèmes d'évolution, C. R. Acad. Sci. I 273 (1971), pp. 624–627.
[89] G. Iooss (ed.), *Bifurcation of maps and applications*, North-Holland Mathematics Studies, Amsterdam, Netherland, Vol. 36, 1979, p. 44.
[90] G. Iooss and R. Lozi, Convection entre deux plaques planes en rotation et effet dynamo résultant d'une bifurcation secondaire, J. Phys. Colloq. 39(Suppl. 8) (1978), pp. C5–15.

[91] O. Jallouli, S. El Assad, M. Chetto, and R. Lozi, Design and analysis of two stream ciphers based on chaotic coupling and multiplexing techniques, Multimed. Tools Appl. 77 (2018), pp. 13391–13417.

[92] W. Just, A. Pelster, M. Schanz, and E. Schöll, Delayed complex systems, Phil. Trans. R. Soc. A368 (2010), pp. 301–513.

[93] H. Kawakami and R. Lozi, *Switched dynamical systems – a switch defines semi-flows on a pair of half planes (in Japanese)*, in *Proceedings of the RIMS, Kokyuroku, Japan Conference*, Vol. 804, 1992, pp. 38–54.

[94] H. Kawakami and R. Lozi, *Switched dynamical systems – dynamics of a class of circuits with switch*, Advanced Series in Dynamical Systems Vol. 11, S. Ushiki, ed., World Scientific Publishing, Singapore, 1992, pp. 39–58.

[95] H.B. Keller, *Numerical solution of bifurcation and nonlinear eigenvalue problems*, in *Applications of Bifurcation Theory*, P. Rabinowitz, ed., Academic Press, New York, 1977, pp. 359–384.

[96] H.B. Keller, Numerical Methods for Two-Point Boundary-Value Problems, Dover, Mineola, New York, 1992.

[97] A-A. Khennaoui, A. Ouannas, S. Bendoukha, G. Grassi, R.P. Lozi, and V.-T. Pham, On fractional-order discrete-time systems: Chaos, stabilization and synchronization, Chaos Solit. Fractals119 (2019), pp. 150–162.

[98] A.A. King, R.F. Costantino, J.M. Cushing, S.M. Henson, R.A. Desharnais, and B. Dennis, Anatomy of a chaotic attractor: Subtle model predicted patterns revealed in population data, Proc. Nat. Acad. Sci. 101(1) (2003), pp. 408–413.

[99] N.E. Klinshpont, E.A. Sataev, and R.V. Plykin, Geometrical and dynamical properties of Lorenz type system, J. Phys. Conf. Ser. 23 (2005), pp. 96–104.

[100] H. Kostial, M. Asche, R. Hey, K. Ploog, B. Kehrer, W. Quade, and E. Schöll, Low temperature breakdown and current filamentation in n-type GaAs with homogeneous and partially ordered Si doping, Semicond. Sci. Technol. 10 (1995), pp. 775–784. https://doi.org/10.1088/0268-1242/10/6/006.

[101] Y. Kuramoto and D. Battogtokh, Coexistence of coherence and incoherence in nonlocally coupled phase oscillators, Nonlinear Phenom. Complex Syst. 5 (2002), p. 380.

[102] S.P. Kuznetsov, Dynamical chaos and uniformly hyperbolic attractors: From mathematics to physics, Phys. Uspekhi 54(2) (2011), pp. 119–144.

[103] L.D. Landau, On the problem of turbulence (in russian), Dokl. Akad. Nauk SSSR 44 (1944), pp. 339–342.

[104] P.T. Landsberg, Thermodynamics, Interscience Publishers, New York, London, 1961.

[105] P.T. Landsberg, Recombination in Semiconductors, Cambridge University Press, Cambridge, 1991.

[106] P.T. Landsberg and A. Pimpale, Recombination-induced nonequilibrium phase transitions in semiconductors, J. Phys. C9 (1976), p. 1243.

[107] P.T. Landsberg, E. Schöll, and P. Shukla, A simple model for the origin of chaos in semiconductors, Physica D 30(1–2) (1988), pp. 235–243.

[108] N.N. Leonov, Point transformation of a straight line into a straight line (in Russian), Izv. Vyssh. Uchebn. Zaved. Radiofiz. 2(6) (1959), pp. 942–956.

[109] N.N. Leonov, A discontinuous point transformation of a straight line into a straight line, Dokl. Akad. Nauk SSSR 3 (1962), pp. 532–535. [*English translation*], Soviet Math. Dokl., original work published in (1962).

[110] C. Letellier, R. Abraham, D.L. Shepelyansky, O.E. Rössler, P. Holmes, R. Lozi, L. Glass, A. Pikovsky, L.F. Olsen, I. Tsuda, C. Grebogi, U. Parlitz, R. Gilmore, L.M. Pecora, and T.L. Carrol, Some elements for a history of the dynamical systems theory, Chaos 31 (2021), p. 053110.

[111] T.-Y. Li and J.A. Yorke, Period three implies chaos, Amer. Math. Mon. 82 (1975), pp. 985–992.

[112] C. Lobry and R. Lozi, *Bifurcations of motifs in families of mixed two vectors fields*, Springer Series in Synergetics Vol. 12, Springer, Berlin, Heidelberg, 1981, pp. 67–70.

[113] E.N. Lorenz, Deterministic nonperiodic flow, J. Atmos. Sci. 20(2) (1963), pp. 130–141.

[114] E.N. Lorenz, Compound windows of the Hénon map, Physica D 237 (2008), pp. 1689–1704.

[115] R. Lozi, *Analyse numérique de certains problèmes de bifurcation*, Ph.D. thesis, University of Nice, France, 1975.

[116] R. Lozi, A computing method for bifurcation boughs of nonlinear eigenvalue problems, Bull. Amer. Math. Soc. 81 (1975), pp. 1127–1130.

[117] R. Lozi, Un attracteur étrange du type attracteur de Hénon, J. Phys. Colloq. 39(C5) (1978), pp. C5-9–C5-10.

[118] R. Lozi, *Strange attractors: A class of mappings of R2 which leaves some cantor set invariant*, International Workshop on Intrinsic Stochasticity, in *Plasmas*, G. Laval and D. Grésillon, eds., Éditions de Physique, Orsay, France, 1979, pp. 373–381.

[119] R. Lozi, Sur un modèle mathématique de suite de bifurcations de motifs dans la réaction de Belousov–Zhabotinsky, C. R. Acad. Sci. Ser. I Math. 294 (1982), pp. 21–26.

[120] R. Lozi, *Dimensional bifurcations between thread and sheet strange attractor*, in *Théorie de l'itération et ses applications*, Vol. 322, R. Thibault, ed., Colloques Internationaux du CNRS, 1982, pp. 145–152.

[121] R. Lozi, *Modèles mathématiques simples et consistants pour l'étude de quelques systèmes dynamiques expérimentaux*, Thèse d'État, University of Nice, France, 1983.

[122] R. Lozi, *Motifs et confineurs organisés dans la réaction de Bélousov-Zhabotinskii et le 'double scroll'*, Les rencontres physiciens–mathématiciens de Strasbourg -RCP25, Tome 41 (1990), Exposé no. 1, 62 p. Available at http://www.numdam.org/item/RCP25_1990__41__1_0.

[123] R. Lozi, Survey of recent applications of the chaotic Lozi map, Algorithms 16(10) (2023), p. 491. https://doi.org/10.3390/a16100491.

[124] R. Lozi and S. Ushiki, Organized confinors and anticonfinors and their bifurcations in constrained 'Lorenz' system, Ann. Télécomm 43 (1988), pp. 187–208.

[125] J.-P. Lozi, O. Garasym, and R. Lozi, *The challenging problem of industrial applications of multicore-generated iterates of nonlinear mappings*, in *Industrial Mathematics and Complex Systems*, Industrial and Applied Mathematics, P. Manchanda, R. Lozi, and A. Siddiqi, eds., Springer, Singapore, 2017, pp. 43–67.

[126] M. Lyubich, Dynamics of quadratic polynomials, I–II, Acta Math. 178 (1997), pp. 185–297.

[127] A.G. Maier, *A structurally stable transformation of a circle into a circle* (in Russian), Uch. Zap. Gork. Univ. 12 (1939), pp. 215–229.

[128] Y.L. Maistrenko, I.M. Sushko, and L. Gardini, About two mechanisms of reunion of chaotic attractors, Chaos Solit. Fractals 9(8) (1998), pp. 1373–1390.

[129] P. Manchanda, R. Lozi, and A.H. Siddqi, *Industrial Mathematics and Complex Systems*, Industrial and Applied Mathematics, Springer, Singapore, Singapore, 2017.

[130] P. Manchanda, R. Lozi, and A.H. Siddqi, *Mathematical Modelling, Optimization, Analytic and Numerical Solutions*, Industrial and Applied Mathematics, Springer, Singapore, Singapore, 2020.

[131] J.E. Marsden, Qualitative methods in bifurcation theory, Bull. Amer. Math. Soc. 84 (1978), pp. 1125–1148.

[132] T. Matsumoto, A chaotic attractor from Chua's circuit, IEEE Trans. Circ. Syst. CAS-31 (1984), pp. 1055–1058.

[133] C. Mira, Chaotic Dynamics. From the One-Dimensional Endomorphism to the Two-Dimensional Diffeomorphism, World Scientific, Singapore, 1987.

[134] C. Mira, L. Gardini, A. Barugola, and J.C. Cathala, Chaotic Dynamics in Two-Dimensional Nonivertible Maps, World Scientific Series on Nonlinear Science Vol. 20, World Scientific, New Jersey, 1996.

[135] M. Misiurewicz, Strange attractors for the Lozi mappings, Ann. New York Acad. Sci. 357 (1980), pp. 348–358.

[136] M. Misiurewicz, Absolutely continuous measures for certain maps of an interval, Publ. Math. IHES53 (1981), pp. 17–51.

[137] M. Misiurewicz, Periodic points of maps of degree one of a circle, Ergod. Theory Dyn. Syst. 2(2) (1982), pp. 221–227.

[138] M. Misiurewicz and W. Szlenk, Entropy of piecewise monotone mappings, Astérisque 50 (1977), pp. 299–310. Full version: Studia Math. 67 (1980), pp. 45–63.

[139] M. Misiurewicz and S. Roth, Constant slope maps on the extended real line, Ergod. Theory Dyn. Syst. 38 (2018), pp. 3145–3169.

[140] M.I. Motova and V.D. Shalfeev, *The school of academician A.A. Andronov and its development at Nizhny Novgorod University (Notes on the 90th anniversary of the school, in Russian)*, N.I. Lobachevsky National Research University, Nizhny Novgorod, 2023, 156 p.

[141] S. Newhouse, J. Palis, and F. Takens, Bifurcations and stability of families of diffeomorphisms, Inst. Hautes Etudes Sci. Publ. Math. 57 (1983), pp. 5–71.

[142] G. Nicolis and I. Prigogine, Self-Organization in Nonequilibrium Systems - From Dissipative Structures to Order Through Fluctuations, John Wiley & Sons, New York, London, Sydney, Toronto, 1977.

[143] I. Omelchenko, Y. Maistrenko, P. Hövel, and E. Schöll, Loss of coherence in dynamical networks: Spatial chaos and chimera states, Phys. Rev. Lett. 106 (2011), p. 234102.

[144] I. Omelchenko, B. Riemenschneider, P. Hövel, Y. Maistrenko, and E. Schöll, Transition from spatial coherence to incoherence in coupled chaotic systems, Phys. Rev. E 85 (2012), p. 026212.

[145] J. Peinke, J. Parisi, O.E. Rössler, and R. Stoop, Encounter with Chaos, Springer, Berlin, Heidelberg, 1992.

[146] Y.B. Pesin, Dynamical systems with generalized hyperbolic attractors: Hyperbolic, ergodic and topological properties, Ergod. Theory Dyn. Syst. 12 (1992), pp. 123–151. https://doi.org/10.1017/s0143385700006635.

[147] J. Petitot (ed.), *Logos et théorie des catastrophes à partir de l'œuvre de René Thom*, in *Actes du colloque international de 1982*, Cerisy La Salle, Patino, 1988.

[148] R.V. Plykin, Hyperbolic attractors of diffeomorphisms (in Russian), Russ. Math. Surv. 35(3) (1980), pp. 109–121.

[149] H. Poincaré, Sur les courbes définies par une équation différentielle, J. Math. Pures Appl. 3e Sér. t.7 (1881), pp. 375–422. t. 8 (1882), pp. 251–296; C. R., XCVIII (1884), pp. 287–289; Jordan J. (4), I (1885), pp. 167–244; II (1886), pp. 151–211.

[150] H. Poincaré, Sur l'équilibre d'une masse fluide animée d'un mouvement de rotation, C. R. A. S. 100 (1885), pp. 346–348.

[151] T. Poston and I. Stewart, Catastrophe Theory and Its Applications, Pitman, Boston, 1978.

[152] M.I. Rabinovich, Pathways to and properties of stochasticity in dissipative systems, Sov. Phys. Usp.26(1) (1983), pp. 186–187.

[153] P. Rabinowitz, *Variational methods for nonlinear eigenvalue problems*, in *Eigenvalues of Non-Linear Problems*, Vol. 67, G. Prodi, ed., Springer, Berlin, Heidelberg, 1974, pp. 139–195.

[154] R.C. Rael, R.F. Costantino, J.M. Cushing, and T.L. Vincent, Using stage-structured evolutionary game theory to model the experimentally observed evolution of a genetic polymorphism, Evol. Ecol. Res. 11 (2009), pp. 141–151.

[155] R.G. Rakhmankulov, The cyclic groups of a single-valued continuous transformation of the circle into the circle (in Russian), Volzhsk. Mat. Sb. 13 (1972), pp. 80–92.

[156] R.G. Rakhmankulov, Coexistence of cycles of a continuous transformations of a circle (in Russian), Izv. Vyssh. Uchebn. Zaved. Mat. 2 (1974), p. 9706.

[157] G. Reeb and D. Bennequin, *Textes de sept conférences données à la 48ème réunion de travail de la RCP n °25 en l'honneur de René THOM*, 11–13 mai 1989.

[158] D. Ruelle, *Dynamical systems with turbulent behavior*, in *International Mathematics Physics Conference*, Lecture Notes in Physics Vol. 80 (1978), Roma, 1977, pp. 341–360.

[159] D. Ruelle, Measure describing a turbulent flow, Ann. N.-Y. Acad. Sci. 357 (1980), pp. 1–9.

[160] D. Ruelle and F. Takens, On the nature of turbulence, Commun. Math. Phys. 20 (1971), pp. 167–192.

[161] E. Rybalova, V. Anishchenko, G. Strelkova, and A. Zakharova, Forced synchronization of a multilayer heterogeneous network of chaotic maps in the chimera state mode, Chaos 29(3) (2019), p. 033134.

[162] E. Rybalova, T. Vadivasova, G. Strelkova, V. Anishchenko, and A. Zakharova, Solitary states and solitary state chimera in neural networks, Chaos 29(7) (2019), p. 071106.

[163] E. Rybalova, G. Strelkova, E. Schöll, and V. Anishchenko, Relay and complete synchronization in heterogeneous multiplex networks of chaotic maps, Chaos 30(6) (2020), p. 061104.

[164] E.V. Rybalova, A. Zakharova, and G.I. Strelkova, Interplay between solitary states and chimeras in multiplex neural networks, Chaos Solit. Fractals 148 (2021), p. 111011.

[165] E.A. Sataev, Invariant measures for hyperbolic maps with singularities, Russ. Math. Surv. 47 (1992), pp. 191–251. https://doi.org/10.1070/rm1992v047n01abeh000864.

[166] E.A. Sataev, Ergodic properties of the Belykh map, J Math. Sci. 95 (1999), pp. 2564–2575. https://doi.org/10.1007/bf02169056.

[167] F. Schlögl, Chemical reaction models for non-equilibrium phase transitions, Z. Phys. 253 (1972), pp. 147–161.

[168] E. Schöll, Current layers and filaments in a semiconductor model with an impact ionization induced instability, Z. Phys. B 48 (1982), pp. 153–165.

[169] E. Schöll, Bistability and nonequilibrium phase transitions in a semiconductor recombination model with impact ionization of donors, Z. Phys. B 46 (1982), pp. 23–30.

[170] E. Schöll, Nonequilibrium Phase Transitions in Semiconductors, Springer, Berlin, 1987.

[171] E. Schöll (ed.), Theory of Transport Properties of Semiconductor Nanostructures, Electronic Materials Series Vol. 4, Chapman and Hall, London, 1998.

[172] E. Schöll, Nonlinear Spatio-Temporal Dynamics and Chaos in Semiconductors, Nonlinear Science Series Vol. 10, Cambridge University Press, Cambridge, 2001.

[173] E. Schöll, *Synchronization in delay-coupled complex networks*, in *Advances in Analysis and Control of Time-Delayed Dynamical Systems*, chap. 4, J.-Q. Sun and Q. Ding, eds., World Scientific, Singapore, 2013, pp. 57–83.

[174] E. Schöll, S.H.L. Klapp, and P. Hövel, Control of Selforganizing Nonlinear Systems, Springer, Berlin, 2016.

[175] E. Schï£¡ï£¡ï£¡ï£¡ll, Synchronization patterns and chimera states incomplex networks: Interplay of topology and dynamics, Eur. Phys. J. Spec. Top. 225 (2016), pp. 891–919.

[176] E. Schöll, Partial synchronization patterns in brain networks, Europhys. Lett. 136 (2021), p. 18001.

[177] E. Schöll and P.T. Landsberg, Semiconductor models for first and second order non-equilibrium phase transitions, Proc. Roy. Soc. A 365 (1979), p. 495. https://doi.org/10.1098/rspa.1979.0030.

[178] E. Schöll and H.G. Schuster (eds.), Handbook of Chaos Control, Wiley-VCH, Weinheim, 2008. Second completely revised and enlarged edition.

[179] E. Schöll, D. Bimberg, H. Schumacher, and P.T. Landsberg, Kinetics of picosecond pulse generation in semiconductor lasers with bimolecular recombination at high current injection, IEEE J. Quantum Electron. 20 (1984), pp. 394–399.

[180] E. Schöll, J. Parisi, B. Röhricht, J. Peinke, and R.P. Huebener, Spatial correlations of chaotic oscillations in the post-breakdown regime of p-Ge, Phys. Lett. A 119 (1987), pp. 419–424.

[181] E. Schöll, P. Hövel, V. Flunkert, and M.A. Dahlem, *Timedelayed feedback control: From simple models to lasers and neural systems*, in *Complex Time-Delay Systems: Theory and Applications*, F.M. Atay, ed., Springer, Berlin, 2010, pp. 85–150.

[182] E. Schöll, A. Zakharova, and R.G. Andrzejak, *Chimera States in Complex Networks*, in *Research Topic, Frontiers in Applied Mathematics and Statistics*, Frontiers Media SA, Lausanne, 2020, Ebook.

[183] F.M. Scudo and J.R. Ziegler, The Golden Age of Theoretical Ecology: 1923–1940, Lecture Notes in Biomathematics Vol. 22, Springer, Berlin, 1978.

[184] N. Semenova, A. Zakharova, E. Schöll, and V. Anishchenko, Does hyperbolicity impede emergence of chimera states in networks of nonlocally coupled chaotic oscillators?, EPL (Euro. Lett.) 112 (2015), p. 40002.

[185] N. Semenova, G. Strelkova, V. Anishchenko, and A. Zakharova, Temporal intermittency and the lifetime of chimera states in ensembles of nonlocally coupled chaotic oscillators, Chaos 27(6) (2017), p. 061102.

[186] A.N. Sharkovsky, Co-existence of cycles of a continuous mapping of the line into itself, Ukr. Math. J. 16 (1964), pp. 61–71.

[187] A.N. Sharkovsky, Coexistence of cycles of a continuous map of the line into itself, Int. J. Bifurc. Chaos Appl. Sci. Eng. 5(5) (1995), pp. 1263–1273. Original work published in 1964.

[188] M.P. Shaw, V.V. Mitin, E. Schöll, and H.L. Grubin, The Physics of Instabilities in Solid State Electron Devices, Plenum Press, New York, 1992.

[189] I.A. Shepelev, S.S. Muni, E. Schöll, and G. Strelkova, Repulsive inter-layer coupling induces anti-phase synchronization, Chaos 31(6) (2021), p. 063116.

[190] L.P. Shilnikov, A case of the existence of a denumerable set of periodic motions, Sov. Math. Dokl. 6 (1965), pp. 163–166.

[191] L.P. Shilnikov, Mathematical problems of nonlinear dynamics: A tutorial, Int. J. Bifurc. Chaos 7(9) (1997), pp. 1953–2001.

[192] S. Smale, Differentiable dynamical systems. I. Diffeormorphisms, Bull. Amer. Math. Soc. 73 (1967), pp. 747–817.

[193] S. Smale, *On how I got started in dynamical systems, 1959–1962*, in *The Mathematics of Time: Essays on Dynamical Systems, Economic Processes and Related Topics*, Springer-Verlag, New York, 1980.

[194] P. Stefan, A theorem of Sarkovskii on the existence of periodic orbits of continuous endomorphisms of the real line, Commun. Math. Phys. 54 (1977), pp. 237–248.

[195] G.I. Strelkova and V.S. Anishchenko, Spatio-temporal structures in ensembles of coupled chaotic systems, Physics-Uspekhi 63(2) (2020), pp. 145–161.

[196] F. Takens, Implicit differential equations: Some open problems, Lect. Notes Math. 535 (1976), pp. 237–253.

[197] R. Thom, Structural Stability and Morphogenesis: An Outline of a General Theory of Models, W. A. Benjamin, New York, 1973. (Translated from the French by David Fowler).

[198] L. Tumash, S. Olmi, and E. Schöll, Effect of disorder and noise in shaping the dynamics of power grids, Europhys. Lett. 123 (2018), p. 20001.

[199] R. Turner, *Positive solutions of nonlinear eigenvalue problems*, in *Eigenvalues of Non-Linear Problems*, Vol. 67, G. Prodi, ed., Springer, Berlin, Heidelberg, 1974, pp. 213–239.

[200] S. Ushiki and R. Lozi, Confinor and anticonfinor in constrained 'Lorenz' system, Jpn. J. Appl. Math.4 (1987), pp. 433–454.

[201] T. Uzer, B. Chirikov, F. Vivaldi, and G. Casati, Obitaries Joseph Ford, Phys. Today 48(10) (1995), p. 88.

[202] C. del Valle and D.J. Nader, Toward the theory of the Yukawa potential, J. Math. Phys. 59 (2018), p. 102103.

[203] C. Vidal, J.-C. Roux, S. Bachelart, and A. Rossi, Experimental study of the transition to turbulence in the Belousov–Zhabotinsky reaction, Ann. N. Y. Acad. Sci. 357 (1980), pp. 377–396.

[204] C. Vidal, S. Bachelart, and A. Rossi, Bifurcations en cascade conduisant à la turbulence dans la reaction de Belousov–Zhabotinsky, J. Phys. 43 (1982), pp. 7–14.

[205] M. Weispfenning, I. Hoeser, W. Böhm, W. Prettl, and E. Schöll, Cyclotron-resonance-induced nonequilibrium phase transition in n-GaAs, Phys. Rev. Lett. 55 (1985), pp. 754–757.

[206] R. Williams, Expanding attractors, Publ. Math. IHES 43 (1974), pp. 169–203.

[207] M. Yamaguti and S. Ushiki, Chaos in numerical analysis of ordinary differential equations, Physica D 3 (1981), pp. 618–626.

[208] A. Zakharova, *Chimera patterns*, in *Networks: Interplay between Dynamics, Structure, Noise, and Delay, Understanding Complex Systems*, Springer, Cham, 2020.

[209] E. Zeeman, Catastrophe Theory: Selected Papers, 1972–1977, Addison-Wesley, Reading, MA, 1977.

[210] Available at https://thestarryeye.typepad.com/weather/2012/02/new-york-weather-highlights-1979.html.

Two more mathematicians and one physicist tell what motivated their careers and how they began

René Lozi, Bo Zheng, Miguel Sanjuán and Guanrong Chen

ABSTRACT
This chapter is a continuation of Chapter 1 dedicated to the early steps of nine mathematicians on their paths to the realm of dynamical systems. A physicist and two mathematicians complete the worldwide panel of scientists chosen for this book, which describes how young people became outstanding researchers and who helped them at the beginning of their careers.

1. Introduction to Chapter 2 by René Lozi

This second chapter is a continuation of Chapter 1 which describes the lives of nine mathematicians from five countries and the paths they took to mathematics, chaos, and discrete dynamical systems. I asked three more outstanding researchers, who were unable to meet the deadline for Chapter 1, to write their personal stories for this chapter so as to complete the worldwide panel of researchers chosen for this book. Since Asia was not represented in Chapter 1, I chose two Chinese mathematicians, and I completed this chapter with a Spanish physicist. This choice was motivated by the nature of dynamical systems that belong to the fuzzy boundary between mathematics and physics. Two of the authors in this chapter, Miguel Sanjuán from Madrid and Guanrong Chen from Hong Kong, published articles in the special issue on chaotic attractors which is the basis of the collection of the three books ([31, 32] and this book). They are outstanding researchers who have been awarded many prizes. I had the pleasure of working with them for many years. The third author, Bo Zheng from Guangzhou, is a specialist in mathematical biology (like Jim Cushing in Chapter 1). She is famous for having co-published as principal author in the journal *Nature* in 2019, an article on experiments involving the release of mosquitoes infected with the bacteria *Wolbachia* in the wild, with the goal of suppressing wild mosquitoes carrying dengue fever.

Therefore, in the two first chapters of this book, there are twelve personal stories, two from China, four from Western Europe (France, Germany, Italy, and Spain), three from Russia, and three from the USA, covering a large part of the world. A third of the contributions of both chapters are by women. This ratio is not entirely satisfactory, but it is significantly higher than the ratio of women to men among mathematicians in the world. The gender gap in mathematical research worldwide is difficult to be exactly determined. However, in a recent publication concerning gender representation on the editorial boards of 435 journals in the mathematical sciences, it was found that only 8.9% of the 13067 editorships are held by women [48]. The balance in this book is near four times higher.

In conclusion, I find it interesting to discover the similarities and differences among the career paths of these mathematicians and physicists and to learn about the outstanding researchers who influenced them, both at the beginning of and later in their careers, about which books were decisive in their choice of this part of mathematics, and about their other interests outside mathematics.

This second chapter is organized as follows: in Section 2, Bo Zheng describes her love for mathematics which she attributes to the patience and guidance of her eldest sister, who detected an interest in mathematics in her when she was very young. While watching her sister doing homework, Bo Zheng would mimic her by drawing geometric shapes (this interest remains in her academic career, as we can notice the geometric arguments and illustrations in some of her publications). In high school, Bo Zheng almost always scored the highest in mathematics in her class, in contrast to other relatively weak performances in other subjects. She graduated from high school in 1998, and she scored 144 out of 150 on the National College Entrance Examination in China in mathematics. Her professors encouraged her to pursue postgraduate studies at Hunan University, the best university in the Hunan Province. Hunan University not only granted her admission but also waived her three-year tuition fees, greatly relieving her family's financial burden.

In 2002, she was admitted as a graduate student at Hunan University, majoring in mathematics. Her family was proud because she was the first graduate student ever in her village. She was particularly fascinated by the course on Ordinary Differential Equations because it was the first time she realized how mathematics could be magnificently applied in real life.

When Bo Zheng was a second-year student, her supervisor, Jianshe Yu, was appointed President of Guangzhou University. Yu transferred his whole research team to Guangzhou. At that time, Bo Zheng was asked to apply Morse index theory to examine the existence and multiplicity of periodic solutions to a discrete system. Her success resulted in her first publication in 2008. Before the publication of the paper, she presented this work in 2006 at *The International Conference on Recent Developments in Differential Equations and Applications* held at Guangzhou University. This was her first academic presentation, which was chaired by Professor Moxun Tang, who became her second mentor in mathematical biology.

Bo Zheng started her academic career at Guangzhou University in March 2008, after she received her Ph.D. degree. Her twin daughters were born in September of the same year. Becoming a mom made her more serene and patient, although it was also the first time that she realized that 1 + 1 was much greater than 2 when taking care of two babies at the same time. Fortunately, she found her secret weapon for soothing her twin daughters: reading math papers or books in English aloud when they cried for unknown reasons. It worked surprisingly well, and she thinks this was the secret to balancing childcare with doing research.

Since 2012, Bo Zheng has started the research in mathematical biology, in 2014, his first paper was entitled "*Establishing mathematical models to study the dynamics of Wolbachia spread in mosquito populations*". In 2015, Guangzhou established the world's largest mosquito factory. At that time Jianshe Yu, Moxun Tang and Bo Zheng collaborated closely.

They were in charge of combining experimental data with mathematical models in order to predict the optimal release ratio for effective suppression.

It was during her visit at Michigan State University in 2015–2016 that Bo Zheng completed three articles, including the one published in *Nature* in 2019. This *Nature* article, still driven by semi-field cage experiments, developed a novel discrete mathematical model to explore effective release strategies for population suppression. The optimal release ratio of sterile males to wild males 5:1, discovered and established through the model analysis, has become an international convention for initial *Wolbachia* release. Ever since this series of studies, she has been publishing intensively in the field of mathematical biology. Her contribution has been recognized by invited and plenary lectures at many international conferences and workshops. In her words, "I found more and more challenging mathematical problems from applications in biological sciences, but I am lucky to have more and more international and interdisciplinary collaborators".

In Section 3, Miguel Sanjuán recalls that all his ancestors hail from Cebrones del Río, a small village in the province of León, in Spain. With his family he spent every summer in the village, where he has vivid recollections of discovering and exploring the nature around him. When he was 10 years old he entered the *Instituto* and studied there until the age of 16. There he was interested in everything, science and humanities, classical and modern languages, history, and geography, as still he is. In his final year of the *Instituto*, which was oriented towards university studies, he was advised to pursue studies in Physical Sciences, Mathematics, or Astronomy. Miguel says that he has always been fascinated by languages, geography, and history, among other subjects. Particularly, this fascination with languages has accompanied him throughout his life. At the *Instituto*, he studied French every year and Latin for two years and self-studied English and German. Later he participated in a summer school to learn Slovak and secured a scholarship from the Finnish Ministry of Education to study Finnish. Later he studied Russian on his own. This fascination with languages also extended to his interest in minority languages; their usage; grammar such as Mirandese, a language derived from Old Leonese; Gallo, a Romance language spoken in the eastern part of Brittany in France; or Aranese, an Occitan language spoken by about 8,000 people in the Valley of Aran in Spain's Pyrenees.

In 1976, Miguel pursued studies in Physical Sciences at the University of Valladolid. Over time, he was particularly drawn to Theoretical Physics, especially Mechanics, Mathematical Physics, Relativity, and Quantum Mechanics. From that year, he keeps the vivid memory of always having aspired to be a researcher. In 1984, he published his first solo research article in the *International Journal of Theoretical Physics* (IJTP) and received a scholarship to pursue a doctorate at the National University at a Distance (UNED) under the supervision of Manuel García Velarde. This scholarship also allowed him to go to the University of Bremen in Germany to work under Ludwig Arnold in the *Institut für Dynamische Systeme* of Bremen.

Some years later, Miguel secured a position at the Polytechnic University of Madrid. At this time, he became deeply interested in deterministic dynamical systems and chaotic systems. In September 1990, he defended his doctoral thesis "Contribution to the analytical study of some properties of nonlinear oscillators", which was a systematic study of the Helmholtz nonlinear oscillator. In the 1995/1996 academic year, he secured a training grant at the Institute for Physical Sciences and Technology (IPST) of the University of Maryland, where he worked under the supervision of Jim Yorke.

In the rest of Section 3, Miguel explains why nonlinear dynamics has captivated his interest and he gives a list of the fundamental books that have shaped his understanding. During these formative years he recalls how his passion for foreign languages has played a significant role, enabling him to access a diverse range of sources beyond those available in English and French. In 1997, he joined the newly created, Universidad Rey Juan Carlos (URJC), where he had the opportunity to create and lead the research group on Nonlinear Dynamics, Chaos, and Complex Systems. He gives also a recollection of his research studies during the last 27 years, the important encounters with outstanding scholars like Jim Yorke, Yoshi Ueda, Edward Lorenz, Leon Chua, Hao Bailin, Wei-Mou Zheng, and many others (he has publications with more than 200 co-authors of 25 different nationalities). In 2002, he had the opportunity to serve as a nominator for the 2003 Japan Prize awarded by the Government of Japan, where Benoît Mandelbrot and Jim Yorke were honored with the award. He was also deeply honored to attend the awards ceremony, where the Emperor of Japan personally presented the prize.

Miguel describes the numerous countries he visited (Bolivia, Brazil, Chile, Colombia, India, Japan, Germany, Lithuania, Mexico, Poland, Peru, Russia, Ukraine, etc.) and those where he has extending stays like China and the USA. In conclusion he declares that his scientific work has been based on two fundamental ideas – interdisciplinarity and internationalization – and affirms that the interdisciplinary methodology of nonlinear dynamics leads us to use similar methods and tools to address numerous research problems across various disciplines. In the intellectual process of interdisciplinary activity, it is not only very important to know how to understand another language but also to be able to translate between different languages.

In 2023, he was honored to join the Royal Academy of Sciences of Spain, with his induction ceremony held in June 2024.

In Section 4, Guanrong Chen recalls that he was born and raised in a rather poor family in Guangzhou, a city in southern China. His parents grew up in a small village in the Zhongshan county of the Guangdong Province. Not being able to finish their elementary school education, his parents moved to Guangzhou to work for survival in the 1930s. Therefore, he did not receive any kind of intellectual assistance in education from them. Nevertheless, he did well in both elementary and middle schools on just about all subjects, particularly in mathematics and Chinese literature. That, in some way, nurtured his self-studying and self-learning abilities with intrinsic motivation. In middle school, driven by curiosity and interest, he started to self-learn mathematics beyond the course curriculum from some interesting books written for middle and high school students. When entered in high school in 1964, he had a chance to present several rather complicated mathematical formulas in trigonometric functions and combinatorics that he derived by himself to his mathematics teacher, whose significantly positive appraisal of his "impressive re-discoveries" greatly encouraged him to look forward to studying higher mathematics, hopefully in college.

Two years later, however, the so-called Chinese Cultural Revolution took place throughout the whole country that lasted for an entire decade, during which all universities in China were closed. Together with several middle school and high school students, he was sent to a mountain village on the inland of Hainan Island in the South China Sea to reclaim wastelands. There, during the following seven years, he lived in grass huts, engaged in hard manual labor, and experienced constant hunger. However, he did not want to waste his life without learning new scientific knowledge. He realized that the only thing he could do was to read books. Fortunately, he had opportunities to bring some books in mathematics and literature, discarded by previous university professors and students, back to the village from Guangzhou, when he was allowed to visit his parents once a year for ten days. He told himself that if he could solve one mathematics exercise problem and read one poem every night, then he might be able to finish a mathematics course and write a number of poems in a year and if so he would not be bored from the world-isolated, rural mountain life.

At the end of 1977, after the end of the Cultural Revolution all universities were re-opened. He passed a higher-level entrance exam in several advanced mathematical subjects and was admitted into the postgraduate program of mathematics in Sun Yat-sen University and gained his Master's degree three years later. Then, he moved on to the USA to study for his Ph.D. degree in applied mathematics under the supervision of Charles K. Chui at Texas A&M University. There he studied mathematical approximation theory and computational methods for complex systems modeling and control. In 1987, moving again, he became Visiting Assistant Professor at Rice University where he started to work with Rui J. P. de Figueiredo on nonlinear control systems. His interest was attracted to chaos theory and bifurcation analysis starting from his first encounter with Leon O. Chua who visited de Figueiredo the same year.

In the remainder of this Section, Guanrong recalls how he used his control theory and engineering background to stabilize chaotic dynamics when they are harmful, using various feedback control techniques, and how in 1998, looking at the classic Lorenz system again, he found what is known now as "Chen system" which is dual to it and has much more complex dynamics. He, together with his colleagues Simin Yu and Jinhu Lu, developed an effective methodology to generate all sorts of multi-scroll and multi-wing chaotic attractors from electronic circuits, using multiple saturated functions.

In 2000, he moved from the University of Houston to City University of Hong Kong where he assumed the position of chair and founded the Research Centre for Complexity and Complex Networks. There he was one of the very first in China to notice and apply the small-world network model and the scale-free network model. Consequently, he led great efforts in the Chinese scientific communities to pursue a research development in complex networks and organized the first national conference in 2004, which has then become an annual conference series on complex networks in China.

Throughout his academic career, he received a number of encouraging awards and recognitions: IEEE Life Fellow, Euler Gold Medal from Russia, Honorary Doctor Degrees by the Saint Petersburg State University, Russia, in 2011 and by the University of Le Havre Normandy, France, in 2014. He received the China National Natural Science Awards for three times, in 2008, 2012 and 2016, respectively, etc. Since 2010, he has succeeded his mentor Leon O. Chua to serve as Editor-in-Chief for the *International Journal of Bifurcation and Chaos*. Now he has officially retired and is currently on a short-term extension of work at City University of Hong Kong. During breaks from teaching and research, he maintains his habit of reading Chinese literature. Perhaps as he looks forward to tomorrow, he will rededicate himself to writing new chapters of his life!

2. My path to dynamics by Bo Zheng

The first person, and also an important one I credit to my love for mathematics, is my eldest sister, Aihui Zheng, who is 11 years older than I and is the first college student in our home village. Since none of my parents or relatives worked in the education system, it was quite a momentous event when our family produced its first university student. My sister detected an interest in mathematics in me when I was very young; while watching her do her homework, I would mimic her by drawing geometric shapes. When I was a fifth-grade elementary school student, my sister graduated from Lingling Normal College (renamed as Hunan University of Science and Engineering in 2004) and became a chemistry teacher at Maogongqiao Middle School in our home village. Despite her major in chemistry, she was incredibly fond of mathematics and decided to pass on her passion to me when she noticed my interest in geometric shapes (this interest remains in my academic career, as you may notice my geometric arguments and illustrations in some of my publications). To help me acquaint myself with the challenging topics, including functions and plane geometry in junior high, she borrowed some mathematics books from other junior high students. While helping me with my homework, she also offered me a head start on the upcoming mathematical materials. My sister was very patient with me. She always gave me enough time to think independently. When I had an idea, she would let me explain my thinking as if I were a teacher. Of course, we didn't have a blackboard at home, so we used chalk to write on the door. Sometimes, as I moved, my thoughts would magically become clearer, and I could even solve problems entirely on my own. More often, however, I would get stuck. Then, my sister would provide a hint. With her guidance, I would engage in further independent thinking. This cycle would continue until the problem was solved. This kind of guidance not only allowed me to appreciate the beauty of logical reasoning in mathematics but also gave me a sense of satisfaction and exhilaration after solving challenging math problems, dispelling any initial frustration I felt when struggling with them. Thus, I was fascinated with mathematics.

With my eldest sister's warm and inspiring companion, my school years were filled with a constant sense of joy at learning new techniques in mathematics and solving difficult mathematical problems. In high school, I almost always scored the highest in mathematics in our class, in contrast to relatively weak performances in other subjects. I graduated from high school in 1998, and I scored 144 out of 150 on the National College Entrance Examination in China (like SAT or ACT in USA) in mathematics. Thus, I was admitted to the Mathematics and Applied Mathematics program (nowadays the College of Mathematics and Statistics) at Jishou University. My academic road started there. Particularly, I was fascinated by the course on Ordinary Differential Equations because it was the first time I realized how mathematics could be magnificently applied in real life. My professors encouraged me to pursue postgraduate studies at Hunan University, the best university in the Hunan Province. Not only I made it, but Hunan University also waived my three-year tuition fees, greatly relieving my family's financial burden.

In September 2002, I became a graduate student in the College of Mathematics and Econometrics at Hunan University (renamed as the School of Mathematics in 2019). My family was proud of me because I was the first graduate student ever in our village. I began working on my thesis under the supervision of Professor Jianshe Yu, Vice President of

Hunan University (and former Chair of the College of Mathematics and Econometrics). In 2003, when Professor Yu was appointed as President of Guangzhou University, the whole research team led by him was transferred there. This was the year when our fruitful collaborations started. At that time, I was advised to read two books, *Discrete Hamiltonian Systems* by Calvin D. Ahlbrandt and Allan C. Peterson [1] and *Critical Point Theory and Hamiltonian Systems* by Jean Mawhin and Michel Willem [35]. I was lucky to take courses from distinguished professors like Jianshe Yu, Zhiming Guo, and Zhan Zhou and also be a member of the brilliant research team applying Critical Point Theory to study discrete systems. Before 2003, there was a big gap between the variational method and its application to discrete Hamiltonian systems and discrete Schrödinger equations in the literature. As Zalman Balanov, Carlos García-Azpeitia, and Wieslaw Krawcewicz pointed out in [7]: *Due to the discrete nature of difference systems, a direct application of the classical methods of Critical Point Theory to this type of systems met a paradigmatic difficulty: essentially, it was not clear how to provide a framework allowing to view a periodic solution/solution to Boundary Value Problems as a critical point of a smooth functional defined on an appropriate (finite-dimensional) vector space.* The breakthrough was made by Jianshe Yu and Zhiming Guo in 2003 [24–26], where the variational structure was first introduced to study the existence of periodic and subharmonic solutions to second-order difference equations. These pioneering studies have opened the door for a systematic study of difference systems, including boundary value problems, homoclinic and heteroclinic orbits, and applications to the discrete Schr¨odinger equation. After nearly 20 years of effort, *The Variational Method for Discrete Systems and Its Applications* was awarded the First Prize in Natural Science by the Guangdong Province in 2020.

It was precisely in 2003 that it was suggested to me that I should try to apply Morse index theory and Conley index theory to a discrete system. Similar to critical point theory, they deal with the existence and multiplicity of periodic solutions but from different angles. Morse index theory is brilliant, while the Conley index turned out to be hard for me. Maybe the main reason was that I was not as lucky as Jim Cushing, who had Charles Conley as his office mate for a while (see Chapter 1, Section 8). This project was challenging but ultimately rewarding, and I got my first publication in this area in 2008 [57]. On July 17–21, 2006, *the International Conference on Recent Developments in Differential Equations and Applications* was held at Guangzhou University. As a Ph.D. student, I delivered my first academic presentation at this conference about *Multiplicity of periodic solutions for second-order discrete Hamiltonian systems with a small forcing term.* I was so nervous that I even cut my hair short (it did not work for me, though) (see Figure 1), and my roommate said I was practicing my presentation in my dreams the night before. Professor Moxun Tang from Michigan State University (MSU) chaired the section where I presented my talk. Thanks to his professional help, the talk went smoothly.

My academic career started at Guangzhou University in March 2008, after I received my Ph.D. degree. My twin daughters were born in September of the same year. Becoming a mom has made me more serene and patient, although it was also the first time that I realized 1 + 1 was much greater than 2 when taking care of two babies at the same time. Fortunately, I found my secret weapon for soothing my twin daughters: reading math papers or books in English aloud when they cried for unknown reasons. It worked surprisingly well, and I think this was the secret to balancing childcare with doing research. In

Figure 1. Left: I was lecturing at the *International Conference on Recent Developments in Differential Equations and Applications* held at Guangzhou University in 2006. **Right:** The photo was taken during the conference's banquet. The second row includes George Sell, Roger Nussbaum, Tibor Krisztin, and Yun Tang. I was sitting next to Elena Braverman in the front.

September 2011, the twins went to kindergarten, which allowed me to focus completely on my research during the daytime. I realized it was time to challenge myself to step out of my comfort zone to explore other research areas.

At that time, mathematical biology had become a highly active field of research, covering a wide range of problems from the molecular level to ecosystem scales. When attending various academic conferences, I often heard many presentations about mathematical models in biology, introducing new mathematical models or using existing ones to explain the dynamical behavior of biological systems. Motivated by this, I emailed Professor Moxun Tang and asked for advice on possible research topics in mathematical biology. I should say this was the bifurcation point in my academic life. Professor Tang was very kind and promptly gave me a new project: establish mathematical models to study the dynamics of *Wolbachia* spread in mosquito populations. Later, I found out the reason that Professor Tang replied so quickly was because he and Professor Zhiyong Xi, a microbiologist from Department of Microbiology, Genetics, & Immunology at Michigan State University (MSU), were neighbors at the time. When he received my email, they were at a family gathering where Professor Xi mentioned that he had just received funding from the Bill & Melinda Gates Foundation to introduce *Wolbachia*-based vector-borne disease control technology to China and planned to establish *The Sun Yat-Sen University-Michigan University Joint Center of Vector Control for Tropical Disease* in Guangzhou. If everything went well, they would be very happy to have mathematicians join their team since mathematical models are necessary to interpret complex experimental data, optimize experimental conditions, and predict experimental outcomes. Along with the email, Professor Tang sent me the paper written by Michael Turelli [49] as well as the first two papers on discrete models for *Wolbachia* spread dynamics [10, 21]. When reading these three articles, almost every word was new to me and sometimes even after figuring out the meaning of each word, I still didn't understand the whole sentence. I remember one time when I told Professor Tang, "This is too much difficult for me". He replied, "'Don't worry. I'll invite the author

Figure 2. Left: The photo was taken during a team meeting at Sun Yat-sen University in 2014. In the front row, Professor Zhiyong Xi is fourth from the right, Professor Moxun Tang is third from the right, and I am second from the right. **Right:** The photo was taken in 2015 during the summer school in Guangzhou University. In the front row, from right to left, Professor Moxun Tang, Professor Michael Turelli, Professor Zhiyong Xi, and me. In the second row, from right to left, Dr. Mugen Huang and Dr. Linchao Hu.

over, and we can discuss it together". Of course, I thought he was joking, but it came true in the Summer of 2015 (see Figure 2). As a reward of three-year effort, in 2014, we published our first paper on this topic [60]. This is the paper that I spent the longest time writing. Professor Tang to whom I owe a lot, in an extremely patient way, guided me to carefully craft and meticulously consider word by word, sentence by sentence. This is a painful yet worthwhile process. It taught me the value of patience, the importance of persistence, and the power of dedication in doing research and writing papers.

In 2015, Guangzhou established the world's largest mosquito factory, with a 3,500 m^2 laboratory producing more than 60 million mosquitoes per week nowadays. In March 2015, *Wolbachia*-infected male mosquitoes were released on Shazai Island and Dadaosha Island, south of Guangzhou City (see Figure 3). Professor Jianshe Yu, Professor Moxun Tang, and I were in charge of combining experimental data to establish a mathematical model to predict the optimal release ratio for effective suppression. Without any hesitation, I dove into this project.

From August 2015 to August 2016, with the support of the China Scholarship Council, I went to MSU for a year-long visit to Professor Moxun Tang and Professor Zhiyong Xi, from whom I have learned a lot about bridging the gap between theoretical models and practical biological applications. Professor Tang and I visited Professor Xi's lab very often. After identifying the special biological questions, we defined the problem in terms that can be quantified and modeled. Before the data were collected through experiments and field trials, we determined what types of data were needed and how they should be collected to support the modeling process. Based on the biological problem and available data, we developed mathematical models and tested them with simulations. Adjustments and refinements were always required when multiple cycles of trial and feedback were necessary for meaningful results. This collaboration was very fruitful, and we completed three articles. One was published in *Ecological Modelling* in 2018 [64] and one in the *Journal of Mathematical Biology* in the same year [61]. The third one below is the most influential. The

Figure 3. In March 2015, *Wolbachia*-infected male mosquitoes were released for the first time in the Nansha district of Guangzhou City.

three-year implementation from 2015 to 2017 in Guangzhou enabled near-elimination of wild-type *Aedes albopictus* field populations. This result was published in *Nature* in July 2019 [65] (see Figure 4). Driven by semi-field cage experiments, we developed the following discrete mathematical model (selected from at least 15 tries of different models)

$$\begin{aligned} &[\text{Egg hatch rate}]\ \xi_i = \frac{b_i\ \xi_0}{b_i + \lambda c_i}, \\ &[\text{Number of wildfemales}]\ a_i = \mu a_{i-1} + 75\cdot\frac{\xi_{i-1}}{\xi_0} + 8, \\ &[\text{Number of wildmales}]\ b_i = \mu b_{i-1} + 75\cdot\frac{\xi_{i-1}}{\xi_0}, \\ &[\text{Number of releasedmales}]\ c_i = \mu c_{i-1} + 75\ r, \end{aligned} \tag{1}$$

for $i = 12, 13, \cdots, 36$, to explore effective release strategies for population suppression. In model (1), ξ_0 is the optimal egg hatch rate of wild population, λ is the relative mating competitiveness of released males to wild males, μ is the survival rate of adult mosquitoes, and r is the ratio of released males to wild males. Model (1) accurately described and predicted target population dynamics in the semi-field cage experiments and supported the notion that the 5:1 "over-flooding" ratio of released males to wild males is sufficient for effective population suppression and/or elimination. Currently, the release ratio 5:1 has become an international convention for initial *Wolbachia* release.

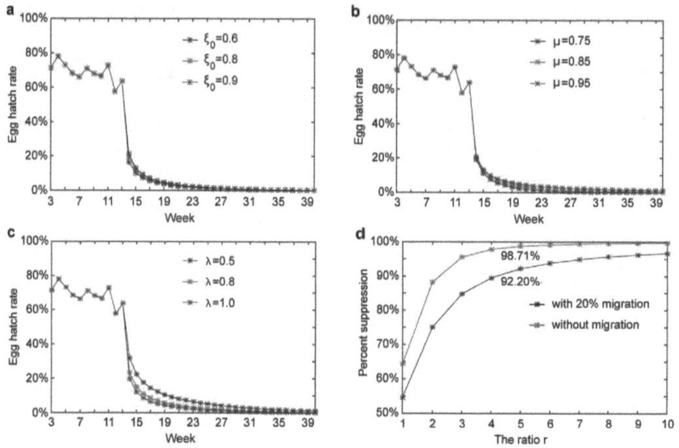

Extended Data Fig. 3 | Sensitivity analysis of the robustness of the 5:1 HC:GUA male release ratio to induce population suppression. **a–c**, The mathematical model of the semi-field cage experiments shown in Fig. 1e, f and described in the Supplementary Information provides an accurate approximation to the semi-field data when $r = 5$ and one of three parameters listed in (5) in the Supplementary Information are varied across a wide range of values. **a**, $R^2 \in [0.9259, 0.9355]$ for $\xi_0 \in [0.6, 0.9]$. **b**, $R^2 \in [0.9301, 0.9329]$ for $\mu \in [0.75, 0.95]$. **c**, $R^2 \in [0.9325, 0.9573]$ for $\lambda \in [0.5, 1]$. **d**, The effect of mosquito migration and the efficiency of population suppression as measured by egg hatch rate. We fixed $\lambda = 0.6$, $\mu = 0.85$, $\xi_0 = 0.80$ and $b_0 = 75$. The 5:1 ratio is sufficient to offset 20% migration with a suppression efficiency 92.20% as compared to 98.71% suppression efficiency without migration.

Figure 4. This figure offers a sensitivity analysis of the robustness of the 5:1 release ratio to induce population suppression.

After completing this field project, our work in mathematics did not stop but instead accelerated owing to our continuous communications with biologists. Reflecting on the entire process of this project, we realized that the mission of the released *Wolbachia*-infected mosquitoes (called sterile mosquitoes hereafter) is to sterilize wild mosquitoes with whom they mate, in the sense that eggs produced by the mating do not hatch. There are three key parameters in designing release strategies: the amount c of each release, the waiting time T between two consecutive releases, and the mating competitiveness between released and wild mosquitoes. The last parameter is mainly determined by the sexual lifespan $\overline{T}$ of the sterile mosquitoes, which characterizes how long the released mosquitoes stay sexually active and energetic. Motivated by these observations, instead of assigning a separate dynamical equation for the released mosquitoes as done in all previous literature (to the best of my knowledge), we introduced an innovative modeling idea that includes the sexually active sterile mosquitoes as a control function, which is nonnegative and depends on the design of release strategies. Based on the new modeling idea, we proposed new mathematical models, *time-switched* models, along with new mathematical questions whose solutions require new mathematical methods (including novel geometric arguments).

For starters of time-switched difference equation models, we refer the readers to

$$x_{n+1} = \begin{cases} \dfrac{(1-s_f)(1+\alpha)x_n}{x_n^2 - \left[1+s_f-\alpha(1-s_f)\right]x_n + 1}, & n \in Z(1+kT, kT+N), \\[2ex] \dfrac{(1-s_f)x_n}{x_n^2-(1+s_f)x_n+1}, & n \in Z(kT+N+1, (k+1)T), \end{cases} \tag{2}$$

for a release strategy that only infected males are supplementarily released to speed up the population replacement [58, 59, 62]. Regarding the time-switched differential equation models, we refer the readers to

$$x'(t)=\begin{cases}\dfrac{ax^2}{x+c}-[\mu+\xi(x+c)]x, & t\in\left[kT,kT+\bar{T}\right),\\ ax-(\mu+\xi x)x, & t\in\left[kT+\bar{T},(k+1)T\right),\end{cases} \tag{3}$$

for the release strategy with $T>\bar{T}$, and

$$x'(t)=\begin{cases}\dfrac{ax^2}{x+(p+1)c}-\left[\mu+\xi\left(x+(p+1)c\right)\right]x, & t\in\left[kT,kT+q\right)\\ \dfrac{ax^2}{x+pc}-\left[\mu+\xi\left(x+pc\right)\right]x, & t\in\left[kT+q,(k+1)T\right)\end{cases} \tag{4}$$

for the release strategy with $T<\bar{T}$, $\bar{T}=pT+q$ with $q\in(O,T)$ [55, 56, 63]. For the above models, we obtained the necessary and/or sufficient conditions for the existence, uniqueness, and stability of periodic solutions, which are challenging questions even without time switching into the model. Also, we initiated our study on time-switched delay differential equation models specified as

$$x't)=\frac{ax^2(t-\tau)}{x(t-\tau)+g(t-\tau)}-\left[\mu+\xi\left(x(t)+g(t)\right)\right]x(t),\quad t\geq 0, \tag{5}$$

where $g(t)$ is the release function, which depends on the implemented release strategy. By introducing the notation of *good solutions*, we counted the exact number of periodic solutions of some special cases of model (4) and analyzed the stability of the periodic solutions. We should mention here that for time-switched models, the number of unsolved problems far exceeds the number of problems that have been solved. For example, the study of planar time-switched systems is almost an unexplored territory due to the lack of effective methods. Currently, we collaborated with Professor Jia Li and Professor Shangbing Ai from the University of Alabama in Huntsville and Professor Jianhong Wu from York University and developed some geometric approaches to study the planar system [2]. However, this seems to be just the beginning of a long path towards fully understanding the complicated dynamics of the model.

In 2023 (see Figure 5 left), as one of ten plenary speakers at the 28th ICDEA conference held in Thailand, I introduced our generalized discrete models

$$x_{n+1}=\frac{(1-\mu)\left(1-s_f\right)\left(1+m_n\right)\left(x_n+f_n\right)}{s_h x_n^2-\left(s_f+s_h+m_n\left(s_f-s_h\right)\right)x_n+1+\left(1-s_f\right)f_n+\left(1-s_h\right)m_n+\left(1-s_f\right)f_n m_n}, \tag{6}$$

with $n = 0, 1, 2$, for *Wolbachia*-release dynamics in mosquito populations based on the first discrete model developed in 1959 [10] (model (6) with $\mu = 0$, $m_n = f_n = 0$) and later in 1978 [21] (model (6) with $m_n = f_n = 0$) and then systematic studies on *Wolbachia* spread dynamics in *Drosophila simulans* since the 1990s by Ary Hoffmann and Michael Turelli [49]. Surprisingly, here is where we come full circle, as these three articles were the very first ones Professor Moxun Tang gave me 13 years ago, back in 2011. Unlike back then when I struggled to understand the articles, I have now introduced a new modeling approach grounded in biological experimentation into these classic models. I love our comprehensive discrete models on this subject because we can formulate conjectures from real-life problems to inspire further studies. These conjectures undoubtedly constitute very challenging mathematical problems, but, more importantly, the answers to these conjectures will provide a much deeper understanding of the full dynamics of the model, which in turn will shed many important insights on the efficacy of control strategies. In May 2024, funded by The National Natural Science Foundation of China (NSFC), I presented our team research at the Workshop of *The Ecology and Evolution of Infectious Diseases* held in London, UK, with experts in a variety of disciplines, including life sciences, mathematics, and social science (see Figure 5 right). This workshop initiated our collaboration with Professor Guanhong Wang from the Chinese Academy of Sciences, Professor Michael C. Wimberly from the University of Oklahoma, and Professor Ravinder Sehgal from San Francisco University.

I am at the crossroads of past accomplishments and future possibilities. Reflecting on the journey that has brought me to this point, I am filled with gratitude for the mentors who guided me, the challenges that shaped me, and the opportunities that allowed me to grow. The road ahead is deterministically uncertain, but I am ready to embrace whatever comes next with the same determination and passion that have carried me this far. As I continue to explore the uncharted territories of science and academia, I am reminded that every new beginning is just another chapter in the ongoing story of discovery, innovation, and the pursuit of excellence.

Figure 5. Left: In 2023, I was lecturing at the 28th ICDEA conference held in Phitsanulok, Thailand, on *Wolbachia-Release Dynamics in Mosquito Populations via Discrete Models*. **Right:** China's Representative at the Workshop of *The Ecology and Evolution of Infectious Diseases* held in London, UK, in 2014.

The author thanks Professor Elena Braverman, Professor Xingfu Zou, Professor Jian Fang, and Professor Huafeng Xiao for their help in photographic materials. Also, many thanks to Professor Jianhong Wu, Professor René Lozi, Professor Saber Elaydi, Professor Jianshe Yu, Professor Moxun Tang, Professor Jia Li, Professor Yuming Chen, Dr. Hongpeng Guo, and my Ph.D. student Huichao Yang for their valuable comments and suggestions, which significantly improved the quality of this article.

3. My journey into nonlinear dynamics and chaos by Miguel A.F. Sanjuán

Exploring how I came to study nonlinear dynamics, chaos, and complex systems is a profound question that traces back to my origins, an idea that has always been close to my intellectual interests. This journey undoubtedly stems from my early fascination with science and the exploration of nature. Although I was born in the city of León in northwest Spain, all my ancestors hail from Cebrones del Río, a small village in the province of León, which is traversed by the River Orbigo, a tributary of the River Esla, the most important and voluminous river in the province. We spent every summer in the village, where I have vivid recollections of discovering and exploring the nature around me, particularly the river, where I spent much time fishing for fish and crayfish as a child, although the hobby lasted much longer (see Figure 6).

Figure 6. With a capture of a trout in the River Orbigo at Cebrones del Río, León.

In those years after primary school, you continued the studies for *Bachillerato* at the *Instituto* for a duration of 7 years. I entered the *Instituto* when I was 10 years old and I finished when I was 16 (see Figure 7). I was interested in everything, science and humanities, classical and modern languages, history, and geography, as still I am.

In the 1970s, the secondary school system in Spain forced us to choose between sciences and humanities for the last three years of study before entering the University. However, it was clear to me that I wanted to pursue a path in the sciences. I vividly remember the awe I felt as a child reading Colin A. Ronan's *Man Probes the Universe* (*Secretos del Cosmos* in his Spanish version) [41], a short history of Astronomy. I remember its price was 25 pesetas, which corresponds to around 0.15 € today. At that time, access to scientific literature was limited, as were my financial resources. Physics and mathematics felt natural to me, complemented by a genuine interest for philosophy. Thus, in my final year of the *Instituto*, which was oriented towards university studies, I was advised to pursue studies in Physical Sciences, Mathematics, or Astronomy. This choice was reflected in the official document at the end of my studies at the *Instituto*.

I have always been fascinated by languages, geography, and history, among other subjects. At the *Instituto*, I studied French every year and Latin for two years. My interest in languages led me to self-study English and German when I was just 14, which I found incredibly enjoyable. This fascination with languages has accompanied me throughout my life. Years later, while working on my Ph.D. thesis, I was invited by the lecturer of Slovak language at the Complutense University of Madrid, who lived in the same student residence as I did, to participate in a summer school to learn Slovak in Bratislava, which was then part of Czechoslovakia. There, I had the opportunity to meet people from around the world who were eager to learn the language or simply seeking an international experience. This experience inspired and motivated me to learn Russian on my own later. In fact, I had read several books by classic authors of Russian literature years earlier, which sparked my interest in the language. Furthermore, I secured a scholarship from the Finnish Ministry of Education to study Finnish, which I pursued at a summer school in Rauma, where I met students from various countries. Subsequently, I decided to travel and explore Scandinavia, and my knowledge of English and German helped me grasp the basics of the three Scandinavian languages. This fascination with languages also extends to my interest in minority languages, their usage, grammar, and, anecdotally, various editions of *Le Petit Prince* that I keep, despite not being at all a book collector (see Figure 8). Some recent curiosities come to mind, such as Mirandese, a language derived from Old Leonese spoken in a small region of Portugal around Miranda do Douro. It is the second official language in Portugal, even though it is spoken by only around 15,000 people. When I hear it in videos on YouTube, it evokes local sounds I remember from my province many years ago. Another interesting language I recently discovered is Gallo, a Romance language spoken in the eastern part of Brittany in France, where Breton, a Celtic language, is spoken in the west. Finally, there is Aranese, an Occitan language spoken by about 8,000 people in the Valley of Aran in Spain's Pyrenees, where it holds official status in Catalonia. These languages and their unique histories continue to deepen my appreciation for linguistic diversity.

Another activity I truly began to enjoy during those years was exploring the mountains in the northern part of my province, particularly the *Picos de Europa* which often inspired many adventures and experiences facing challenges and exploring nature.

Figure 7. This was my ID at the *Instituto* when I was only 12.

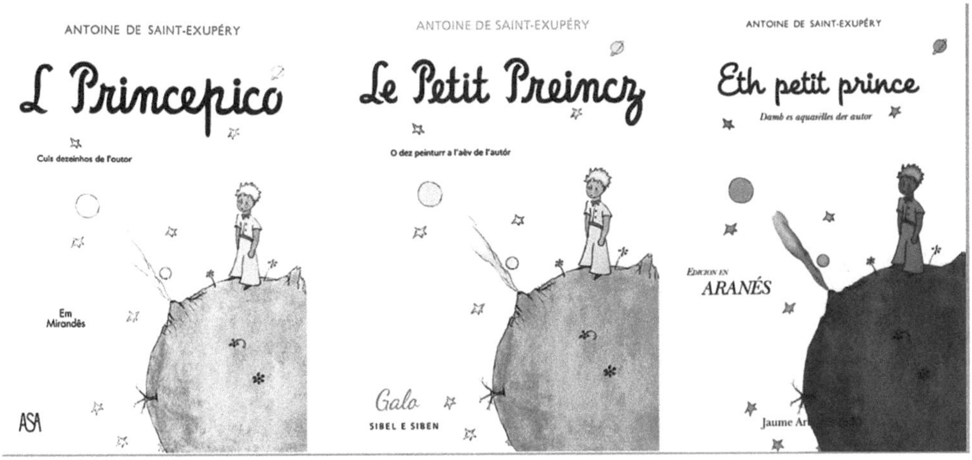

Figure 8. Three versions of *Le Petit Prince* of the French author Antoine de Saint-Exupéry. From left to right, Mirandese, Gallo and Aranese.

Besides languages and geography, history and the origins of things have been always a source of motivation for me. I have developed a genuine curiosity to have a knowledge of the world, the geography, the peoples, the languages, the habits and traditions, etc. I always have enjoyed much reading geographical maps. Incidentally, at the first international scientific conference I attended, which it was in Spain, I had the pleasure of meeting Benjamin Widom, renowned for his work on critical exponents with Kenneth G. Wilson (1982 Physics Nobel Prize) and Michael E. Fisher, though I wasn't aware of this at the time. I remember discussing about the USA and its geography with him, which surprised him greatly, since I had never been there at the moment. Similarly, it is noteworthy how surprised Chinese colleagues are when they see what I know of their provinces. This understanding certainly helps to create connections between people and different cultures.

In 1976, I moved from my hometown in León in the northwest of Spain to the neighbor city of Valladolid to pursue my studies in Physical Sciences at the University of Valladolid, as it offered ample mathematics and the possibility, if desired, to eventually study Astrophysics. I ended up specializing in Fundamental Physics, abandoning the idea of following the specialty of Astrophysics in the last two years of the undergraduate studies, which would have required relocating once again to another University. Over time, I realized I was particularly drawn to Theoretical Physics, especially Mechanics, Mathematical Physics, Relativity, and Quantum Mechanics. I learned that I had a genuine interest more in abstract and general ideas than the more specific ones. From those years, I keep the vivid memory of always having aspired to be a researcher. This ambition has been both a great advantage and a challenge, as difficulties inevitably arise. Moreover, my passion for learning, my curiosity, and my habit of asking many questions and trying to answer them have remained a constant source of knowledge and learning.

My first research project was my Master's degree in physics "Anisotropic contractions of the Poincaré group" [42] (see Figure 9), where Mariano Santander was my advisor, and it was focused on ultrarelativistic groups, where I became acquainted with the Poincaré Group, the Galileo Group, Lie groups, and algebras and group theoretical methods in physics. Defending my work publicly was highly satisfying since I noticed that the efforts done during so much time were rewarded. At the end of the work, I got an intuitive idea connecting the nine Cayley-Klein plane geometries with the kinematics of the group of motions that led me to publish my first solo research article in the *International Journal of Theoretical Physics* (IJTP) in 1984 [43].

Circumstances directed my interest towards statistical mechanics and nonlinear science, as I continually sought alternative paths and novel questions in physics. I recall seeing numerous advertisements in American journals like *Physics Today* about the establishment of nonlinear science centers, including Los Alamos, among others at the beginning

Figure 9. A photo for the official graduation picture of the Physics course in 1981.

of the 1980s. I started to be interested in this sort of phenomena that were totally new and intriguing to me.

This set the stage for me to receive a scholarship in 1984 to pursue a doctorate at the National University at a Distance (UNED) under the supervision of Manuel García Velarde. This opportunity fulfilled one of my dreams: to go to Germany, specifically Bremen, to work under Ludwig Arnold in the *Institut für Dynamische Systeme* of the University of Bremen. During this period, the first fractals exhibition was held at Sparkasse Bremen with Heinz-Otto Peitgen, and Dietmar Saupe, who later wrote the remarkable book on fractals and chaos [39, 40].

It was also during this time that I encountered the book *Synergetics* by Herman Haken [27], which captivated a group of students while in Bremen and inspired us to study it independently. It was a noteworthy initiative on our part, as we decided on our own to dive into the material and present our findings in seminars. I took responsibility for preparing deterministic dynamical systems and particularly the Van der Pol nonlinear oscillator. This was a true challenge. Some years later, I secured a position at the Polytechnic University of Madrid. It was a difficult period with a high teaching load that certainly slowed down my doctoral studies. I became deeply interested in deterministic dynamical systems and chaotic systems. I sought analytical and theoretical methods, leading me to study the Melnikov method and research the Helmholtz nonlinear oscillator, a simple quadratic oscillator though asymmetrical. It was an interesting example that allowed me to use many known methods to understand its dynamics, compute its solutions via Jacobi elliptic functions, and analyze the chaotic properties with the Melnikov method. I explored a great deal of literature, and most of the work was done independently.

Years later, I decided it was time to complete my thesis and began drafting all the work I had done. Before finalizing it, I was offered a project on the integrability of coupled Lorenz systems, where I saw the potential to involve cubic nonlinearities, and mastering techniques of integrability of dynamical systems. My thesis work "Contribution to the analytical study of some properties of nonlinear oscillators" [44] was a systematic study of the Helmholtz nonlinear oscillator, incorporating almost all the techniques available in the literature, in addition to the Lorenz coupled system and integrability.

I mentioned earlier how much I have enjoyed exploring the mountains in the northern part of my home province. In fact, I remember finalizing my Ph.D. thesis while hiking in the mountains of La Uña (León), on my way from La Vecilla to Cain in the Valdeón Valley of Picos de Europa. It was very early in the morning, and a thick fog enveloped the landscape. In that moment, I found inspiration for how to structure my thesis based on all the material I had collected. After that episode, writing it up was simply a matter of sitting down and getting it done.

In September 1990, I defended my doctoral thesis. On October 3, 1990, I climbed Peña Bermeja (2393 m) in the Macizo del Cornión, Valley of Valdeón, Picos de Europa, in the province of León (see Figure 10). This ascent took place just a few days after my Ph.D. defense. I had attempted the climb earlier, but apparently I felt stronger and more prepared to reach the summit after having finished with my Ph.D. thesis.

I had the chance to attend the NATO *Advanced Study Institute Summer School on Chaos, Order, and Patterns* held at Centro di Cultura Scientifica "A. Volta" in Como, Italy, that was organized by Giulio Casati, in the Summer of 1990, where I had the privilege to meet most

Figure 10. At the top of Peña Bermeja (2393 m) in the Valdeón Valley of Picos de Europa mountains on October 3, 1990, a few days after my Ph.D. thesis defense.

of the chaos scientists that I have known through reading the literature. This caused me a big impression. A crucial reading was also the book *Chaos* [22] by James Gleick where all the names contributing to the chaos revolution appear inside, so that I was able to recognize most of them. A pivotal moment was my meeting there with Jim Yorke,[1] with whom I established a lasting relationship. He provided me with a copy of the *Dynamics* software, which he has been writing for years out of his research work in dynamical systems and this allowed me to begin my initial explorations into numerical studies of chaotic dynamical systems. Some years later, he published a book [36] with the software and a manual to use it properly. This represented a significant advancement because, suddenly, numerical methods enabled me to see things that analytical methods could not. After Como, I attended another international meeting in Rota, Cádiz, where many well-known researchers discussed the relationship between turbulence and chaos.

Immediately after having defended my Ph.D. thesis, I decided I wanted to experience a postdoctoral position and sought for opportunities for this aim. At the same time, I obtained a position as an Associate Professor and some years later I had the great opportunity to secure a training grant, allowing me to spend the 1995/1996 academic year at the IPST of the University of Maryland working under the supervision of Jim Yorke (see Figure 11). I also collaborated with Judy Kennedy, Celso Grebogi, and Ed Ott. The book by Ed Ott on *Chaos in Dynamical Systems* [37] had just been published, and Jim Yorke was preparing his *Chaos* [3] book that I had the honor to participate in the revisions preview to its publication. These experiences profoundly impacted my professional and personal life.

1 Jim Yorke is the usual name of James A. Yorke.

Figure 11. In my office at the University of Maryland in 1996.

Understanding why nonlinear dynamics has captivated my interest requires an exploration of its broad implications and the profound connections it fosters with various fields of study. Nonlinear dynamics focus on systems that are not governed by linear equations often leads to complex, unpredictable behavior that can be described as chaotic. This complexity resonates deeply with me, particularly due to its connections with uncertainty and randomness, which are central themes in quantum mechanics.

In quantum mechanics, the fundamental unpredictability and probabilistic nature of particles challenges classical notions of determinism and inspires a fascination with how similar principles manifest in other complex systems.

Moreover, nonlinear dynamics raises intriguing philosophical questions about the nature of reality and our ability to describe it. The unpredictability and sensitive dependence on initial conditions inherent in chaotic systems prompt reflections on the limits of human knowledge and the nature of causality. These philosophical inquiries offer a profound perspective on how we understand and interpret the universe, bridging scientific inquiry with deeper existential considerations. My enthusiasm for nonlinear dynamics extends to the broader spectrum of dynamical systems and their applications across various scientific disciplines. The principles of nonlinear dynamics are not confined to any single field but are rather universal, appearing in physics, mathematics, engineering, biology, and astrophysics, among others. This universality underscores how dynamical phenomena, from the behavior of fluid flows and weather systems to population dynamics and cosmic evolution, can be described and understood through the lens of nonlinear dynamics. Ultimately, nonlinear dynamics provides a cohesive framework for viewing the world as an interconnected whole. It encourages a holistic perspective, where the intricate interrelationships and feedback loops within systems are recognized and explored. This comprehensive approach allows for a deeper appreciation of the complexity and richness of natural phenomena, reinforcing my commitment to understanding and contributing to this fascinating field.

It is crucial to acknowledge the fundamental books that have shaped my understanding during these formative years. My passion for foreign languages has played a significant role in this journey, enabling me to access a diverse range of sources beyond those available in English and French. During that period, literature on the foundations of nonlinear dynamics and chaos theory was sparse and often highly specialized, scattered primarily across research journals. My familiarity with the Russian literature due to having learnt Russian was also helpful to browse the contents of several books existing in this language.

Among the most influential books where I learned a great deal, particularly many aspects related to my doctoral thesis, is the book by J. Guckenheimer and P. Holmes [23] that helped me understand the basic ideas of the Melnikov method, which was one of the few analytical methods available to determine the threshold at which chaos appears in a given dynamical system, particularly in nonlinear oscillators, which was my area of interest. One of the first books that provided a broad overview of certain aspects related to chaos, although oriented more towards fluid physics, was the French book by Bergé-Pomeau-Vidal, published in 1984 [9]. A book that I thoroughly enjoyed and that had a profound influence was Bruce West's essay [54], as it contributed to understanding and awareness of the role of linearity and nonlinearity in science. The theoretical ideas presented in Tabor's book [47] and the comprehensive work by A.J. Lichtenberg and M.A. Lieberman [30] were also instrumental in my academic development.

I remember learning many fundamental aspects of elliptic functions, crucial for studying exact solutions of nonlinear oscillators, from Derek Lawden's book [29], with whom I maintained certain correspondence. A particularly noteworthy text from that era was Robert Devaney's *An Introduction to Chaotic Dynamical Systems* [19], though primarily a mathematics book, it provided critical conceptual insights into chaos theory.

In 1986, I came across the book *Chaos in dynamischen Systemen* [28] by Albrecht Kunick and Willi-Hans Steeb in German. I decided to translate it completely for myself as an exercise to improve my German skills and to gain a broad and systematic knowledge of the contents described in the book about chaotic dynamical systems.

These foundational texts, along with my linguistic skills and diligent study, have been pivotal in shaping my knowledge and approach to nonlinear dynamics and chaos theory.

During those years, I also developed a keen interest in various disciplines, which enabled me to conduct research using techniques from physics and nonlinear dynamics a few years later. It comes to my recollection, the interest in neurosciences. This is especially noteworthy after the announcement of the 2024 Nobel Prize in Physics awarded to John Hopfield and Geoffrey Hinton "for foundational discoveries and inventions that enable machine learning with artificial neural networks". Just a few years after my Ph.D. defense, I acquired Daniel Amit's book, *Modeling Brain Function: The world of attractor neural networks* [4], written shortly after John Hopfield's seminal paper. It is difficult to have even figured out at that time that his work would go on to deserve the 2024 Nobel Prize in Physics.

Upon returning from Maryland, I had the strong desire to find a place where I could continue my research career and develop all the work I desired to do in nonlinear dynamics. In 1996, I learned about the creation of a new university, Universidad Rey Juan Carlos

(URJC), by the Government of the Community of Madrid, located in the southern part of Madrid. In 1997, I joined the university with the mission of organizing the activities around Physics, where I had the opportunity to create and lead the research group on Nonlinear Dynamics, Chaos, and Complex Systems. The early years were filled with excitement and hard work, as everything had to be built from the ground up. After several years, I earned a full professorship in 2002.

Since 1997, I started attending the SIAM conferences on Dynamical Systems in Snowbird, where over the years I met most renowned scholars working in various areas of nonlinear dynamics. For many years, this conference was incredibly exciting, arguably the most important in the field, where you could meet virtually all practitioners involved in nonlinear dynamics.

At the beginning of the century I was eager to broaden my international horizons and get to know Asia, so that I started looking for opportunities to visit Japan and China primarily. In 2002, I received a JSPS fellowship from the Japanese government allowing me to visit the University of Tokyo to work with Kazuyuki Aihara, whom I had met in 1998 in a conference on dynamical systems and statistics organized at the Newton Institute at Cambridge University. This was a substantial change, leading to continuous trips to Japan until recently (see Figure 12).

The same year 2002, I was participating in *Dynamics Days Asia-Pacific* in Hangzhou, China. It was my first visit to China. Numerous renown scientists were present in that meeting.

In 2002, I had the opportunity to serve as nominator for the 2003 Japan Prize, that year devoted to the Science and Technology of Complexity awarded by the Government of Japan, where Benoît Mandelbrot and Jim Yorke were honored with the award. I was also

Figure 12. At the *Dynamics Days Asia-Pacific* held in Nara, Japan, in 2008 with Yoshi Ueda and Kazuyuki Aihara.

deeply honored to attend the awards ceremony, where the Emperor of Japan personally presented the prize (see Figure 13).

During these years, I also visited Potsdam for a brief stay with Juergen Kurths and participated in various international meetings and scientific projects. Around the early 2000s, I began supervising my first doctoral theses, and to date, I have supervised more than 20 Ph.D. students [34]. My strong desire was to establish a school of nonlinear dynamics within the field of physics, in which I was deeply engaged.

In 2004, I had the extraordinary opportunity to visit Lithuania, where I met scientists working in nonlinear dynamics such as Kestutis Pyragas from at the time Semiconductor Institute in Vilnius, known from his work in control of chaos and Minvydas Ragulskis from Kaunas Technological University (KTU) leading to lasting collaborations with KTU, where I have imparted many seminars to the students.

In the Summer of 2005, I participate in a Gordon Conference on Nonlinear Science in Maine, where among other scientists I had the opportunity to meet Edward Lorenz (see Figure 14).

During the same decade, I started receiving postdocs. The first one was Hongjun Cao from Beijing Jiatong University, China, and later Arsen Janoev from the State University of Moscow. Additionally, Guohei Tanaka and Martin Siewe Siewe from the University of Tokyo, Japan, and from the University of Yaoundé, Cameroon, followed for short visits. In 2007, I had the chance to serve at Martin Siewe Siewe's doctoral thesis defense in Yaoundé (see Figure 15).

In 2005, I returned to China and presented seminars at Tsinghua University and Beijing Jiaotong University, both in Beijing. In 2008, I visited India for the first time. It was at the occasion of the International Conference on Recent Developments in Nonlinear Dynamics of the School of Physics of the Bharathidasan University at Tiruchirappalli, Tamil Nadu, India, where the 60th anniversary of Muthusamy Lakshmanan was celebrated, and new important collaborations started. I formed extraordinary collaborations with Indian

Figure 13. At the occasion of the Japan Prize ceremony in Tokyo, Japan, in 2003. On the left with Benoît Mandelbrot and a Japanese scientist. On the right with Jim Yorke and his daughter, Bekra.

colleagues such as Shanmuganathan Rajasekar, with whom I have co-authored numerous articles and books. Several visits to Trichy and Kolkata followed. I met Leon Chua in Cargèse, France, in 2008, the same week that the news about the memristor[2] had been published in *Nature* (see Figure 16). It was a new occasion where I encountered many other notable scientists working in nonlinear dynamics, and the series of conferences *International Workshop-School on Chaos and Dynamics in Biological Networks* at the "Institut d'Études Scientifiques de Cargèse", Corsica, France, were organized by Maurice Courbage until 2014 every two years and I had the opportunity to be invited.

In those years, I began serving on the Editorial Boards of scientific journals such as DCDS-B, CNSNS, and IJBC. It has been an extraordinary experience from which I have learned a great deal.

In the first semester of the school year 2010/2011, I was Visiting Research Professor at Beijing Jiaotong University in Beijing invited by Hongjun Cao under the "Key Invitation Program for Top-Level Experts" from the "State Administration of Foreign Experts Affairs" of the Chinese government. During this period I had the opportunity to visit several prestigious universities in China, including Peking University, Tsinghua University, Beihang University, the Institute of Mathematics and the Institute of Theoretical Physics of the Chinese Academy of Sciences, Fudan University in Shanghai, and City University of Hong Kong, where I delivered numerous research seminars. Precisely at the Institute of Theoretical Physics of the Chinese Academy of Sciences, I had the opportunity to meet Profs. Hao Bailin and Wei-Mou Zheng [6], well-known physicists for their pioneering work in chaos in China (see Figure 17).

In the years that followed, I embarked on numerous extended stays in China, which have continued up to the present day. Besides numerous visits to Fudan University in Shanghai, I imparted numerous courses and seminars in other Chinese universities. These experiences have not only enriched my professional and personal life but have also provided valuable insights and opportunities for collaboration in my field. I was appointed Honorary Professor at several universities in China, including the Institute of Computational Physics and Complex Systems of Lanzhou University, Sichuan University of Science and Engineering at Zigong, and Huaqiao University at Xiamen, Fujian, as well as a member of the International Advisory Committee of the Institute of Systems Science of the same university.

In 2011, I organized a special session of the Physics and Control conference that was held in my hometown, León, Spain, where Jim Yorke, Judy Kennedy, Celso Grebogi, and many foreign colleagues attended. It was certainly a special occasion to meet in the city I was born and raised so many colleagues from all over the world.

In 2014, the degree of *Doctor Honoris Causa* was awarded to Jim Yorke at URJC, where I was acting as his sponsor (see Figure 18). In 2015, Jim Yorke visited Spain again where he participated as a Plenary Speaker as well as Jesús María Sanz Serna, Ex-president of the Royal Academy of Sciences of Spain and other relevant scientists at the *International*

2 A memristor is a non-linear two-terminal electrical component relating electric charge and magnetic flux linkage, completing a theoretical quartet of fundamental electrical components which also comprises the resistor, capacitor and inductor. It was described and named in 1971 by Leon Chua.

Figure 14. With Edward Lorenz in Maine in the Summer of 2005.

Conference on Nonlinear Dynamics and Complexity in La Manga, in the southeast of Spain, organized by Albert Luo and collaborators.

During my new stay at the IPST at the University of Maryland from 2017 to 2018, supported by a Fulbright Scholar grant, I had the opportunity to continue collaborating with Jim Yorke on a new problem of hetero-chaos in higher dimensional systems. This period also allowed me to connect with many friends and colleagues and give numerous seminars in universities in the Washington area of the USA. It was particularly important that in this occasion I was with my whole family.

Other areas where I have developed my efforts in developing nonlinear dynamics and chaos concern the organization of scientific conferences in Spain with the goal to advance nonlinear dynamics and chaos theory within the field of physics and occasionally in collaboration with mathematicians. Numerous activities have been developed at URJC, including nonlinear dynamics courses over the years, doctoral courses, and organizing a Master's program in nonlinear dynamics since 2009. Furthermore, numerous public outreach lectures by invitation at many institutions.

Besides numerous publications, the first in *Reviews of Modern Physics* was a significant challenge. Other publications followed in *Physics Reports* and later in *Reports on Progress in Physics*. And the last one, an article written for *Physics Today*, appeared in the November issue of 2024 [18]. Meriting special mention are the books I have written on different topics on nonlinear dynamics with Manuel de León, with Celso Grebogi and with Shanmuganathan Rajasekar.

Figure 15. Martin Siewe Siewe's doctoral defense in Université of Yaoundé I, Cameroon, in 2007.

Figure 16. During the Cargèse conference in 2008 with Leon Chua.

Figure 17. A visit to Bailin Hao (on the right hand side) and Wei-Mou Zheng (on the left hand side) at the Institute of Theoretical Physics of the Chinese Academy of Sciences in Beijing, China, during my stay in 2011.

Linked to the international dimension of my research, I can mention the several visits to countries in South America like Chile, Peru, Mexico, Bolivia, and Brazil. I was also invited to impart courses on topics related to nonlinear dynamics and chaos to students in Mexico, Bolivia, and Colombia.

A special mention goes to Eastern Europe, where I have maintained scientific relationships with Czech Republic (see Figure 19), Poland, Russia, and Ukraine through various visits and collaborations. In the conference *Dynamical Systems – Theory and Applications* (DSTA 2019), organized by Jan Awrejcewicz in the Lodz University of Technology (Politechnika Łódzka) in Lodz, Poland, I was honored with a special session celebrating my 60th anniversary where many colleagues and friends were present. Later, a research monograph [5] dedicated to my 60th anniversary, compiling numerous papers by colleagues and friends, was published. Furthermore, since 2019, I have been invited to the *Nonlinearity* conferences that Branko Dragović started to organize in Belgrade, Serbia, and in 2024, I was honored to be elected as a member of the Serbian Academy of Nonlinear Sciences.

Due to numerous international collaborations, I was named a member of the Academy of Sciences in Lithuania in 2008. In 2015, I was appointed as a corresponding member of the Royal Academy of Sciences of Spain (RAC[3]), and in 2017, I became a member of

3 RAC: Real Academia de Ciencias Exactas, Físicas y Naturales de España.

Academia Europaea–The Academy of Europe. Most recently, in 2023, I was honored to join the Royal Academy of Sciences of Spain, with my induction ceremony held in June 2024. The main goal of my acceptance speech was to offer a broad overview of the field of Nonlinear Dynamics, Chaos and Complexity [45] that has been the field to which I have dedicated my scientific efforts.

My experiences with learning foreign languages and the interdisciplinary work I have done on complex systems have led me to reflect on the connections between interdisciplinarity and language. Many scientists have pointed out that one of the main difficulties for true interdisciplinarity, intrinsic to the sciences of complexity, is the lack of a common language. In this sense, the problem is very similar to the knowledge of languages. When one masters several languages, one is able not only to express oneself in various ways but also to translate. In the intellectual process of interdisciplinary activity, it is not only very important to know how to understand another language but also to be able to translate different languages. Each individual discipline has its own methods and its own traditions, and this also makes it very difficult to deal with problems from other disciplines and, above all, to interpret them correctly. The best way to deal with interdisciplinary problems is to be anchored in a discipline, otherwise one may be in the absence of true references. However, it would be very convenient for academic training to pay greater attention to interdisciplinary dialogue, so that one can learn various forms of expression, as is done in language learning.

As I have already outlined, for more than thirty years, I have maintained a fruitful collaboration with Jim Yorke, sharing ideas through numerous joint projects and multiple meetings. Additionally, besides the Honorary Doctorate by URJC, where I had the honor of serving as his sponsor, he has been elected as a foreign member of the Royal Academy of Sciences of Spain.

My scientific work has been based on two fundamental ideas: interdisciplinarity and internationalization. The interdisciplinary methodology of nonlinear dynamics leads us to use similar methods and tools to address numerous research problems across various disciplines. Furthermore, my research has a significant international dimension with

Figure 18. With Jim Yorke at the ceremony of his Honorary Doctorate at the Universidad Rey Juan Carlos in 2014, and a later visit in 2015 Toledo, Spain.

Figure 19. With René Lozi and Ron Chen in Ostrava, Czech Republic, in 2013.

more than two hundred collaborators of twenty-five different nationalities among my publications.

This interdisciplinary research covers a wide range of thematic areas such as dynamical systems and fractal structures, with developments on ideas such as basin entropy and Wada basins. Applications in celestial mechanics and astrophysics, such as fractal structures in the shadows of binary black holes, and the predictability of chaotic dynamics in galactic models. Applications in physics involve chaos and quantum entanglement, chaotic dynamics and fractal structures in cold atom experiments, both classical and relativistic chaotic scattering, and plasma physics problems. Methods of control of chaotic systems, such as the partial control method and the phase control method. Noise and stochastic phenomena in physical, mechanical, or biological dynamic systems. Machine learning algorithms for classifying attractor basins, as well as explorations of chaotic systems in Evolutionary Game Theory. Nonlinear mechanics, dynamics of nonlinear oscillators, and the development of vibrational resonance phenomena and other nonlinear resonances. Modeling of biological systems: developments in computational neuroscience; physics and modeling of cancer dynamics through dynamic systems; and ecological complexity models; genetic networks; and control of infectious diseases. Finally, the fundamentals of nonlinear dynamics and chaos theory, such as the analysis of so-called indecomposable continua related to the topological structure of chaos, and the concept of hetero-chaos as a new paradigm in the study of high-dimensional dynamic systems. References to all these works can be easily found in Google Scholar [46].

This has been a brief overview of my scientific journey and my personal adventure in the fascinating field of research of nonlinear dynamics and chaos. I would like to conclude by conveying a profound message of optimism and hope. I am deeply convinced that, given the nature and youth of the discipline, an extremely bright and promising future awaits us. There is still much work to be done and numerous objectives, challenges, and difficulties to be faced as well as new opportunities to advance knowledge.

4. My academic life through a winding path by Guanrong Chen

Ever since my childhood, I have always been and still am a forward-looking person, with little interest in looking back to the winding path of my life and academic career. But when beckoned to reflect, I concede that I have accumulated many small stories that are perhaps worth telling.

I was born in 1948 and raised in a rather poor family in Guangzhou, a city in southern China. My parents grew up in a small village in the Zhongshan county of the Guangdong Province, near the home of Sun Yat-sen (1866–1925), who was the founding president of the Republic of China. Not being able to finish their elementary school education, my parents moved to Guangzhou to work for survival in the 1930s. Obviously, I did not receive any kind of intellectual assistance in education from them. Nevertheless, I did well in both elementary and middle schools on just about all subjects, particularly in mathematics and Chinese literature. That, in some way, nurtured my self-studying and self-learning abilities with intrinsic motivation. In middle school, driven by curiosity and interest, I started to self-learn mathematics beyond the course curriculum from some interesting books written for middle and high school students by, for example, the famous Chinese mathematician Loo-Keng Hua[4] (1910–1985), who published a series of elementary mathematics treatises for the youths that stimulated and guided me to pursue further studies of mathematics. Hua and Shiing-Shen Chern[5] (1911–2004) were considered the greatest mathematicians of modern China in their time. While Hua lived in China, Chern was Professor at the University of California at Berkeley in the USA.

I entered high school in 1964 and had a chance to present several rather complicated mathematical formulas in trigonometric functions and combinatorics that I derived by myself to my mathematics teacher, whose significantly positive appraisal of my "impressive re-discoveries" greatly encouraged me to look forward to studying higher mathematics, hopefully in college.

Two years later, however, the so-called Chinese Cultural Revolution took place throughout the whole country that lasted for an entire decade, during which all universities in China were closed. Of course, my dream of going to college was shattered, as were those of the rest of a whole generation of youth. Together with several middle school and high school students, I was sent to a mountain village in the inland of Hainan Island in the South China Sea to reclaim wastelands. There, during the following seven years, we lived in grass huts, engaged in hard manual labor, and experienced constant hunger.

Unlike the other young fellows in the village who completely gave up their hopes for the future and passively spent their time in playing poker and chess games after work

4 Hua's main contributions to number theory include the Goldbach-Waring problem (on formulating every integer as the sum of powers of primes). A number of important mathematics results, such as Brauer-Cartan-Hua theorem, Hua's lemma, Hua's inequality and Hua's identity in Jordan algebra, are named after him.

5 Chern, commonly regarded as one of the greatest mathematicians of the twentieth century, is the "father of modern differential geometry".

Figure 20. I was sent to a mountain village in the inland of the Hainan Island to work for seven years (1968–1975), where I lived in a hut and self-studied mathematics at night (recent photos taken by some unknown tourists).

every day, I did not want to waste my life without learning new scientific knowledge. Since there were absolutely no conditions for anyone to learn physics or engineering that require experimentation, I realized that the only thing I could do was to read books. Fortunately, we were allowed to visit our parents once a year for ten days each time, which gave me opportunities to bring some books in mathematics and literature back to the village from Guangzhou, since I knew that reading books in these two subjects is most time-consuming. Motivated by my strong interests in mathematics and Chinese classic poetry, I brought back from every home visit some mathematics books discarded by previous university professors and students, along with some collections of classic poems. I told myself that if I could solve one mathematics exercise problem and read one poem every night, then I might be able to finish a mathematics course and write a number of poems in a year and if so I would not be bored from the world-isolated, rural mountain life.

Driven by my determination to accomplish that goal, I lit up a small kerosene lamp every night in the hut to study mathematics. When I felt tired of mathematics, I read and wrote poems.[6] I studied alone because no one was interested in joining me to learn "useless mathematics" that had already been abandoned by the revolution. However, I strongly felt that learning mathematics was an enjoyable way to enrich my tiring and boring life in the mountain village. Figure 20 resembles a scenario of my living environment, with recent photos taken by some unknown tourists.

I started learning Calculus from a college textbook with exercises and solutions and made a clear plan of reading the textbook page by page and trying to solve one exercise problem each day. Moving forward steadily, I spent more than a year to basically complete the self-study of Calculus, of course leaving some difficult exercise problems unsolved.

6 See concluding remark.

I then moved on to learn Linear Algebra. In this way, through the seven years in the mountain village, I essentially completed teaching myself the first courses of Calculus, Linear Algebra, Ordinary Differential Equations, and Theoretical Mechanics. Meanwhile, I also studied English with some help from an enthusiastic elementary school teacher who was sent from the South China Normal University in Guangzhou to work in the same village. I started from the English alphabet, following a couple of discarded middle school textbooks. But listening and speaking English were out of the question. Yet by the time when I left the island, I could read simple English articles with a dictionary.

In 1975, near the end of the Cultural Revolution, all of us young fellows found our ways to return home. At that time, my father filed an excuse of being seriously sick and requested from the authority for me to replace him as a porter in a railroad station in Guangzhou, at my return from the island. In Guangzhou, while I worked in the railroad station during the weekdays, I was able to continue learning mathematics by myself in the evenings and the weekends.

For my self-study, I was fortunate to be introduced by a family friend to two mathematics professors, Mingjun Chen (1934–2008) and Yuesheng Li (1930–2024), of the Sun Yat-sen University in Guangzhou (see Figure 21). During those days, they were my highly respected mentors, who had helped me to become a self-taught "amateur mathematician". In particular, my beloved mentor, Mingjun Chen, hosted my frequent visits to his home and often treated me to lunches with his family after giving me free tutorials. I also paid occasional visits to Yuesheng Li, who was a pioneer and leader in computational mathematics in China and specialized in spline function theory and numerical analysis. In fact, I visited them almost every weekend, and they were happy to provide me with free tutorials, explaining some basic concepts and solving some difficult exercise problems. This was timely in my favor, as the university was still closed and they had no teaching duties during that period. Within three years, I had partially completed learning the basics of Probability Theory, Statistics, Partial Differential Equations, and Functional Analysis.

Figure 21. Mingjun Chen (1934–2008) and Yuesheng Li (1930–2024).

In 1976, the Cultural Revolution officially ended and all universities were re-opened by the end of 1977. I took and passed a stringent nationwide entrance exam on several high school subjects and thereby was admitted to the undergraduate program of the South China University of Technology in Guangzhou. Three months later, in early 1978, I took and passed a higher level entrance exam in several advanced mathematical subjects and was consequently admitted into the postgraduate program of mathematics at Sun Yat-sen University. There, I studied computational mathematics, with an emphasis on spline function theory and numerical analysis under the supervision of Yuesheng Li. In the first two years, I took some courses on Advanced Linear Algebra, Real Analysis, and Partial Differential Equations.

Three years later, I gained my Master's degree. Then, through a rather tedious process I moved on to the USA to study for my Ph.D. degree in applied mathematics at Texas A&M University. There, I studied mathematical approximation theory and computational methods for complex systems modeling and control, under the supervision of Charles K. Chui (see Figure 22), who was a world-leading expert in these fields and, since 1990, one of the pioneers of wavelets and a leading authority in computational and applied harmonic analysis. Continuously driven by my innate desire of making up for the loss of ten years' time, I studied tirelessly on several advanced subjects in applied mathematics as well as systems engineering. Meanwhile, I also started to do research on complex dynamical systems modeling, control, analysis, and computation. There was no hurry for me to graduate, though, partially because I might have to return to China to work thereafter. Admittedly, instead of submitting my thesis earlier, I started to write, together with my Ph.D. advisor, several papers and books, resulting in a textbook on optimal estimation, entitled *Kalman Filtering with Real-Time Applications* [16] and a research monograph on optimal systems reduction, entitled *Discrete* H^∞ *Optimization* [17].

Figure 22. With Charles K. Chui (photo taken in 2016).

It happened one day in the spring of 1987, just before I defended my Ph.D. thesis, I received a phone call from Rui J. P. de Figueiredo (1929–2013) at Rice University. Although we did not know each other, he asked if I would be interested in working with him as a postdoc fellow, based on his reading of a couple of my papers. I consequently became Visiting Assistant Professor at Rice for three years, where I started to work with Rui on nonlinear control systems. Rui is an Indian born in Goa, a Portugal colony, therefore inheriting a Portuguese name. He had great leadership in the IEEE Circuits and Systems Society and was a leading expert in the fields of nonlinear systems, signal processing, and circuits theory (see Figure 23). At Rice, together with Rui, we completed and later published a research monograph entitled *Nonlinear Feedback Control Systems—An Operator Theory Approach* [20].

After my postdoctoral research at Rice, my focus turned to nonlinear science and engineering, encompassing a wide spectrum of research topics in nonlinear control systems and complex dynamical analysis. My interest was attracted to chaos theory and bifurcation analysis since the late 1980s, starting from my first encounter with Leon O. Chua from the University of California at Berkeley, who visited Rui at Rice in 1987. I immediately recognized that Chua's circuit is an excellent platform for me to work on regarding chaos theory.

With control theory and systems engineering background, I first worked on chaos control by means of stabilizing chaotic dynamics when they are harmful, using various feedback control techniques. Those mathematical and engineering methodologies differ from the parameter-variation approach that was popular in the physics community, typically the Ott-Grebogi-Yorke perturbation method for "controlling chaos" [38]. After several years of pursuit, I summarized the research progress of the field in a large-sized book,

Figure 23. With Rui J. P. de Figueiredo (photo taken during the IEEE ISCAS in 2008).

written with my Ph.D. student Xiaoning Dong, entitled *From Chaos to Order: Methodologies, Perspectives and Applications* [13].

In the early 1990s, I also worked on engineering applications of chaos, specifically using chaos to benefit secure communication, data encryption, and liquid mixing. For those applications, we expect to generate or sustain chaos by means of control, which I called anti-control of chaos or chaotification. One day in 1998, I looked at the classic Lorenz system again,

$$\begin{cases} \dot{x} = a(y-x), \\ \dot{y} = cx - xz - y, \\ \dot{z} = xy - bz, \end{cases} \tag{7}$$

where a, b, and c are real parameters. As is well known, when $a = 10$, $b = 8/3$, and $c = 28$, the system is chaotic, with an attractor, as shown in Figure 24 (a). With the parameters a, b, and c not in the range of chaos, I applied my anti-control technique to add a linear controller

$$u = \alpha x + \beta y + \gamma z \tag{8}$$

to the right-hand side of the second equation of the Lorenz system (7), where α, β, γ are real constants to be determined. Then, under the Shilnikov condition for having chaos, I determined these three coefficients such that the resultant controlled system is chaotic. It turned out that one simple choice is $\alpha = -a$, $\beta = c+1$, $\gamma = 0$, yielding the so-called "Chen system", named by others afterward, described by

$$\begin{cases} \dot{x} = a(y-x), \\ \dot{y} = (c-a)x - xz + cy, \\ \dot{z} = xy - bz, \end{cases} \tag{9}$$

where a, b, and c are the Lorenz parameters. When $a = 35$, $b = 3$, and $c = 28$, the system is chaotic, with an attractor, as shown in Figure 24 (b), which was immediately verified and visualized by my postdoctoral fellow Tetsushi Ueta.

The Chen system appears to be similar in formulation, yet nevertheless is not mathematically equivalent to the Lorenz system. In fact, in the sense of algebraic structures it is a dual system to the Lorenz system and it has much more complex dynamics. Soon after, with my Ph.D. student Jinhu Lu, we coined another chaotic system bridging the two dual chaotic systems. Later, together with my colleague Sergej Celikovsky, based on the three closely related chaotic systems we constructed a large family of generalized Lorenz systems [11].

Meanwhile, with my colleague Simin Yu and also Jinhu Lu, we developed an effective methodology to generate all sorts of multi-scroll and multi-wing chaotic attractors from electronic circuits, using multiple saturated functions, as demonstrated by the example shown in Figure 25 [33], for which we can control the number, location, and orientation of the scrolls and wings of the chaotic attractor.

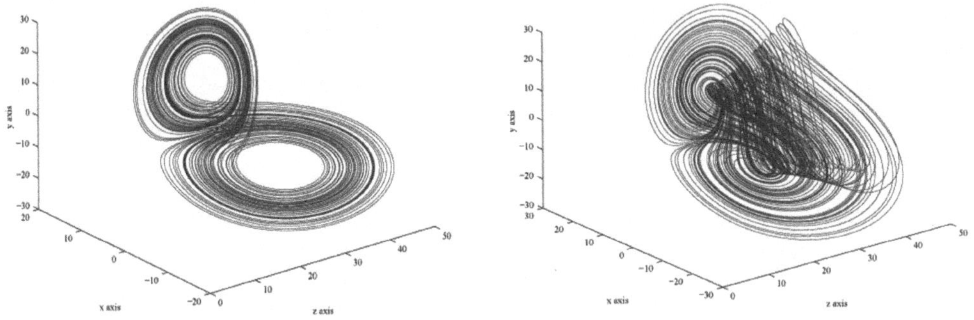

Figure 24. Phase portraits of the two chaotic systems: (a) Lorenz attractor and (b) Chen attractor.

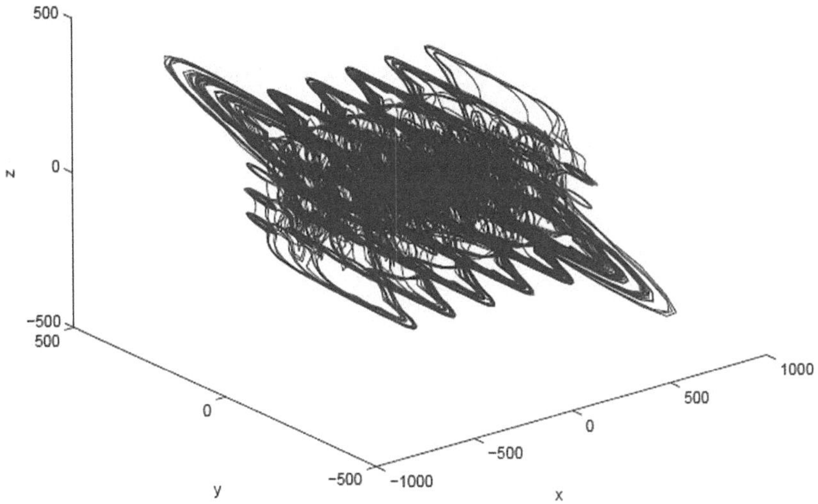

Figure 25. A chaotic attractor with multi-scrolls in three directions.

For linear maps, on the other hand, in collaboration with colleagues Dejian Lai and especially Yuming Shi, my anti-control technique provides a unified way to rigorously generate chaos in the sense of Li-Yorke or in the sense of Devaney via simple and implementable feedback controls [14].

Recently, my Ph.D. student Xiong Wang and I, along with several pivotal collaborators, found chaos from some first-order three-dimensional quadratic polynomial systems that have no equilibria [51] (see Equation (10) and Figure 26 (a)), or have only stable equilibria [50] (see Equation (11) and 26 (b)), or have infinitely many equilibria that form a curve or a surface or even a subspace, summarized in our edited book (with Nikolay V. Kuznetsov) entitled *Chaotic Systems with Multistability and Hidden Attractors* [52]. One example is

$$\begin{cases} \dot{x} = y, \\ \dot{y} = z, \\ \dot{z} = -y + 3y^2 - x^2 - xz - a, \end{cases} \tag{10}$$

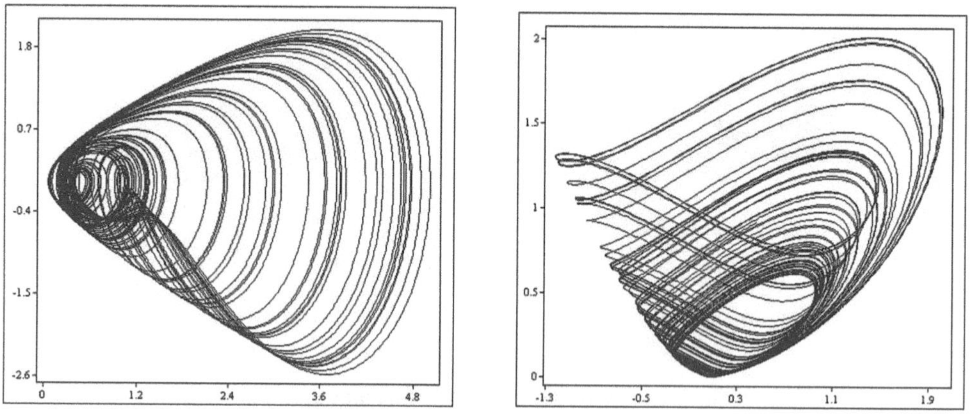

Figure 26. Chaotic attractors: (a) System (10) and (b) System (11).

which, for any real parameter $a > 0$, has no equilibrium point, but interestingly it has a chaotic attractor, as shown in Figure 26 (a). Another example is

$$\begin{cases} \dot{x} = yz - a, \\ \dot{y} = x^2 - y, \\ \dot{z} = 1 - 4x, \end{cases} \tag{11}$$

which, with $a = 0.005$ or $a = -0.05$, has only one stable equilibrium point, but surprisingly it has a chaotic attractor, as shown in Figure 26 (b). Both systems (10) and (11) can be verified to be chaotic by all measures such as positive Lyapunov exponents and fractional dimensions. This great collaborative research effort widens up a research direction on non-hyperbolic chaotic systems.

I moved to the City University of Hong Kong from the University of Houston in year 2000, to work as Chair Professor and now Hong Kong Shun Hing Education and Charity Fund Chair Professor in Engineering, where I founded the Research Centre for Complexity and Complex Networks. In my endeavour to extend the study of synchronization from two coupled chaotic oscillators to a network of chaotic oscillators, I started to look for efficient complex network topologies. I was one of the very first in China to notice and apply the small-world network model [53] and the scale-free network model [8]. I consequently led great efforts in the Chinese scientific communities to pursue a research development in complex networks and organized the first national conference in 2004, which has then become an annual conference series on complex networks in China (see Figure 27).

With my colleagues Xiaofan Wang, Xiang Li, and Zhisheng Duan, we initiated the pinning control approach to achieving complex network synchronization, resulting in a unified pinning control methodology, and established the first criterion for network synchronization, both published in 2002. We put all these earlier advancements together into a textbook type of research monograph entitled *Introduction to Complex Networks—Models, Structures and Dynamics* [15]. Recently, I also published a textbook entitled *Nonlinear Systems: Stability, Dynamics and Control* [12]. Figure 28 shows the two book covers.

Figure 27. Chinese forums on complex networks (from the first in 2004 to the twentieth in 2024).

Throughout my academic career, I received a number of encouraging awards and recognitions: I was elected IEEE Fellow in 1997 and am now Life Fellow. I was awarded the 2011 Euler Gold Medal from Russia and conferred Honorary Doctor Degrees by the Saint Petersburg State University, Russia, in 2011 and by the University of Le Havre Normandy, France, in 2014. I received the China National Natural Science Awards for three times in 2008, 2012, and 2016. Moreover, I became a member of Academia Europaea in 2014 and a fellow of The World Academy of Sciences in 2015. Since 2010, I have succeeded my mentor Leon O. Chua (see Figure 29) to serve as the Editor-in-Chief for the *International Journal of Bifurcation and Chaos*. These are honors both bestowed and earned through sheer diligence, dedication, and dogged persistence of academic endeavors, under a long-term unconditional strong support from my wife Helen and our two lovely daughters, Julie and Leslie.

In retrospect, I did not wish to inherit a legacy from the ten-year deprivation of my youth marked by helplessness or lack of control over my life. Instead, by relentlessly pursuing my goals, with unwavering belief in myself, and drawing on my strong discipline for self-study of new subjects and self-exploration of new research directions, I have recovered a decade of my life that was lost to the Chinese Cultural Revolution. I have since turned that loss into a personal abundance that has launched a fulfilling career and enriched an entire academic life.

Finally, as a closing remark, I have officially retired and am currently on a short-term extension of work at the same City University of Hong Kong. When having breaks during teaching and research work, I maintain my habit of reading Chinese literature books. But I have not been writing Chinese classic poems for years, leaving a collection of my old

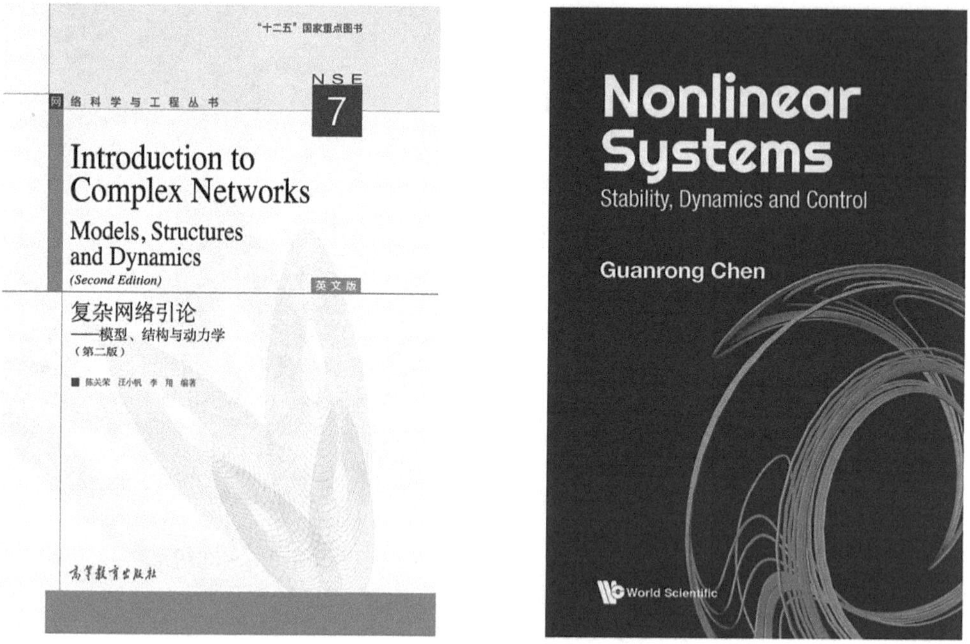

Figure 28. Two recent books: (a) A textbook on complex networks (2012; 2nd ed. 2015) [15] and (b) A textbook on nonlinear systems (2023) [12].

Figure 29. With Leon O. Chua (photo taken in 2012).

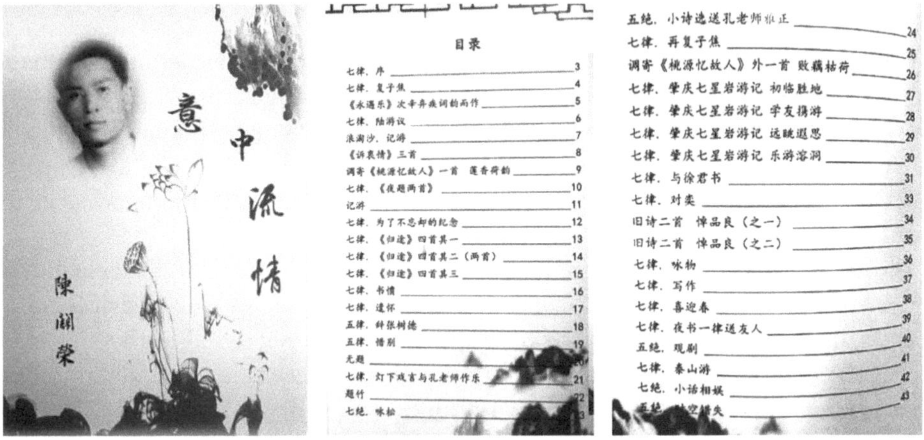

目录

七律. 序 3
七律. 复子焦 4
《永遇乐》次辛弃疾词韵而作 5
七律. 陆游议 6
浪淘沙. 记游 7
《诉衷情》三首 8
调寄《桃源忆故人》一首 莲香荷韵 9
七律. 《夜题两首》 10
记游 11
七律. 为了不忘却的纪念 12
七律. 《归途》四首其一 13
七律. 《归途》四首其二（两首） 14
七律. 《归途》四首其三 15
七律. 书愤 16
七律. 遣怀 17
五律. 钟张树德 18
五律. 惜别 19
无题 20
七律. 灯下戏言与孔老师作乐 21
题竹 22
七绝. 咏松 23
五绝. 小诗选送孔老师雅正 24
七律. 再复子焦 25
调寄《桃源忆故人》外一首 败藕枯荷 26
七律. 肇庆七星岩游记 初临胜地 27
七律. 肇庆七星岩游记 学友携游 28
七律. 肇庆七星岩游记 远眺遐思 29
七律. 肇庆七星岩游记 乐游溶洞 30
七律. 与徐君书 31
七律. 对奕 33
旧诗二首 悼品良（之一） 34
旧诗二首 悼品良（之二） 35
七律. 咏物 36
七律. 写作 37
七律. 喜迎春 38
七律. 夜书一律送友人 39
五绝. 观剧 40
七律. 泰山游 41
七绝. 小话相娱 42
五绝. 空错失 43

Figure 30. A self-printed collection of my Chinese classic poems. On the left is the cover page, in the middle and on the right is the table of contents

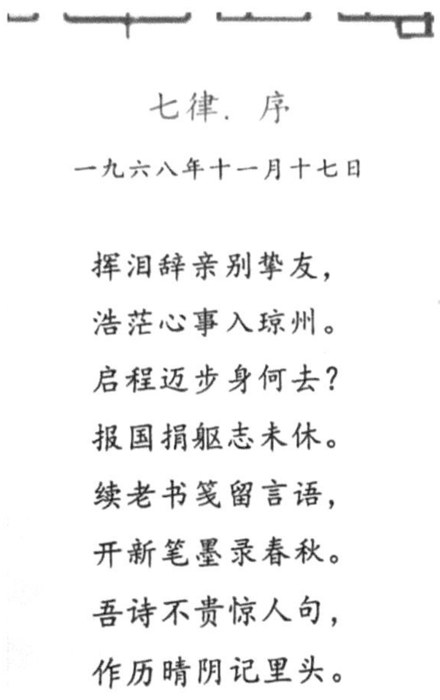

七律. 序

一九六八年十一月十七日

挥泪辞亲别挚友，
浩茫心事入琼州。
启程迈步身何去？
报国捐躯志未休。
续老书笺留言语，
开新笔墨录春秋。
吾诗不贵惊人句，
作历晴阴记里头。

Figure 31. My very first poem (see translation in foot note).

poems in the bookshelf (see Figures 30, 31).[7] Perhaps as I look forward to tomorrow, I will rededicate myself to writing new chapters of my life!

7 In Figure 30, on the left is the cover page, in the middle and on the right is the table of contents. In Figure 31, is my very first poem:
A New Start (November 17, 1968)
Saying goodbye to my family and friends with tears, my heart is filled with worries.
Where will I be going?
I am not afraid of sacrificing my life for my country.
I will continue to write my life history on the old papers, but to write it with a new pen.
I will not value my poems with surprising sentences; they record my life in the sunny and cloudy days.

5. Conclusion

In the conclusion of Chapter 1, we asked whether there were similarities in the paths and ways in which nine long-standing mathematicians specializing in discrete dynamical systems chose their academic careers. With two additional mathematicians and one physicist, we have broadened our panel, covering the world more extensively, including China and Spain. We have to ask ourselves whether our conclusions are still valid or whether they need to be revised.

The first remark was that for the nine mathematicians of Chapter 1, mathematics was always present from their youth, to some extent. In this chapter, we can see that Guanrong Chen in middle school, driven by curiosity and interest, started to self-learn mathematics beyond the course curriculum from some interesting books written for middle and high school students by, for example, the famous Chinese mathematician Loo-Keng Hua. After, in 1964, he entered high school, he had a chance to present several rather complicated mathematical formulas in trigonometric functions and combinatorics that he derived by himself to his mathematics teacher, whose significantly positive appraisal of his "impressive re-discoveries" greatly encouraged him to look forward to studying higher mathematics, hopefully in college. This is reminiscent of what said by Michal Misiurewicz, Jim Michael Cushing, and René Lozi and the decision to study mathematics taken by Laura Gardini when she was 12 years old. In similar way, Bo Zheng, under the kind of guidance of her eldest sister, felt a sense of satisfaction and exhilaration after solving challenging mathematical problems. Her school years were filled with a constant sense of joy at learning new techniques in mathematics and solving difficult mathematical problems.

In Chapter 1, it is also highlighted that mathematics was for some in competition with other topics. Eckehard Schöll recalled that as a schoolboy he was interested in a very broad spectrum of subjects, ranging from Latin, English, French, German via History, Music, Art to Mathematics and Physics. Galina Strelkova was very attracted to medicine and she also seriously studied music, as did Lyudmila Efremova who made attempts to write music. Similarly, it can be noted that Guanrong Chen was attracted to Chinese literature and Miguel Sanjuán at the *Instituto* was interested in everything, science and humanities, and particularly classical and modern languages, history, and geography. Besides the study of French and Latin there, his interest in languages led him to self-study English and German when he was just 14, which he found incredibly enjoyable. He also participated in a summer school to learn Slovak and studied Finnish secured by a scholarship from Finnish Ministry of Education. Later he studied Russian. He has also shown interest in minority languages, such as Mirandese, Gallo, and Aranese, among others.

A third remark in the conclusion of Chapter 1 is that electronics research inspired several authors like Belykh who visited the Electronics Laboratory at Danish Technical University (DTU) in Lyngby or Schöll who gave talks at the International Conference on the Physics of Semiconductors (ICPS) in Edinburgh, at the German Physical Society (DPG) Annual Spring Meeting in Münster and at the International Conference on Hot Carriers in Semiconductors (HCIS) in Montpellier. More generally, both physics and mathematics were his privileged domains of research. After finishing his British Ph.D. in Mathematics at Southampton, he moved to Aachen and prepared another Doctoral Thesis (Dr. rer. nat.) in Physics, which he completed in 1981. We can compare his career with those of Sanjuán who pursued studies in Physical Sciences at the University of Valladolid, which resulted in his specializing in Fundamental Physics. Over time, he realized that he was particularly drawn to Theoretical Physics, especially Mechanics, Mathematical Physics, Relativity, and Quantum Mechanics. Later his interests were directed towards statistical mechanics and nonlinear science (deterministic dynamical systems and chaotic systems) leading him to study the Melnikov method and research the Helmholtz nonlinear oscillator, which is more in the domain of mathematics.

Guanrong Chen completed his Ph.D. degree in applied mathematics at Texas A&M University. There, he studied tirelessly mathematical approximation theory and computational methods for complex systems modeling and control and several advanced subjects in applied mathematics as well as systems engineering. He also began research on complex dynamical systems modeling, control, analysis, and computation. After his postdoctoral research at Rice University, his focus turned to nonlinear science and engineering, encompassing a wide spectrum of research topics in nonlinear control systems and complex dynamical analysis. Control theory uses mathematical tools, however, in a different scope than mathematics.

Therefore, one can see that there is no strict border between mathematics, physics, and control theory, especially in the domain of dynamical systems. As said by Sanjuán, "The principles of nonlinear dynamics are not confined to any single field but are rather universal, appearing in physics, mathematics, engineering, biology, and astrophysics, among others".

The case of biology is especially interesting in its links with mathematics. In Chapter 1, Jim Cushing recalls that period of his professional life, from the late 1970s to the early 1990s, was an active one that involved many studies and publications in theoretical ecology. The overall goal of his work during this period was to investigate the extent to which basic principles of classical theoretical ecological dynamics holdup (or do not hold up) when demographic structure within a population is included in the models – principles such as logistic equilibration, predator-prey oscillations, competitive exclusion, and so on. Then he met biologist Robert Costantino, an expert on flour beetles. With him, his former postdoc Robert Desharnais and Brian Dennis, an ecologist and statistician, they formed the *Beetle Team*, later joined by Shandelle Henson and Aaron King. For years they developed and utilized predictive models for the dynamics of a biological population (namely, the flour beetle), which would be validated by carefully designed, controlled, and replicated laboratory cultures. Their beetle chaos project accomplished much more than just a tight model-data fit and a demonstration of biological chaos. Numerous other dynamic phenomena were explained by the model, often with jaw-dropping accuracy. Some of these models predicted phenomena were unknown to them prior to the experiments and some even seemed impossible or highly unlikely to the biologists and yet were validated by experiments.

One can compare the work accomplished by this *Beetle Team* with the research of Bo Zheng when she was involved in research on modeling the spread of *Wolbachia*-infected mosquitos for disease control. In 2015, the city of Guangzhou established the world's largest mosquito factory, with a 3,500m^2 laboratory producing more than 60 million mosquitoes per week nowadays. In March 2015, *Wolbachia*-infected male mosquitoes were released on Shazai Island and Dadaosha Island, south of Guangzhou City. Bo Zheng with

Professor Jianshe Yu and Professor Moxun Tang were in charge of combining experimental data to establish a mathematical model to predict the optimal release ratio for effective suppression of wild mosquitoes. After identifying the special biological questions, they defined the problem in terms of that can be quantified and modeled. Before the data were collected through experiments and field trials, they determined what types of data were needed and how they should be collected to support the modeling process. They developed a discrete mathematical model (selected from at least 15 tries of different models) to explore effective release strategies for population suppression. Reflecting on the entire process of this project, they realized that the mission of the released *Wolbachia*-infected mosquitoes which are sterile is to sterilize wild mosquitoes with whom they mate, in the sense that eggs produced by the mating do not hatch. They found that the 5:1 "over-flooding" ratio of released males to wild males is sufficient for effective population suppression. This ratio has become an international convention for initial *Wolbachia* release. The study of dynamics of population of beetle and mosquito gives us two examples of strong and direct link between mathematics and biology.

Another key common point highlighted for the nine mathematicians of Chapter 1 is that many outstanding researchers had a great influence at the beginning and during their careers. They met them, either in person or by reading their books and several times interacted with them. These mathematicians also interacted with each other, forming an intriguing social network worldwide, across all borders of nationality and languages. This is also the case for Chen, Sanjuán, and Zheng. The names of Jim Yorke and Leon Chua which are cited in Chapter 1 are also present in this Chapter. Saber Elaydi met Yorke at the fifth ICDEA, which was held in Temuco, Chili, in January 2000. Sanjuán worked under the supervision of Yorke during the 1995/1996 academic year at the IPST of the University of Maryland. In 2002, he had the opportunity to serve as a nominator for the 2003 Japan Prize awarded by the Government of Japan, where Benoît Mandelbrot and Jim Yorke were honored with the award. Finally he established a lasting relationship with him. Leon Chua is cited by Belykh who collaborated with him at UC Berkeley and by Lozi who met him for the first time in 1986 and had a very long cooperation that lasted more than 38 years. In this chapter Sanjuán said that he met Chua in Cargèse, France, in 2008, the same week that the news about the memristor had been published in *Nature*. The first encounter of Chen with Chua was during the visit of Chua to Rui de Figueiredo at Rice University in 1987. Chen immediately recognized that Chua's circuit is an excellent platform for him to work on regarding chaos theory. Later in 2010 up to now, he has succeeded to him to serve as the Editor-in-Chief for the *International Journal of Bifurcation and Chaos*. During the career of Sanjuán one can note many renowned scientists who have directly worked with him like Manuel García Velarde, Ludwig Arnold, Celso Grebogi, Ed Ott, Kazuyuki Aihara, Muthusamy Lakshmanan, Hao Bailin, Wei-Mou Zheng, and Judy Kennedy. He met Edward Lorenz (the father of the butterly effect) during a Gordon Conference on Nonlinear Science in Maine in 2005. He mentions that attending the NATO *Advanced Study Institute Summer School on Chaos, Order, and Patterns* held at Centro di Cultura Scientifica "A. Volta" in Como, Italy, in the Summer of 1990, he had the privilege to meet most of the chaos scientists that he had known through reading the literature. He highlights also that "my research has a significant international dimension with more than two hundred collaborators of twenty-five different nationalities among my publications".

Bo Zheng had the privilege to be mentored by Jianshe Yu, President of the Guangzhou University, and after to be guided by Moxun Tang and Zhiyong Xi from Michigan State University.

In 1975, near the end of the Cultural Revolution, Guanrong Chen was fortunate to be introduced by a family friend to two mathematics professors, Mingjun Chen and Yuesheng Li, of the Sun Yat-sen University in Guangzhou. During those days, they were his highly respected mentors, who had helped him to become a self-taught "amateur mathematician". In particular, his beloved mentor, Mingjun Chen, hosted his frequent visits to his home and often treated him to lunches with his family after giving him free tutorials. He also paid occasional visits to Yuesheng Li, who was a pioneer and leader in computational mathematics in China and specialized in spline function theory and numerical analysis. Some years after, when he moved to Texas A&M University, his mentor was Charles K. Chui, an expert in approximation theory and computational control. In 1987, Rui de Figueiredo at Rice University asked him to collaborate for three years as a postdoctorate fellow.

The influence of outstanding scholars extends to the books that Chen, Sanjuán, and Zheng read early in their careers. Sanjuán mentions many of the fundamental books that have shaped his understanding during his formative years: *Synergetics* by Hermann Haken; *Chaos* by James Gleick; *Chaos in Dynamical Systems* by Ed Ott; *Nonlinear Oscillations, Dynamical Systems, and Bifurcations of Vector Fields* by J. Guckenheimer and P. Holmes; the French book *L'ordre dans le chaos: Vers une approche déterministe de la turbulence* by Bergé, Pomeau, and Vidal, published in 1984. A book that he thoroughly enjoyed and that had a profound influence was Bruce West's *An Essay on the Importance of Being Nonlinear*. He says that the theoretical ideas presented in Tabor's book *Chaos and Integrability in Nonlinear Dynamics: An Introduction* and the comprehensive work *Regular and Stochastic Motion* by A. J. Lichtenberg and M. A. Lieberman were also instrumental in his academic development. He highlights that a particularly noteworthy text from that era was Robert Devaney's *An Introduction to Chaotic Dynamical Systems*; though primarily a mathematics book, it provided critical conceptual insights into chaos theory. He translated completely the German book *Chaos in dynamischen Systemen* by Albrecht Kunick and Willi-Hans Steeb. After his Ph.D. defense, he acquired Daniel Amit's book, *Modeling Brain Function: The World of Attractor Neural Networks*. He remembers also having learned many fundamental aspects of elliptic functions from Derek Lawden's book *Elliptic Functions and Applications*. He had the honor to participate in the revisions preview of *Chaos: An Introduction to Dynamical Systems* by K. T. Alligood, T. Sauer, and Jim Yorke, when he was working with the last author in IPST at Maryland. Finally, Sanjuán mentions the book of Jim Yorke and H. E. Nusse *Dynamics: Numerical Explorations, Accompanying Computer Program Dynamics*, published with the software *Chaos* and a manual to use it properly.

Bo Zheng recalls that, when she began working on her thesis under the supervision of Jianshe Yu, she was advised to read two books, *Discrete Hamiltonian Systems* by Calvin D. Ahlbrandt and Allan C. Peterson and *Critical Point Theory and Hamiltonian Systems* by Jean Mawhin and Michel Willem.

As for Guanrong Chen, it is clear that the most influential books he read were from the period in his life when he was sent to a mountain village in the interior of Hainan Island in the South China Sea and decided to study mathematics on his own. Although Chen does not cite these books, he says that their subjects were calculus, linear algebra, ordinary differential equations, and theoretical mechanics.

As a final note on the early careers of the twelve mathematicians interviewed in Chapters 1 and 2, one can remark that three of them experienced historical upheavals. Vladimir Belykh moved from an elite military boarding school to a high school due to massive reorganizations of the Soviet army, Saber Elaydi had a difficult life in a refugee camp in the Gaza Strip, and Guanrong Chen was deprived of a university education during the ten years of the Cultural Revolution. Nevertheless, their love of mathematics allowed them to overcome these difficult times.

In conclusion of this chapter, we can note as Miguel Sanjuán says that his scientific work has been based on two fundamental ideas, namely interdisciplinarity and internationalization, with respect to which he states

> "interdisciplinary methodology of nonlinear dynamics leads us to use similar methods and tools to address numerous research problems across various disciplines. Furthermore, my research has a significant international dimension with more than two hundred collaborators of twenty-five different nationalities among my publications."

These remarks can be applied to all the researchers of these Chapters.

References

[1] C. D. Ahlbrandt and A. C. Peterson, *Discrete Hamiltonian Systems: Difference Equations, Continued Fractions, and Riccati Equations*, Dordrecht: Kluwer Academic Publishers (1996), 376 pp. https://doi.org/10.1007/978-1-4757-2467-7

[2] S. Ai, J. Li, J. Yu and B. Zheng, *Stage-structured models for interactive wild and periodically and impulsively released sterile mosquitoes*, Discrete Cont. Dyn–B. 27(6) (2022), pp. 3039–3052. https://doi.org/10.3934/dcdsb.2021172

[3] K. T. Alligood, T. Sauer and J. A. Yorke, *Chaos: An Introduction to Dynamical Systems*, Springer, New York (1996), 603 pp. https://doi.org/10.1007/b97589

[4] D. J. Amit, *Modeling Brain Function: The World of Attractor Neural Networks*, Cambridge University Press, Cambridge (1989), 487 pp. https://doi.org/10.1017/ CBO9780511623257

[5] J. Awrejcewicz, S. Rajasekar and M. Ragulskis, eds. *Recent Trends in Chaotic, Nonlinear and Complex Dynamics*, World Scientific, Singapore (2021), 560 pp. https://doi.org/10. 1142/11872

[6] H. Bailin and W.-M. Zheng, *Applied Symbolic Dynamics and Chaos*, Peking University-World Scientific, Singapore, Second edition (2018), 544 pp. https://doi.org/10.1142/ 10887

[7] Z. Balanov, C. García–Azpeitia and W. Krawcewicz, *On variational and topological methods in nonlinear difference equations*, Commun. Pur. Appl. Anal. 17(6) (2018), pp. 2813–2844. https://doi.org/10.3934/cpaa.2018133

[8] A.-L. Barabasi and R. Albert, *Emergence of scaling in random networks*, Science 286 (1999), pp. 509–512. https://doi.org/10.1126/science.286.5439.509.

[9] P. Bergé, Y. Pomeau and Ch. Vidal, *L'ordre dans le chaos: Vers une approche d´eterministe de la turbulence*, Editions Hermann, Paris (1984), 352 pp.

[10] E. Caspari and G. S. Watson, *On the evolutionary importance of cytoplasmic sterility in mosquitoes*, Evolution 13 (1959), pp. 568–570. https://doi.org/10.1111/j.1558-5646.1959.tb03045.x

[11] G. Chen, *Generalized Lorenz Systems Family*. https://doi.org/10.48550/arXiv.2006. 04066

[12] G. Chen, *Nonlinear Systems Stability, Dynamics and Control*, World Scientific, Singapore (2023), 236 pp. https://doi.org/10.1142/13488

[13] G. Chen and X. Dong, *From Chaos to Order, Methodologies, Perspectives and Applications*, World Scientific, Singapore (1998), 776 pp. https://doi.org/10.1142/3033
[14] G. Chen and Y. Shi, *Introduction to anti-control of discrete chaos: theory and applications*, Phil. Trans. R. Soc. A 364 (2006), pp. 2433–2447. https://doi.org/10.1098/rsta. 2006.1833
[15] G. Chen, X. F. Wang and X. Li, *Introduction to Complex Networks: Models, Structures and Dynamics*, Higher Education Press, Beijing, (2012); with the 2nd edition (2015), 392 pp.
[16] C. K. Chui and G. Chen, *Kalman Filtering with Real-Time Applications*, Springer (1984, with 5th edition 2017), 247 pp. https://doi.org/10.1007/978-3-319-47612-4
[17] C. K. Chui and G. Chen, *Discrete H^∞ Optimization*, Springer (1989, with 2th edition 1997), 260 pp. https://link.springer.com/book/10.1007/978-3-642-59145-7
[18] Daza, A. Wagemakers and M. A. F. Sanjuán, *Multistability and Unpredictability*, Physics Today 77(11) (2024), pp. 44–50. https://link.springer.com/book/10.1063/pt.kcxv.poxf
[19] R. L. Devaney, *An Introduction to Chaotic Dynamical Systems*. 2nd Edition, Addison-Wesley, New York (1989), 336 pp.
[20] R. J. P. de Figueiredo and G. Chen, *Nonlinear Feedback Control Systems – An Operator Theory Approach*, Academic Press, San Diego (1993), 312 pp.
[21] P. Fine, *On the dynamics of symbiote–dependent cytoplasmic incompatibility in* Culicine *mosquitoes*, J. Invert. Path. 31 (1978), pp. 10–18. https://doi.org/10.1016/0022-2011(78)90102-7.
[22] J. Gleick, *Chaos: Making a New Science*, Abacus, New York, N.Y., U.S.A. (1988), 352 pp.
[23] J. Guckenheimer and P. Holmes, *Nonlinear Oscillations, Dynamical Systems, and Bifurcations of Vector Fields*, Springer-Verlag, New York (1983), 462 pp. https://doi.org/10.1007/978-1-4612-1140-2
[24] Z. Guo and J. Yu, *The existence of periodic and subharmonic solutions for second-order superlinear difference equations*, Sci. China Ser. A 46(4) (2003), pp. 506–515. https://doi.org/10.1007/BF02884022
[25] Z. Guo and J. Yu, *The existence of periodic and subharmonic solutions to subquadratic second-order difference equations*, J. London Math. Soc. 68 (2003), pp. 419–430. https://doi.org/10.1112/S0024610703004563
[26] Z. Guo and J. Yu, *Periodic and subharmonic solutions for superquadratic discrete Hamiltonian systems*, Nonlinear Anal. 55(7-8) (2003), pp. 669–683. https://doi.org/10.1016/j. na.2003.07.019
[27] H. Haken, *Synergetics: An Introduction*, Springer, Berlin, 3rd ed. (1983), 390 pp. https://doi.org/10.1007/978-3-642-88338-5
[28] A. Kunick and W.-H. Steeb, *Chaos in dynamischen Systemen*. Bibliographisches Institut, Mannheim (1986), 172 pp.
[29] D. F. Lawden, *Elliptic Functions and Applications*, Springer New York, New York (1989), 336 pp.
[30] J. Lichtenberg and M. A. Lieberman, *Regular and Stochastic Motion*, Springer (1983), 499 pp.
[31] R. Lozi, L. Efremova, M.- S. Abdelouahab, and S. El Assad, *Lozi, Hénon and Belykh Chaotic Attractors: New Results Fifty Years On*, CRC Press, Oxon, UK (2025) https://doi.org/10.1201/9781003568643
[32] R. Lozi, L. Efremova, and M. Pluháček, *Recent Improvements in the Theory of Chaotic Attractors*, CRC Press, Oxon, UK (2025) https://doi.org/10.4324/9781003568513
[33] J. Lu and G. Chen, *Generating multiscroll chaotic attractors: theories, method and applications*, Int. J. Bif. Chaos 16(04) (2006), pp. 775-858. https://doi.org/10.1142/ S0218127406015179
[34] *Mathematics Genealogy Project. Miguel A. F. Sanjuán*. https://genealogy.math.ndsu.nodak.edu/id.php?id=118446
[35] J. Mawhin and M. Willem, *Critical Point Theory and Hamiltonian Systems*, New York: Springer-Verlag (1989), 278 pp. https://doi.org/10.1007/978-1-4757-2061-7
[36] H. E. Nusse and J. A. Yorke, *Dynamics: Numerical Explorations, Accompanying Computer Program Dynamics*, Applied Mathematical Sciences 101, Springer-Verlag, New York (1994), 485 pp. https://doi.org/10.1007/978-1-4684-0231-5
[37] E. Ott, *Chaos in Dynamical Systems*, Cambridge University Press (1993), 2nd Edition (2012), 385 pp. https://doi.org/10.1017/CBO9780511803260

[38] E. Ott, C. Grebogi and J. A. Yorke, *Controlling chaos*, Phys. Rev. Lett. 64 (1990), pp. 1196–1199; Erratum, Phys. Rev. Lett. 64, p. 2837 (1990). https://doi.org/10.1103/ PhysRevLett.64.1196

[39] H.-O. Peitgen and D. Saupe, eds., *The Science of Fractal Images*, Springer-Verlag, New York, (1988), 312 pp. https://doi.org/10.1007/978-1-4612-3784-6

[40] H.-O. Peitgen, H. Jürgens and D. Saupe, *Chaos and Fractals: New Frontiers of Science* (1992), 864 pp. https://doi.org/10.1007/b97624

[41] C. A. Ronan. *Man Probes the Universe: The Story of Astronomy*, Natural History Press, Garden City, New York, (1964), 156 pp. Spanish version: *Secretos del Cosmos*, Biblioteca Bàsica Salvat de libros RTV, Barcelona, (1969), 180 pp.

[42] M. A. F. Sanjuán. *Anisotropic contractions of the Poincaré group*, Master Thesis, Universidad de Valladolid, Valladolid, Spain, 28 November 1981.

[43] M. A. F. Sanjuán. *Group contraction and the nine Cayley–Klein geometries*, Intern. J. of Theor. Physics 23 (1984), pp. 1–14. https://doi.org/10.1007/BF02080668.

[44] M. A. F. Sanjuán. *Contribution to the analytical study of some properties of nonlinear oscillators*. Ph. D. Thesis. Universidad Nacional de Educacíon a Distancia, Madrid, Spain, 24 September 1990.

[45] M. A. F. Sanjuán. *Nonlinear Dynamics, Chaos, and Complexity: Interdisciplinarity in the Sciences*, Royal Academy of Sciences of Spain (2024) (In Spanish). https://rac.es/ ficheros/doc/ c30b5a5ee5a9cb1e.pdf

[46] M. A. F. Sanjuán. Google Scholar. https://scholar.google.es/citations?user= 6Amfr24AAAAJ\&hl=en

[47] M. Tabor. *Chaos and Integrability in Nonlinear Dynamics: an Introduction*, John Wiley and Sons, New York (1989), 384 pp.

[48] C. M. Topaz and S. Sen, *Gender representation on journal editorial boards in the mathematical sciences*, PLoS One 11(8): e0161357 (2016). https://doi.org/10.1371/journal. pone.0161357.

[49] M. Turelli, *Cytoplasmic incompatibility in populations with overlapping generations*, Evolution 64(1) (2010), pp. 232–241. https://doi.org/10.1111/j.1558-5646.2009.00822.x.

[50] X. F. Wang and G. Chen, *A chaotic system with only one stable equilibrium*, Commun. Nonlinear Sci. Numer. 17(3) (2012), pp. 1264–1272. https://doi.org/10.1016/j.cnsns.2011.07.017.

[51] X. F. Wang and G. Chen, *Constructing a chaotic system with any number of equilibria*, Nonlinear Dynamics 71(3) (2013), pp. 429–436. https://doi.org/10.1007/ s11071-012-0669-7.

[52] X. Wang, N. Kuznetsov and G. Chen, *Chaotic Systems with Multistability and Hidden Attractors*, Springer, Berlin, Heidelberg (2021), 672 pp. https://doi.org/10.1007/ 978-3-030-75821-9.

[53] D. Watts and S. Strogatz, *Collective dynamics of 'small-world' networks*, Nature 393 (1998), pp. 440–442. https://doi.org/10.1038/30918.

[54] B. J. West. *An Essay on the Importance of Being Nonlinear*, Lecture Notes in Biomathematics, Springer, Berlin, Heidelberg (1985), 210 pp. https://doi.org/10.1007/978-3-642-93306-6.

[55] J. Yu, *Existence and stability of a unique and exact two periodic orbits for an interactive wild and sterile mosquito model*, J. Differ. Equations 269(12) (2020), pp. 10395–10415. https://doi.org/10.1016/j.jde.2020.07.019.

[56] J. Yu and J. Li, *Global asymptotic stability in an interactive wild and sterile mosquito model*, J. Differ. Equations 269(7) (2020), pp. 6193–6215. https://doi.org/10.1016/j.jde.2020.04.036.

[57] J. Yu and B. Zheng, *Multiplicity of periodic solutions for second-order discrete Hamiltonian systems with a small forcing term*, Nonlinear Anal–Theory 69(9) (2008), pp. 3016–3029. https://doi.org/10.1016/j.na.2007.08.069.

[58] J. Yu and B. Zheng, *Modeling* ***Wolbachia*** *infection in mosquito population via discrete dynamical models*, J. Differ. Equ. Appl. 25(11) (2019), pp. 1549–1567. https://doi.org/10.1080/10236198.2019.1669578.

[59] B. Zheng, J. Li and J. Yu, *On discrete dynamical model on* ***Wolbachia*** *infection frequency in mosquito populations*, Sci. China Math. 65(8) (2022), pp. 1749–1764. https://doi.org/10.1007/s11425-021-1891-7.

[60] B. Zheng, M. Tang and J. Yu. *Modeling* ***Wolbachia*** *spread in mosquitoes through delay differential equations*, SIAM J. Appl. Math. 74(3) (2014), pp. 743–770. https://doi.org/10.1137/13093354.

[61] B. Zheng, M. Tang, J. Yu and J. Qiu. ***Wolbachia*** *spreading dynamics in mosquitoes with imperfect maternal transmission*, J. Math. Biol. 76(1-2) (2018), pp. 235–263. https://doi.org/10.1007/s00285-017-1142-5.

[62] B. Zheng and J. Yu, ***Wolbachia*** *spread dynamics in mosquito populations in cyclic environments*, J. Differ. Equ. Appl. 30 (2024), pp. 252–268. https://doi.org/10.1080/ 10236198.2023.2279628.

[63] B. Zheng, J. Yu and J. Li. *Modeling and analysis of the implementation of the* ***Wolbachia*** *incompatible and sterile insect technique for mosquito population suppression*, SIAM J. Appl. Math. 81(2) (2021), pp. 718–740. https://doi.org/10.1137/20M1368367.

[64] B. Zheng, J. Yu, Z. Xi and M. Tang, *The annual abundance of dengue and Zika vector* ***Aedes albopictus*** *and its stubbornness to suppression*, Ecol. Model. 387 (2018), pp. 38–48. https://doi.org/10.1016/j.ecolmodel.2018.09.004.

[65] X. Zheng, D. Zhang, Y. Li, C. Yang, Y. Wu, X. Liang et al. *Incompatible and sterile insect techniques combined eliminate mosquitoes*, Nature 572 (7767) (2019), pp. 56–61. https://doi.org/10.1038/s41586-019-1407-9.

Optimal placement of sensor and actuator for controlling the piecewise linear Chua circuit via a discretized controller

Christophe Letellier, Ludovico Minati and Jean-Pierre Barbot

ABSTRACT

Controlling the dynamics of chaotic systems is a task which is often addressed in an empirical way, particularly for placing sensors and actuators. Here, we show that selecting the measured variable and placing the actuator can be guided by considering the observability and controllability symbolic coefficients and applying the notion of flatness. This approach is here demonstrated on the piecewise linear Chua circuit, whose specific features are leveraged in constructing a discretized controller with a switch mechanism and optimally placed sensor and actuator. The feedback linearization is compared to a homogeneous and a passivity-based control laws, the flat control laws being more efficient than the others. It is thus shown that the proposed flat control law by feedback linearization is very efficient. The continuous time and discretized Chua circuit, governed by differential and difference equations, respectively, are treated. Most likely, these results could be extendable to a large group of natural and experimental systems.

1. Introduction

Chaotic systems, owing to their sheer unpredictability and critical sensitivity to initial conditions, attract considerable interest from physicists and control theorists. Foundational works in this area have hinged around the Ott–Grebogi–Yorke (OGY) method, predicated upon the notion that infinite unstable periodic orbits are contained within chaotic attractors, and thus, at least in principle, one can maintain the system near the desirable limit cycle by applying a suitable weak perturbation at each cycle [70]. Countless other approaches have been proposed throughout the years, including the injection of continuous signals, adaptive control, and various assumptions about the variables accessible to the operator for observation and perturbation [12]. In the 1990s, the notions of observability and controllability, central to the formulation of any control system, have been recently extended to the concept of flatness [26]. Successful feedback linearization was demonstrated for diverse chaotic systems, namely, the Rössler, the driven van der Pol, and the

Hénon-Heiles systems [50]. Here, the approach is applied to controlling the Chua circuit [63,64], which has the particularity to be piecewise linear and, as such, is a little bit more tricky to treat. The Chua circuit is, by virtue of its minimal complexity and remarkable potential to generate diverse dynamics, is commonly taken as paradigmatic for chaos in many natural and artificial systems, such as electronic oscillators [27,60]. Its double-scroll attractor – here meant in its chaotic version as initially discussed [64] – is often taken as a hallmark of chaotic dynamics [81], yet, for suitable configurations, this circuit is capable of generating an incredible variety encompassing hundreds of attractor shapes [11]. When multiple Chua circuits are coupled, the emergence of complex dynamical patterns and associated transitions reminiscent of those detected in neural activity can be readily elicited [23,68]. The Chua circuit can therefore arguably be taken as a generic model for universal nonlinear dynamics across diverse natural and artificial systems.

Controlling the Chua circuit is not a new task. Chen and Dong controlled the dynamics produced by a Chua circuit to a limit cycle by using a linear feedback control built with three sensors and three actuators placed on each of the three governing equations: they finally obtained a control using a sensor on y and an actuator placed on $\dot{y}$ [20,21]. Using a feedback control with a single actuator placed on $\dot{z}$, Hartley and Mossayebi controlled the dynamics of the Chua circuit with a cubic nonlinearity to a desired state with a single sensor: the most robust controller was obtained by measuring variable x and placing the actuator on $\dot{z}$ [35]. Saito and Mitsubori stabilized in a periodic orbit the double-scroll dynamics produced by a two-dimensional piecewise linear electronic circuit with two sensors and two actuators [74]. Inspired by Alvarez-Ramírez who controlled the Lorenz system with one sensor (measuring variable y) and one actuator (a first-order differential compensator) applied to $\dot{y}$ [4], Hwang and coworkers controlled the Chua circuit with a cubic nonlinearity using one sensor (measuring variable x) and one actuator applied to $\dot{z}$ [38]. Barone and Singh controlled a Chua circuit with a cubic nonlinearity by using one sensor (measuring x) and one actuator applied on parameter α [8]. More recently, Guo and Liu controlled the piecewise linear Chua circuit by using one sensor (measuring the variable x) and one actuator applied on $\dot{x}$ [34]. In most of these works, the aim was to use a single sensor and a single actuator, but the configurations actually used for controlling the system were never justified. All these control laws were based on an empirical placement of sensors and actuators.

In this work, we aim to propose some guidelines for placing sensors and actuators based on a structural analysis of the system to control and, then, to design some flat control law using this optimal placement, with a special attention devoted to the feedback linearization. The specificity of this work is related to the special case of piecewise linear systems, for which the application of a flat control law is not so straightforward. Indeed, though piecewise linear circuits are straightforward to design [19,22], they are not necessarily trivial to control [43]. We here focus on the use of a flat-control law for the piecewise linear Chua circuit. To rend this study more realistic, we used a discretized control scheme for which it is important to investigate the effect of the delay induced for computing the derivatives on the efficiency of the control law, a point which was not considered in our previous work [50]. The remainder part of this paper is organized as follows. In Section 2, we introduce the well-known dual concepts of observability and controllability, alongside the less-known concept of flatness, which stands as the key point of our approach for selecting an optimal placement for sensors and actuators. A graphical approach combined with symbolic observability, controllability and flatness coefficients are also introduced as useful concepts for treating simple as well as complex systems. Section 3 is devoted to the Chua circuit, its dynamics and its control using a flat law. Section 4 discusses the design of a flat discretized controller. The results were then compared to those provided by some other simple laws, in particular with different placements for the sensor and the actuator. Finally, Section 5 discusses the findings and their relevance.

2. From observability to flatness

Let us consider the nonlinear system

$$\dot{\boldsymbol{x}} = \boldsymbol{F}(\boldsymbol{x}, \boldsymbol{u}), \quad \boldsymbol{x} \in \mathbb{R}^n, \boldsymbol{u} \in \mathbb{R}^{n_{\text{in}}}, \tag{1}$$

$$\boldsymbol{\xi} = \boldsymbol{h}(\boldsymbol{x}), \quad \boldsymbol{\xi} \in \mathbb{R}^{n_{\text{out}}}, \tag{2}$$

where $\boldsymbol{F}(\boldsymbol{x}, \boldsymbol{u}) = \boldsymbol{f}(\boldsymbol{x}) + \boldsymbol{g} \cdot \boldsymbol{u}$ represents the equations governing the controlled dynamics, $\boldsymbol{x} \in \mathbb{R}^n$ is the state vector, $\boldsymbol{u} \in \mathbb{R}^{n_{\text{in}}}$ is the control input sine we here only consider autonomous systems $\dot{\boldsymbol{x}} = \boldsymbol{f}(\boldsymbol{x})$ (meaning that there is no explicit dependency on time and on any other driving term), $\boldsymbol{h}$ is the measurement function returning the measurement (output) vector $\boldsymbol{\xi}$ from the original state $\boldsymbol{x}$, which, in turn, is used for feeding the input $\boldsymbol{u}$ designed for controlling the system $\boldsymbol{f}$; n_{in} is the number of actuators and n_{out} is the number of sensors. When the problem is formulated in this manner, $\boldsymbol{u}$ designates the control law. Strictly speaking, the observability should be determined once the input, namely, the control law, is designed since the input may affect the observability of the controlled system [31]. Nevertheless, when the actuator are placed to the dual of the sensor providing a global observability, the input does not affect the observability, that is, the observability can be determined directly from the autonomous system ($\boldsymbol{u} = \boldsymbol{0}$).

2.1. Observability

Assuming $n_{\text{out}} = 1$ and $n = 3$ without loss of generality, observability may be defined as follows [44].

Definition 2.1: An autonomous dynamical system $\boldsymbol{f}$ is said to be state-observable at time t, if and only if every initial state $\boldsymbol{x}_0 = \boldsymbol{x}(t_0)$ can be uniquely determined from the knowledge of the output $\boldsymbol{\xi}(\tau)$ for $t_0 \leq \tau \leq t$.

The pair $[\boldsymbol{f}, \boldsymbol{\xi}]$ is globally observable when the observability matrix

$$\mathcal{O}_{\xi^n}(\boldsymbol{x}) = \left| \begin{array}{l} \mathrm{d}h(\boldsymbol{x}) \\ \mathrm{d}\mathcal{L}_f h(\boldsymbol{x}) \\ \mathrm{d}\mathcal{L}_f^2 h(\boldsymbol{x}), \end{array} \right. \tag{3}$$

where

$$\mathcal{L}_f^j h(\boldsymbol{x}) = \frac{\partial \mathcal{L}_f^{j-1} h(\boldsymbol{x})}{\partial \boldsymbol{x}} \boldsymbol{f}(\boldsymbol{x}), \tag{4}$$

and $\mathcal{L}_f^0 h(\boldsymbol{x}) = h(\boldsymbol{x})$, is full rank ($\forall \boldsymbol{x} \in \mathbb{R}^3$) [36]. In this paper, we only consider a three-dimensional reconstructed state space, that is, one whose dimension is equal to that of the original state space. Consequently, when h returns one of the variables, the pair $\lceil \boldsymbol{f}, \xi \rceil$ is *globally observable* if and only if $\mathcal{O}_{\xi^3}$ is full rank: here ξ^3 signifies that the first three derivatives of the measured variable ξ (the first one being the variable itself) are used to reconstruct a three-dimensional space. For the dynamicist, the observability matrix $\mathcal{O}_{\xi^3}(\boldsymbol{x})$ is the jacobian matrix of the coordinate transformation $\varphi(\boldsymbol{x})$ between the original state space $\mathbb{R}^3(\boldsymbol{x})$ and the reconstructed one $\mathbb{R}^3(\boldsymbol{\xi})$ where $\boldsymbol{\xi} = (\xi, \dot{\xi}, \ddot{\xi})$ [52]. An output is here said to be flat when it provides a global observability.

The rank condidition of the observability matrix computed from the governing equations of the system without setting the values of the parameters [36] provides a yes-or-no measure: this denotes *structural observability*. In practice, one finds that the observability can be graded from 0 (non-observable) to 1 (globally observable) using suitable coefficients [47,53]. Aguirre introduced a numerical coefficient computed along the trajectory described by the system [2]: this is a form of *dynamical observability*. Another approach can be considered as an intermediate between the structural observability and the dynamical one, since it is based on the structure of the governing equations but returns a coefficient within a unit interval: this is known as the *symbolic observability* [10,49]. A review on these different concepts can be found in Ref. [3]. Here, we note that the structural observability can be quantified through the symbolic observability coefficient η_{ξ^3} [49,55], which is such that $\eta_{\xi^3} \in [0, 1]$, being 1 for a globally observable pair $\lceil \boldsymbol{f}, \boldsymbol{\xi} \rceil$ and null for an everywhere non-observable pair. The symbolic observability coefficients are computed as follows. First, the symbolic jacobian matrix $\tilde{\mathcal{J}}_f$ of the system $\boldsymbol{f}$ is obtained by replacing each constant non-null element J_{ij} with 1, each non-constant non-null polynomial element J_{ij} with $\bar{1}$, and each element J_{ij} with $\bar{\bar{1}}$ when the jth variable is present in the denominator or with $\bar{1}$ otherwise. The rationale underlying this notation is to discriminate the terms which are increasing the complexity of the determinant (as designated with $\bar{1}$ and $\bar{\bar{1}}$) since it was shown that the observability decreases when the complexity of the determinant increases [47]. There is no need to compute algebraically the observability matrix $\mathcal{O}_{\xi^3}$ and the symbolic observabity matrix $\tilde{\mathcal{O}}_{\xi^3}$ can be computed symbolically [10]: this is a relevant advantage because $\mathcal{O}_{\xi^n}$ ($n \geqslant 5$) cannot be always computed, and thus, high-dimensional systems can be treated [55].

The first row of $\tilde{\mathcal{O}}_{\xi^n}$ is such that $\tilde{\mathcal{O}}_{1j} = 1$ when the jth variable is measured, and 0 otherwise. The kth row is the next symbolic derivative $\tilde{\mathcal{L}}_f^k h$, which is obtained as follows. Each element of the ith row of $\tilde{J}$ of the symbolic jacobian matrix $\tilde{\mathcal{J}}_f$ is left multiplied by the vector $\boldsymbol{V} = (\tilde{\mathcal{O}}_{l1}, \ldots, \tilde{\mathcal{O}}_{ln})^{\mathrm{T}}$ where l designates the $(k-1)$th row of $\tilde{\mathcal{O}}_{\xi^n}$. The symbolic product $\otimes$ is defined as [10]

$$\left|\begin{array}{l} 0 \otimes a = 0 \\ 1 \otimes a = a \\ \bar{1} \otimes a = a \text{ for } a \in \{\bar{1}, \bar{\bar{1}}\} \\ \bar{\bar{1}} \otimes a = \bar{\bar{1}} \text{ for } a \neq 0. \end{array}\right. \tag{5}$$

Here too, the underlying rationale is to track the highest nonlinearity involved in each term of the determinant. Thus, $\tilde{\mathcal{J}} \otimes \boldsymbol{V}$ leads to a $n \times n$ matrix which is reduced to a row by summing the elements of each column using symbolic addition [10]

$$
\left| \begin{array}{l}
0 \oplus a = a \\
1 \oplus a = a \text{ for } a \neq 0 \\
\bar{1} \oplus a = a \text{ for } a \notin \{0,1\} \\
\bar{\bar{1}} \oplus a = \bar{\bar{1}}.
\end{array} \right. \tag{6}
$$

The determinant of the so-constructed symbolic observability matrix is computed using the symbolic addition and multiplication. For instance, for a two-dimensional system, the symbolic observability matrix $\tilde{\mathcal{O}}_{\xi^2}$ is a 2×2 matrix whose symbolic determinant is

$$
\text{Det } \tilde{\mathcal{O}}_{\xi^2} = \tilde{O}_{11} \otimes \tilde{O}_{22} \oplus \tilde{O}_{21} \otimes \tilde{O}_{12} \tag{7}
$$

(there is no substraction in these symbolic computation). In the expression of this determinant, the numbers N_1, $N_{\bar{1}}$ and $N_{\bar{\bar{1}}}$ of occurrences of 1, $\bar{1}$, and $\bar{\bar{1}}$, are counted. In the special case where $N_1 = 0$, $N_{\bar{1}} \neq 0$ and/or $N_{\bar{2}} \neq 0$, we impose to N_1 the value of 1. The symbolic observability coefficient η_ξ is therefore equal to

$$
\eta_{\xi^n} = \frac{N_1}{D} + \frac{N_{\bar{1}}}{D^2} + \frac{N_{\bar{\bar{1}}}}{D^3}, \tag{8}
$$

where $D = \max(1, N_1) + N_{\bar{1}} + N_{\bar{\bar{1}}}$. By construction, $0 \leq \eta_{\xi^n} \leq 1$ and $\eta_{\xi^n} = 1$ when the determinant is only made of constant terms, which implies that Det $\mathcal{O}_{\xi^3}(\boldsymbol{x}) \neq 0$, $\forall \boldsymbol{x} \in \mathbb{R}^n$.

2.2. Controllability

The dual of observability is controllability [57,59,78]. Let us start with a definition for accessible points.

Definition 2.2: Let $\boldsymbol{x}_0$ be an initial state at $t = 0$. A point $x_a \in \mathbb{R}^3$ is accessible when there exists a control u and a time $t > 0$ such that $x(t) = x_a$.

Then, controllability may be defined as follows [36].

Definition 2.3: Let $\mathcal{A}(\boldsymbol{x}_0)$ be the set of points accessible to the system (1)–(2) from $\boldsymbol{x}_0 \in \mathbb{R}^n$. Let $M \subset \mathbb{R}^n$ be a set of states over which the system $\boldsymbol{G}$ is defined. The system is said to be *locally* controllable at $\boldsymbol{x}_0$ if $\mathcal{A}(\boldsymbol{x}_0) = M$, and *globally* controllable if $\mathcal{A}(\boldsymbol{x}) = M$, $\forall \boldsymbol{x} \in M$.

When $n_{\text{in}} = 1$, the input vector $\boldsymbol{g}$ (here, a three-dimensional vector) is associated with a control law $u(\xi, \dot{\xi}, \ddot{\xi})$ designed from the measurement vector $\boldsymbol{\xi}$ and applied to the dynamical system according to

$$
\dot{\boldsymbol{x}} = \boldsymbol{f}(\boldsymbol{x}) + \boldsymbol{g}(\boldsymbol{x}) \cdot u(\xi, \dot{\xi}, \ddot{\xi}), \tag{9}
$$

where, in the present cases, $g_i = [\ \delta_{i,1} \quad \delta_{i,2} \quad \delta_{i,3}\ ]^{\mathrm{T}}$ when the control law is applied on the jth component of the vector field. Here, $\boldsymbol{g}(\boldsymbol{x})$ is therefore a constant vector which has a single non-zero comment (equal to 1). The scalar function u is dependent on the measured variable ξ and its successive derivatives $\dot{\xi}$ and $\ddot{\xi}$. By contrast, in the most general case, $\boldsymbol{g}(\boldsymbol{x})$ is a vector field that depends on the state $\boldsymbol{x}$. As we introduced for an observability matrix, the controllability matrix can be defined as [39]

$$\mathcal{C}_g = \left[\mathrm{ad}_f^0 \boldsymbol{g} \ \mathrm{ad}_f^1 \boldsymbol{g} \ \ldots \ \mathrm{ad}_f^{n-1} \boldsymbol{g} \right] (\boldsymbol{x}), \tag{10}$$

where $\mathrm{ad}_f \boldsymbol{g} = [\boldsymbol{f}, \boldsymbol{g}]$ is the adjoint action of $\boldsymbol{f}$ on the smooth vector field $\boldsymbol{g}$ with, by definition,

$$\begin{aligned} \mathrm{ad}_f^k \boldsymbol{g}(\boldsymbol{x}) &= \left[\boldsymbol{f}, \mathrm{ad}_f^{k-1} \boldsymbol{g} \right] (\boldsymbol{x}) \\ &= \mathcal{L}_f \, \mathrm{ad}_f^{k-1} \boldsymbol{g}(\boldsymbol{x}) - \mathcal{L}_g \, \mathrm{ad}_f^{k-1} \boldsymbol{f}(\boldsymbol{x}), \end{aligned} \tag{11}$$

and

$$\mathrm{ad}_f^0 \boldsymbol{g}(\boldsymbol{x}) = \boldsymbol{g}(\boldsymbol{x}). \tag{12}$$

It is always possible to increase n_{in} so as to reach a global controllability [36,59], but in this paper, we limit ourselves to the case $n_{\mathrm{in}} \leqslant n$. A singular controllability manifold,

$$\mathcal{M}_g^{\mathcal{C}} \equiv \left\{ \boldsymbol{x} \in \mathbb{R}^3 \mid \mathrm{Det}\, \mathcal{C}_g = 0 \right\}, \tag{13}$$

can be defined as the domain of the state space over which the local controllability is lost. As we did for observability, the pair $[\boldsymbol{f}, \boldsymbol{g}]$ is said to be globally controllable when the controllability matrix is full rank or, equivalently, when $\mathcal{M}_g^{\mathcal{C}} = \emptyset$.

The algorithm for symbolically computing the controllability matrix is not yet developed. The symbolic controllability matrix is therefore obtained from the analytical one using the same rule as for getting the symbolic jacobian matrix. Then, the symbolic determinant is computed and a symbolic controllability coefficient $\zeta_{\dot{s}}$ is computed using the same procedure as for the symbolic determinant of the symbolic observability matrix. The interpretation of $\zeta_{\dot{s}}$ is as for the symbolic observability coefficients: there is a global controllability when $\zeta_{\dot{s}} = 1$, and a local controllability when $0 < \zeta_{\dot{s}} < 1$.

2.3. Flatness

From Fliess and co-workers, the definition of a zero-flat three-dimensional system is as follows [26,67].

Definition 2.4: The dynamical system (1)–(2) is zero-flat if

(1) The state $\boldsymbol{x}$ can be rewritten as a function of the output and its derivatives, that is, as $\boldsymbol{x} = F_{\mathrm{state}}(\xi, \dot{\xi}, \ddot{\xi})$ where ξ is the measured variable;
(2) The control input can be rewritten as a function of the output and its derivatives that is, as $u = F_{\mathrm{in}}(\xi, \dot{\xi}, \ddot{\xi}, \dddot{\xi})$.
(3) There is an output function h and an input vector $\boldsymbol{g}$ such that [37,42]

$$\begin{cases} \mathcal{L}_g \mathcal{L}_f^k h = 0 \ (k = 0, 1) \\ \mathcal{L}_g \mathcal{L}_f^2 h \neq 0. \end{cases} \tag{14}$$

Initially, the introduction of flatness was algebraic and the role of the singularities was considered as irrelevant since it is associated with a set having null probability [26,56]. In practice, however, those singularities are sufficient to render the control law inefficient: it is therefore relevant to consider them.

From Definition 2.4, two types of problems can be investigated. The most studied problem is to find a flat-output (2) from the governing Equation (1) [6,14,76]. The aim is to select an output ensuring a global observability (the first condition of Definition 2.4) of the state space. This ensures that the system can be rewritten as

$$\begin{cases} \dot{\xi}_1 = \xi_2 \\ \dot{\xi}_2 = \xi_3 \\ \dot{\xi}_3 = \mathcal{L}_f^3\, h(\boldsymbol{x}) + \mathcal{L}_g\, \mathcal{L}_f^2\, h(\boldsymbol{x}) \cdot u, \end{cases} \tag{15}$$

where ξ is the measured variable, and

$$u = \frac{-\mathcal{L}_f^3 h(\boldsymbol{x}) + v}{\mathcal{L}_g\, \mathcal{L}_f^2\, h(\boldsymbol{x})}, \tag{16}$$

is the *feedback control law* [40]. The resulting linearized system,

$$\begin{cases} \dot{\xi}_1 = \xi_2 \\ \dot{\xi}_2 = \xi_3 \\ \dot{\xi}_3 = v, \end{cases} \tag{17}$$

is a chain of integrators. Note that

$$\boldsymbol{\Theta}(\boldsymbol{x}) = \begin{vmatrix} \xi \\ \dot{\xi} \\ \ddot{\xi} \end{vmatrix} \tag{18}$$

does not depend on the control $\boldsymbol{u}$ insofar as the condition (14) (see below) is verified. Then the change of coordinate is (at least) a local diffeomorphism [39]. The control v can be a feedback linearization control law $v = -\boldsymbol{K}\,(\boldsymbol{\xi} - \boldsymbol{\xi}_{\rm d}) + \dddot{\xi}_{\rm d}$, where $\boldsymbol{K}$ is the covector of gains, and $\boldsymbol{\xi}_{\rm d} = (\xi_{\rm d}, \dot{\xi}_{\rm d}, \ddot{\xi}_{\rm d})$ is the vector of desired state. This vector may follow, for instance, a periodic orbit to control the system to that orbit (see Section 3). This system is linear since it can be rewritten as $\dot{\boldsymbol{\xi}} = K\,\boldsymbol{\xi}$, with

$$K = \begin{bmatrix} -\kappa_1 & -\kappa_2 & -\kappa_3 \end{bmatrix}, \tag{19}$$

where the κ_is are chosen so that all eigenvalues of K have negative real part. For simplicity, in this work, we only consider the cases where all eigenvalues are equal. Note that in the form (15) and when $u = 0$, the system (1) is rewritten in a 'jerk' form, that is, in a differential embedding spanned by one of its variables and its successive derivatives. When the condition (14) is fulfilled, each new derivative of the measured variable ξ brings a new independent dimensions to the reconstruct the state space: $\mathcal{L}_g\, \mathcal{L}_f^0\, h = 0$ and $\mathcal{L}_g\, \mathcal{L}_f^1\, h = 0$

ensure that the first two new dimensions are independent, and $\mathcal{L}_g \mathcal{L}_f^2 h \neq 0$ that the system can be actually written in the form (17). This condition is the key point for designing a flat input-output. Such a problem is sometimes intractable [56,61], that is, in spite of a variable providing a global observability and a derivative providing a global controllability, condition (14) is not fulfilled. Nevertheless, this is most often possible for the well-known chaotic systems with one variable providing a global observability [48]. Fulfilling this condition guarantees that there is no residual dynamics (zero-dynamics) [41] after feedback linearization [40].

The second and less common problem is to find a sufficiently smooth vector field $g_i(\boldsymbol{x})\, u_i \in \mathbb{R}^n$ with $i \in \{1, .., n_{\text{out}}\}$ (in other words, a flat input [67,75,80]) from the three-dimensional system

$$\dot{\boldsymbol{x}} = \boldsymbol{f}(\boldsymbol{x}), \quad \boldsymbol{x} \in \mathbb{R}^3, \tag{20}$$

$$\xi = h(\boldsymbol{x}), \quad \xi \in \mathbb{R}, \tag{21}$$

such that the system

$$\dot{\boldsymbol{x}} = \boldsymbol{f}(\boldsymbol{x}) + \boldsymbol{g}_i(\boldsymbol{x})\, u, \quad \boldsymbol{x} \in \mathbb{R}^3, \tag{22}$$

$$\xi = h(\boldsymbol{x}), \quad \xi \in \mathbb{R}, \tag{23}$$

is a zero-flat system with a single sensor and a single actuator [26]. This results from the fact that (21) is not a function of the input, and that condition (14) is fulfilled.

According to the duality between observability and controllability developed by Kalman [45] and Lin [57] for linear systems, the placement of the sensor has its dual, that is, it has some implications for the placement of the actuator. In other words, for each sensor, there is an optimal placement for the actuator, and *vice versa*. Notably, the duality between observability and controllability leads to the fact the number of actuators must be equal to the number of sensors [13,72]. Moreover, if the pair $\lceil \boldsymbol{f}, \boldsymbol{\xi} \rceil$ is globally observable, then there is a dual $\boldsymbol{g}$ of $\boldsymbol{\xi}$ such that the pair $\lceil \boldsymbol{f}, \boldsymbol{g} \rceil$ is globally controllable, and *vice versa*. Thus, one may say that the triplet $\lceil \boldsymbol{f}, \boldsymbol{\xi}, \boldsymbol{g} \rceil$ is differentially zero-flat.

It is possible to introduce a symbolic flatness coefficient as follows.

Definition 2.5: Symbolic flatness coefficients are defined as

$$\varphi_{v^3,\dot{w}} = \left| \begin{array}{ll} \eta_{v^3} \cdot \zeta_{\dot{w}} & \text{if the condition (14) is fulfilled} \\ 0 & \text{otherwise.} \end{array} \right. \tag{24}$$

where η_{v^3} is the symbolic observability coefficient computed from the symbolic observability matrix $\mathcal{O}_{v^3}$ and $\zeta_{\dot{w}}$ is the symbolic controllability coefficient computed from the symbolic controllability matrix $\mathcal{C}_{\dot{w}}$. By definition, $\varphi \in [0, 1]$.

All of this can be performed with the help for a fluence graph which is a graph of directed links which connect at nodes; it allows representing how a signal flows within a system [62]: in other words, when a variable x_i affects the derivative $\dot{x}_j$, there is a link from the ith node to the jth one. When a variable affects a derivative within a nonlinear term, the link is dashed. Then, since nonlinearity is at the source of singularities which should be avoided, we prune the fluence graph from its dashed links [54]. By using the strongly connected components of the pruned fluence graph, that is, the subgraphs whose every nodes is reachable from every other node, we retained those which are without outgoing link [54,58], thus corresponding to the *root strongly connected components* (rSCC). We know that at least one sensor must be placed in each rSCC [58]. This graphical approach is very useful for preselecting the possible placements of sensor and actuator in complex systems [66]. For low-dimensional systems, when combined with the '*observability path*' [50], that is, a path made of linear links visiting only once each variable, which very often provides the optimal placement for sensor and actuator: the sensor is at the end of the observability path, and the actuator is at its beginning.

The symbolic coefficient then help to check the placement suggested with the graphical approach [55]. If necessary, only the required analytical computations can be performed for rigorous result.

Our procedure works as follows.

(1) Determine the optimal placement of sensors with the help of the graphical approach and of the symbolic observability coefficients;
(2) Determine the optimal placement of actuators, being the dual of the measured variable, with the observability path and check it with the largest symbolic controllability coefficients;
(3) Check whether condition (14) is fulfilled;
(4) Design a control law as exemplified in Equation (16), for instance.

In this paper, we will focus to control law based on feedback linearization control law, however, other types of control laws such as feedforward controller [1], adaptative control based on neural network [16], or open-loop control [5], can also be used.

3. Flatness for the Chua circuit

The Chua circuit is governed by the piecewise linear system [63,64]

$$\begin{cases} \dot{x} = \alpha\left[y - x - \Psi(x)\right] \\ \dot{y} = x - y + z \\ \dot{z} = -\beta y, \end{cases} \tag{25}$$

where

$$\Psi(x) = m_1 x - (m_0 - m_1)\,\frac{|x+1| - |x-1|}{2}, \tag{26}$$

which can be rewritten as

$$\Psi(x) = \left| \begin{array}{ll} (2m_1 - m_0)\,x & \text{if } |x| \leqslant 1 \\ m_1 x + (m_1 - m_0) & |x| > +1. \end{array} \right. \tag{27}$$

This is a three-dimensional system involving a piecewise linear function. This system is therefore discontinuous for the singular manifold

$$\mathcal{M}_s \equiv \left\{(y,z) \in \mathbb{R}^2 \mid |x| = 1\right\}. \tag{28}$$

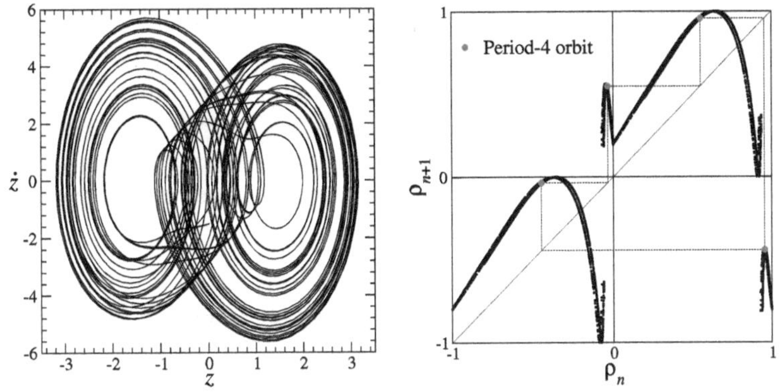

(a) State portrait (b) First-return map

Figure 1. Double-scroll chaotic attractor (a) produced by the Chua circuit (25). The first-return map to the two-component Poincaré section – drawn as two thick lines – is made of ten branches (b). A period-4 orbit extracted from the chaotic attractor (red disks) is plotted in the first-return map. Parameter values: $\alpha = 9$, $\beta = \frac{100}{7}$, $m_0 = -\frac{8}{7}$, and $m_1 = -\frac{5}{7}$.

This system (25) is equivariant under an inversion symmetry, that is, $\boldsymbol{f}(-\boldsymbol{x}) = -\boldsymbol{f}(\boldsymbol{x})$. The dynamics produced by this circuit is extremely rich and can be quasi-periodic, chaotic and even toroidal chaotic [11,24,46], as such, it is an excellent metaphor of a considerable variety of natural and artificial systems, spanning physics, biology and engineering. Parameters β, m_0, and m_1 are generally fixed, while α is elected as a bifurcation parameter. The trajectory produced by the system (25) with appropriate parameter values delineates a double-scroll attractor (Figure 1(a)), known to hallmark the Chua circuit.

The Chua circuit has three singular points, namely,

$$\mathrm{S}_0 = \left| \begin{array}{l} x_0 = 0 \\ y_0 = 0 \\ z_0 = 0 \end{array} \right. \quad \text{and} \quad \mathrm{S}_\pm = \left| \begin{array}{l} x_\pm = \mp \dfrac{m_0 - m_1}{1 + m_1} = -x_\mathrm{p} \\ y_\pm = 0 \\ z_\pm = \pm \dfrac{m_0 - m_1}{1 + m_1}. \end{array} \right. \tag{29}$$

When the parameters are set to $\alpha = 9$, $\beta = \frac{100}{7}$, $m_0 = -\frac{8}{7}$, and $m_1 = -\frac{5}{7}$, the singular point S_0 at the origin of the state space is associated with the eigenvalues

$$\left| \begin{array}{l} \lambda_{1,2} = 0.035 \pm 3.499\, i \\ \lambda_3 = -7.499; \end{array} \right. \tag{30}$$

it is a saddle-focus point with a two-dimensional unstable manifold. The two symmetry-related singular points $\mathrm{S}_\pm$ are characterized by the eigenvalues

$$\left| \begin{array}{l} \lambda_{1,2} = 0.185 \pm 3.047\, i \\ \lambda_3 = -3.942; \end{array} \right. \tag{31}$$

they are both saddle-focus points.

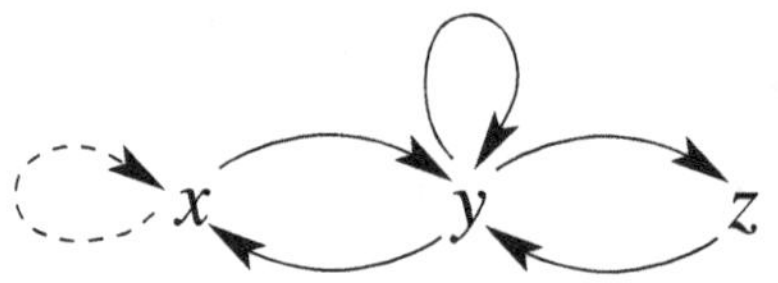

Figure 2. Fluence graph of the Chua circuit (25). Linear and nonlinear links are drawn with solid and dashed lines, respectively.

An adequate Poincaré section $\mathcal{P}$ is made of two components, $\mathcal{P}_1$ and $\mathcal{P}_2$, one in each scroll (Figure 1(b)). It may be defined as

$$\mathcal{P} \equiv \left\{ \left(y_n, z_n \right) \in \mathbb{R}^2 \mid x_n = \pm x_{\text{p}}, \dot{x}_n \gtrless 0 \right\}, \tag{32}$$

where $\boldsymbol{x}_n$ is the state of the nth intersection of the trajectory with the Poincaré section. The first-return map is then built by using the discrete time series $\{y_n\}$, for which y_ns are normalized over the unit interval such as $[0, 1]$ for $\mathcal{P}_1$ and $[-1, 0]$ for $\mathcal{P}_2$, the normalized variable ρ_n being oriented from the centre to the periphery of the attractor [73]. The resulting first-return map is plotted in Figure 1(c): it has 10 branches. Six of them (2×3) are associated with a reinjection of the trajectory in the scroll from where it originates, while the remaining 4 (2×2) are associated with transition from one scroll to the other. A more extended characterization of double-scroll attractors can be found in Refs. [32,33,51]. A ten symbols dynamics is thus used to encode periodic orbits.

The fluence graph of the Chua circuit is drawn in Figure 2. From the pruned fluence graph (dashed lines removed as detailed in [54]), one concludes that there is a single root strongly connected component (defined as a subgraph in which every variable is reachable from every other variable without outgoing arrow) comprising all the variables. There is two possible observability paths, namely $x \to y \to z$ and $z \to y \to x$. Since it is always better to measure a variable without nonlinear link, the sensor is placed at the variable z, that is, the former observability path is retained. The dual of the variable z is thus the derivative $\dot{x}$, where the actuator should be placed.

3.1. Observability

The Jacobian matrix of the Chua circuit is

$$\mathcal{J} = \begin{bmatrix} -\alpha(1 + \psi_x) & \alpha & 0 \\ 1 & -1 & 1 \\ 0 & -\beta & 0 \end{bmatrix}, \tag{33}$$

where

$$\psi_x = \left| \begin{array}{ll} m_1 & \text{if } |x| > 1 \\ 2m_1 - m_0 & \text{otherwise} \end{array} \right. \tag{34}$$

is the term due to the piecewise linear switch which implicitly depends on the variable x. Since ψ_x depends on x, it should be considered as a non-constant term, behaving as a term inducing a singular observability manifold

$$\mathcal{M}_s^{\mathcal{O}} \equiv \left\{(y, z) \in \mathbb{R}^2 \mid |x| = 1\right\}, \tag{35}$$

which matches with the manifold associated with the switching mechanism. The trace of this Jacobian matrix is

$$\mathrm{Tr}(\mathcal{J}) = -(\alpha + 1) - \alpha \left| \begin{array}{ll} m_1 & \text{if } |x| > 1 \\ 2m_1 - m_0 & \text{otherwise,} \end{array} \right. \tag{36}$$

meaning that, when $m_0 = -\frac{8}{7}$ and $m_1 = -\frac{5}{7}$ as set in this work, the system is always dissipative.

When the sensor is placed at the variable z, the observability matrix is

$$\mathcal{O}_{z^3} = \begin{bmatrix} 0 & 0 & 1 \\ 0 & -\beta & 0 \\ -\beta & \beta & -\beta \end{bmatrix}. \tag{37}$$

Its determinant is $\mathrm{Det}\, \mathcal{O}_{z^3} = -\beta^2$. In agreement with the previous analysis, the optimal placement of the sensor is on the variable z provides a global observatbility of the state space.

The previous analysis is confirmed by computing the symbolic observability coefficients as detailed in Section 2. The symbolic Jacobian matrix is

$$\tilde{\mathcal{J}} = \begin{bmatrix} \bar{1} & 1 & 0 \\ 1 & 1 & 1 \\ 0 & 1 & 0 \end{bmatrix}. \tag{38}$$

where the element $\tilde{J}_{11}$ is set to $\bar{1}$ to take into account that ψ_x depends on the x-value, that is, that it behaves as a non-constant term depending on x.

According to the procedure defined in the previous section, when the sensor is placed at the variable z, the symbolic observability matrix is

$$\tilde{\mathcal{O}}_{z^3} = \begin{bmatrix} 0 & 0 & 1 \\ 0 & 1 & 0 \\ 1 & 1 & 1 \end{bmatrix}. \tag{39}$$

Its symbolic determinant is thus $\mathrm{Det}\, \tilde{\mathcal{O}}_{x^3} = 1 \otimes 1 \otimes 1$, that is, only made of three constant terms, leading to $N_1 = 3$, $N_{\bar{1}} = 0$ and $N_{\bar{\bar{1}}} = 0$. Using the expression (8) for the symbolic observability coefficients, one gets $\eta_{x^3} = \frac{3}{3} + \frac{0}{9} = 1$. The symbolic observability coefficient η_{y^3} is explicitly treated in Appendix A.

3.2. Controllability

It is now needed to check whether the derivative $\dot{x}$, the dual of the variable z, provides a global controllability. When the input vector is $\boldsymbol{g} = [\ 1 \quad 0 \quad 0\ ]^{\mathrm{T}}$, the controllability matrix is

$$\mathcal{C}_{\dot{x}} = \begin{bmatrix} 1 & \alpha(1+\psi_x) & \alpha+\alpha^2(1+\psi_x)^2 \\ 0 & -1 & -1-\alpha(1+\psi_x) \\ 0 & 0 & -\beta \end{bmatrix}. \tag{40}$$

Its determinant is Det $\mathcal{C}_{\dot{x}} = \beta$: the pair $[\boldsymbol{f}, g_{\dot{x}}]$ is, therefore, globally controllable.

Symbolic controllability matrices are here obtained from the analytical ones by replacing non-zero constant element by 1, and non-constant ones by $\bar{1}$, leading to the symbolic controllability matrix

$$\tilde{\mathcal{C}}_{\dot{x}} = \begin{bmatrix} 1 & \bar{1} & \bar{1} \\ 0 & 1 & \bar{1} \\ 0 & 1 & 1 \end{bmatrix}. \tag{41}$$

Since Det $\tilde{\mathcal{C}}_{\dot{x}} = 1 \otimes 1 \otimes 1$, we get $\zeta_{\dot{x}} = 1$: the controllability is thus global.

We thus checked that the best strategy to control the Chua circuit is to apply the input to $\dot{x}$ while variable z is measured, as suggested by the observability path.

3.3. *Flatness*

Flatness is related to global observability and global controllability to avoid possible difficulties due to singularities in the observability and controllability manifolds. As an additional necessary condition to design a flat control law, the condition (14) leads to

$$\begin{cases} h = z & \mathcal{L}_g h = 0 \\ \mathcal{L}_f h = -\beta y & \mathcal{L}_g \mathcal{L}_f h = 0 \\ \mathcal{L}_f^2 h = -\beta x + \beta y - \beta z & \mathcal{L}_g \mathcal{L}_f^2 h = -\beta \neq 0, \end{cases} \tag{42}$$

that is, it is fulfilled and the symbolic flatness coefficient $\varphi_{z^3,\dot{x}} = 1$. We identified the optimal placement for the sensor and the actuator. All the results – including those not detailed here – are summarized in Table 1. Note that in our brief review of works devoted to the control of a Chua circuit (Section 1), none was found to use this placement. On the contrary, oftentimes, the sensor and the actuator are placed on a given variable v and on its derivative $\dot{v}$, respectively. Only Hartley and Mossayebi placed a sensor on x and an actuator on $\dot{z}$ [35], but this is the opposite of our selected configuration. In all the works, the placements were not justified and appeared to be the result of an empirical selection.

4. Controlling the Chua circuit

4.1. *A continuous scheme*

We start by considering the case in which the derivatives are known exactly. The governing equations are integrated with a fourth-order Runge–Kutta scheme. The sensor is placed on the variable z and the actuator on the derivative $\dot{x}$. A classical feedback linearization is

Table 1. Symbolic observability, controllability, and flatness coefficients for different combinations of sensor and actuator positioning for the piecewise linear Chua circuit (25). When the actuator is placed on $\dot{y}$, none of the variables allows fulfilling condition (14). The flatness coefficient φ is equal to the product $\eta_{\xi^3} \times \zeta_{\dot{v}}$ only when the condition (14) is fulfilled, otherwise it is null.

	Observability	Controllability	Condition	Flatness	
x	$\eta_{x^3} = 0.78$	$\zeta_{\dot{z}} =$ **1.00**	fulfilled	$\varphi_{x^3,\dot{z}} = 0.78$	
y	$\eta_{y^3} = 0.84$	$\zeta_{\dot{y}} = 0.84$	Not fulfilled	$\varphi_{\xi^3,\dot{y}} = 0.00$	$\xi = \{x, y, z\}$
z	$\eta_{z^3} =$ **1.00**	$\zeta_{\dot{x}} =$ **1.00**	fulfilled	$\varphi_{z^3,\dot{x}} =$ **1.00**	

thus [40]

$$u = -\kappa_g \frac{\mathcal{L}_f^3 h + \kappa_1 (z - z_d) + \kappa_2 (\dot{z} - \dot{z}_d) + \kappa_3 (\ddot{z} - \ddot{z}_d)}{\mathcal{L}_g \mathcal{L}_f^2 h} \tag{43}$$

where z_d is the measured variable z for the desired state, $\dot{z}_d$ and $\ddot{z}_d$ being its successive derivatives and

$$\mathcal{L}_f^3 h = \left| \begin{array}{ll} -\alpha\beta(m_1+1)z - (\alpha m_1 + \beta)\dot{z} & \\ \quad + [\alpha(m_1+1)+1]\ddot{z} - \alpha\beta(m_0 - m_1) & x < -1 \\ \alpha\beta(m_0 - 2m_1 - 1)z + [\alpha(m_0 - 2m_1) & \\ \quad -\beta]\dot{z} + [\alpha(m_0 - 2m_1 - 1) - 1]\ddot{z} & \text{if } |x| < 1 \\ -\alpha\beta(m_1+1)z - (\alpha m_1 + \beta)\dot{z} & \\ \quad + [\alpha(m_1+1)+1]\ddot{z} + \alpha\beta(m_0 - m_1) & x > +1 \end{array} \right.$$

is used for cancelling the x-dynamics to be replaced with a linear combination of the three derivatives of the measured variable z. Obviously, when the desired state is a singular point, the derivatives $\dot{z}_d$, and $\ddot{z}_d$ are null. The κ_i ($i \in \{1, 2, 3\}$) are chosen as [29]

$$\begin{cases} \kappa_1 = -\Lambda_1\Lambda_2\Lambda_3 \\ \kappa_2 = \Lambda_1\Lambda_2 + \Lambda_2\Lambda_3 + \Lambda_3\Lambda_1 \\ \kappa_3 = -(\Lambda_1 + \Lambda_2 + \Lambda_3), \end{cases} \tag{44}$$

where Λ_i ($i \in \{1, 2, 3\}$) are the desired eigenvalues of the matrix

$$A + BK = \begin{bmatrix} 0 & 1 & 0 \\ 0 & 0 & 1 \\ 0 & 0 & 0 \end{bmatrix} + \begin{bmatrix} 0 \\ 0 \\ 1 \end{bmatrix} \begin{bmatrix} -\kappa_1 \\ -\kappa_2 \\ -\kappa_3 \end{bmatrix} = \begin{bmatrix} 0 & 1 & 0 \\ 0 & 0 & 1 \\ -\kappa_1 & -\kappa_2 & -\kappa_3 \end{bmatrix} \tag{45}$$

where A is the jacobian matrix of the linear part of the canonical form (15), B is the matrix defining on which derivative is applied the control law (the last one in the case of a feedback linearization), and K is the vector of gains used to stabilize the control. There are various methods to determine the eigenvalues and assign them to the controlled loop system. The two most common ones, which are not too different, are the pole placement and the root locus [17]. Roughly speaking, the task is to determine the eigenvalues to rend stable the

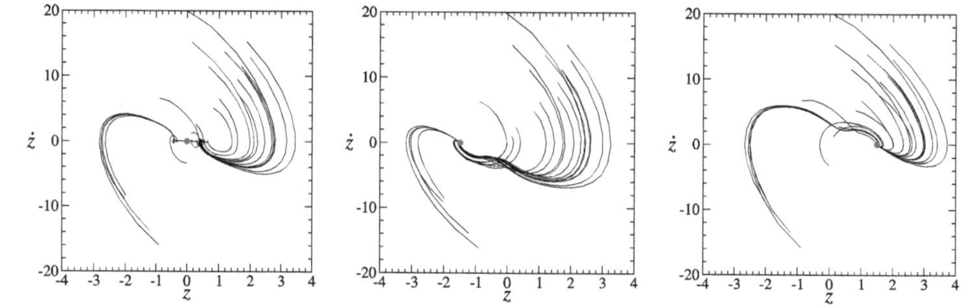

(a) Towards S_0: $\Lambda_i = -1$ (b) Towards S_-: $\Lambda_i = -5$ (c) Towards S_+: $\Lambda_i = -5$

Figure 3. Dynamics of the Chua circuit (25) with the control law (43) when the desired states are the three singular points, respectively. Parameter values: $\alpha = 9$, $\beta = \frac{100}{7}$, $m_0 = -\frac{8}{7}$, and $m_1 = -\frac{5}{7}$. Duration of the trajectory: 85 s. Gains κ_i are obtained from Λ_i according to Equation (44). $\kappa_g = 1$.

behaviour of a closed loop system. The global gain κ_g allows to adjust the amplitude of the input u; it is by default such as $\kappa_g = 1$ but it can be varied to optimize the control as discussed below.

When $z_d = 0$, corresponding to the singular point S_0 as the desired state, starting from 20 initial conditions for the three variables randomly chosen over the range $[-1.5, +1.5]$ with $\Lambda_i = -1$, the controlled trajectory consistently converges to S_0 as shown in Figure 3(a). When, for instance, $z_d = z_+$, the controlled trajectory does not always converge to S_+, but may instead converge to S_-: in such cases, it is sufficient to increase Λ_i to 5 to reach the desired state via the controlled trajectory for every initial condition (Figure 3(b,c)).

A less trivial example consists of controlling the chaotic dynamics onto one of its unstable periodic orbits. An unstable period-4 orbit was extracted from the attractor using a close-return technique [52]. Its four periodic points are plotted in the first-return map (Figure 1(c)). If one labels the ten monotone branches using integers whose parity is defined by the slope sign of the branch (as exemplified in Ref. [52]), and using an explicit notation to exhibit the symmetry between the two scrolls, one may have the natural order

$$\overline{0} < \overline{1} < \overline{2} < \overline{4} < \overline{5} < 0 < 1 < 2 < 4 < 5$$

where $\overline{0}$ designates the increasing branch of the left scroll (negative quadrant), which is symmetric under the inversion symmetry of the increasing branch 0 in the right scroll, and so on. Here, the overbars reflects the symmetry inherent in the system. The orbital sequence of the extracted period-4 orbit is thus $(5\overline{0}\overline{5}0)$, meaning that the orbit crosses the Poincaré section in branch 5, $\overline{0}$, $\overline{5}$, and 0, successively, as plotted in Figure 1(c). It can be seen that the controlled trajectory is quite close to the period-4 orbit but presents some short drifts after crossing the singular manifold $\mathcal{M}_s$ (blue disks in Figure 4): the switching plane is therefore source of perturbations in the control process. This is nevertheless a rather successful control of the chaotic regime onto a periodic orbit [remember that the uncontrolled circuit produces the chaotic trajectory plotted in Figure 1(a)].

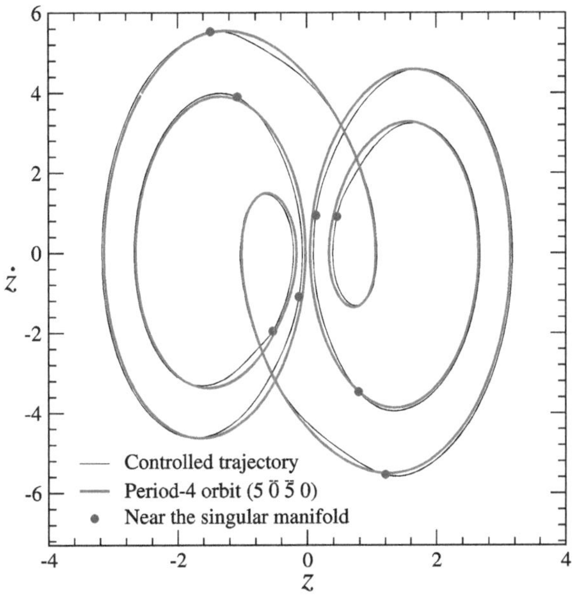

Figure 4. Controlled trajectory with the flat input-output (43) to the unstable period-4 orbit extracted from the chaotic behaviour (on which is developed the uncontrolled trajectory) shown in Figure 3(a) produced by the Chua uncontrolled circuit. Parameter values: $\Lambda_i = -10$, $\kappa_g = 1$, and others as in Figure 3.

4.2. A discretized scheme

Rigorously, a piecewise linear system cannot be integrated with a Runge–Kutta scheme without identifying accurately the time at which the trajectory crosses the switching plane and to consider the switch in computing the corresponding step. The Chua circuit is integrated using an Euler scheme as

$$\begin{cases} x_k = x_{k-1} + \alpha \left[y_{k-1} - x_{k-1} - \Psi(x_{k-1}) \right] \delta t \\ y_k = y_{k-1} + \left(x_{k-1} - y_{k-1} + z_{k-1} \right) \delta t \\ z_k = z_{k-1} - \beta y_{k-1} \delta t \end{cases} \tag{46}$$

where $\Psi(x)$ is defined by Equation (27) and $\delta t = 0.0005$ s. These equations are difference equations. Although in Figure 1(c), the first-return map does not show any difference between an explicit Euler scheme and a 4th-order Runge–Kutta scheme, strictly speaking, the latter is not allowed for the system (25). This restriction is due to the fact that for $|x| = 1$, $\Psi(x)$ is not differentiable and, consequently, higher-derivative terms as involved in a Runge–Kutta scheme cannot be determined. In the other hand, a discretized system offers a simple base for investigating the effect of discretization in practical processes. Thus, we now consider that only the measured variable z is known as a discretized time series at the sampling time τ_s. As we will see, the discretization of the measured signal necessarily induces a delay in the application of the control law which may have a significant effect on its effectiveness. The first step towards applying the control law (43) (which depends on the location in the state space) is thus to detect the switching mechanism by using a reconstructor from these measurements. From Equation (25), the first step is to compute the derivative of z as

$$\dot{z}_{k-1} = \frac{z_k - z_{k-2}}{2\,\tau_s}, \tag{47}$$

where z_k designates the value of variable z at time $t = k\,\tau_s$, and then to assess variable y at time $(k-1)\,\tau_s$ as

$$y_{k-1} = -\frac{\dot{z}_{k-1}}{\beta}. \tag{48}$$

The derivative of y is then computed using a centred Euler scheme

$$\dot{y}_{k-2} = \frac{y_{k-1} - y_{k-3}}{2\,\tau_s} \tag{49}$$

at time $(k-2)\,\tau_s$. Using the second equation of the Chua model (25), one finally gets

$$x_{k-2} = \dot{y}_{k-2} + y_{k-2} - z_{k-2}. \tag{50}$$

Variable x is thus retrieved from the measurements of variable z with a delay of $2\,\tau_s$ and it is possible to detect the switch. To apply the control law (43), it remains necessary to compute the second derivative of z, that is,

$$\ddot{z}_{k-2} = \frac{\dot{z}_{k-1} - \dot{z}_{k-3}}{2\,\tau_s}. \tag{51}$$

The final time-delay with which the control is applied is, therefore, $2\tau_s$. In [7,18,71], real-time discretization techniques based on explicit or implicit Euler schemes are proposed to avoid delay-related problems. A survey can be found in [65].

We apply the feedback linearization (43) as follows. We consider that the sampling time τ_s for discretizing the time series of the variable z is also the clock for applying the control law (43); in a practical scenario, this could correspond to the sampling rate of the analog-to-digital and digital-to-analog converters of a microprocessor-based controller. The sampling period τ_s is set in terms of δt, the time step used for the numerical integration of the Chua model (25), and the control law is evaluated once per τ_s (here $\tau_s = 10\,\delta t$). To assess the quality of the control, we computed the mean control error,

$$\overline{\epsilon} = \frac{1}{k_{\text{end}} - k_{\text{start}}} \sum_{k=k_{\text{start}}}^{k_{\text{end}}} \sqrt{\frac{(z_k - z_d)^2}{\max|z_d|} + \frac{(\dot{z}_k - \dot{z}_d)^2}{\max|\dot{z}_d|} + \frac{(\ddot{z}_k - \ddot{z}_d)^2}{\max|\ddot{z}_d|}} \tag{52}$$

where $k_{\text{start}} = k_{\text{init}} + k_{\text{transient}}$ wherein k_{init} is the discrete time at which the control law starts to be applied and $k_{\text{transient}} = 40,000 = 2.0 \lesssim T_0$ exceeds the duration of the transient regime after which the controlled trajectory becomes globally time invariant (we used $k_{\text{end}} - k_{\text{start}} = 130,000 = 65\,T_0$).

To have a global view of the response of the system to the control law when the global gain κ_g is varied, we used the simple component Poincaré section (for the sake of computational parsimony), defined as

$$\mathcal{P}_s = \left\{(s_n, \ddot{s}_n) \in \mathbb{R}^2 \mid \dot{s}_n = 0, \ddot{s}_n > 0\right\}. \tag{53}$$

We started with the default value $\kappa_g = 1$, and browsed the Λ_i-values for providing the smallest error: we retained $\Lambda_i = -15$. Then we computed a bifurcation diagram as a function of the gain κ_g to detect the range over which the controlled trajectory is with the right

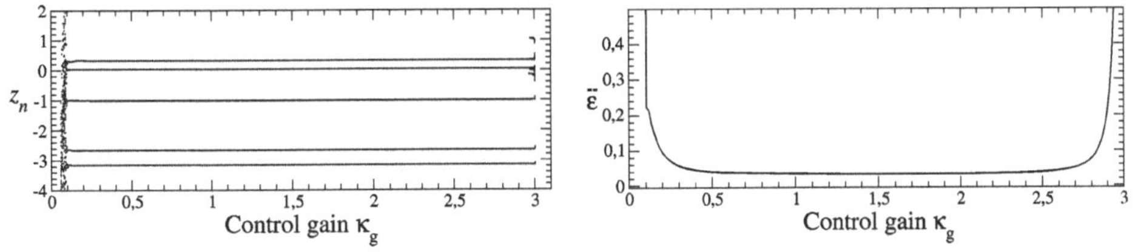

(a) Feedback linearization (43) with a sensor on z and an actuator on $\dot{x}$: $\Lambda_i = -15$

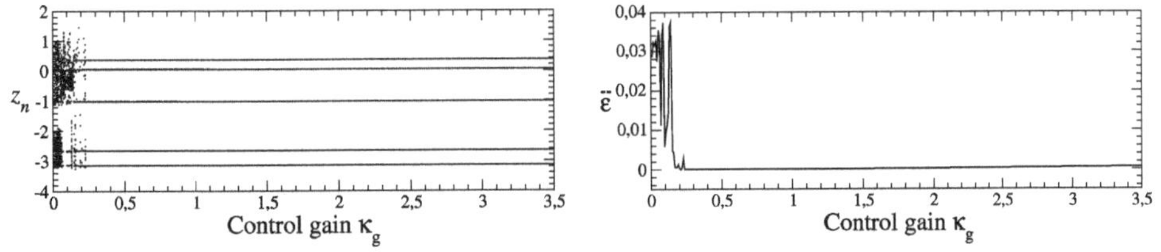

(b) Homogeneous law (56) with a sensor on z and an actuator on $\dot{x}$

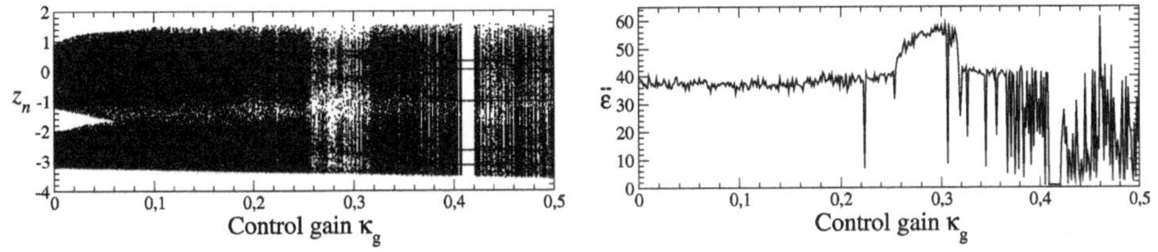

(c) Passivity-based law (57) with a sensor on x and an actuator on $\dot{x}$

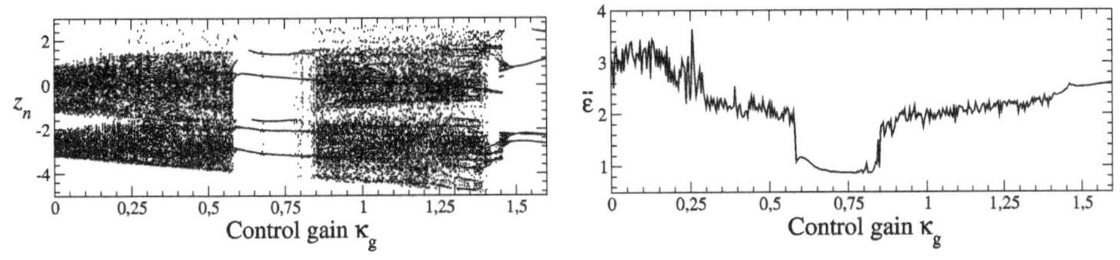

(d) Passivity-based law (57) with a sensor on z and an actuator on $\dot{z}$

Figure 5. Bifurcation diagram and mean error $\bar{\epsilon}$ as a function of the control gain κ for the Chua circuit controlled with the homogeneous control law (a) and the passivity-based control law (b) and (c), respectively. Parameter values: $\chi = 0.3$, $\Lambda_i = -15$, $\tau_d = 2\,\tau_s$, $\tau_u = \tau_s$ and others as in Figure 3.

period (Figure 5(a)). In the present case, the diagram confirmed that the retained control parameter κ_g does not affect too much the result and corresponds to the smallest error ($\bar{\epsilon} = 0.035$).

We determined that the smallest mean control error, $\bar{\epsilon} = 0.035$, is obtained for $\Lambda_i = -15$: the corresponding bifurcation diagram is plotted in Figure 5(a) alongside the corresponding mean control error (this will be discussed in details below). When a realistic controller is considered, there are two aspects which can degrade the efficiency of the control: (i) the delay τ_d due to the computation of the derivatives and to the processing of the data and, (ii) the duration τ_u before which a new value of the input u can be computed. These two delays are expressed in terms of the sampling tme τ_s which is here equal to $10\,\delta t$. By construction, when a centred Euler scheme is used to compute derivatives, $\tau_d \geqslant 2\,\tau_s$.

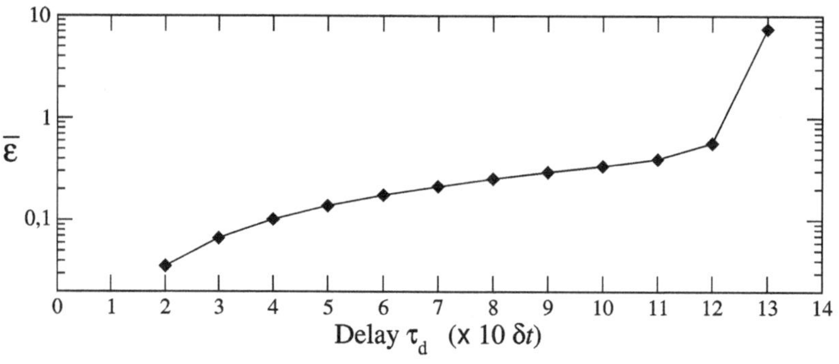

Figure 6. Evolution of the mean control error $\bar{\epsilon}$ as a function of the delay τ_d induced by the discretization or by the device for other reasons. The delay is incremented in terms of the sampling time $\tau_s = 10\,\delta t$.

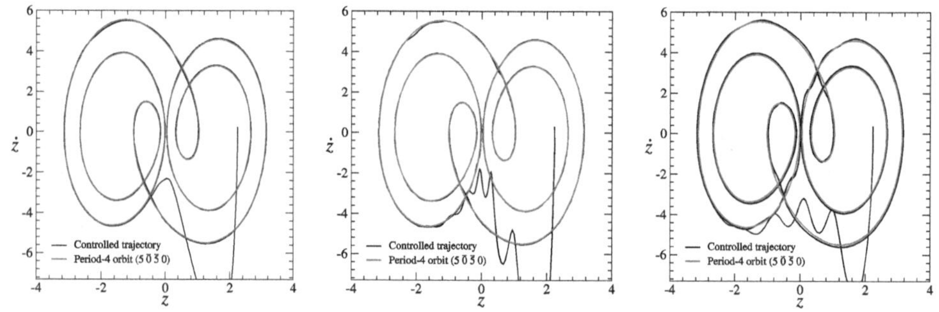

(a) $\tau_u = \tau_s$: $\Lambda_i = -15$, $\bar{\epsilon} = 0.035$ (b) $\tau_u = 2\,\tau_s$: $\Lambda_i = -15$, $\bar{\epsilon} = 0.078$ (c) $\tau_u = 3\,\tau_s$: $\Lambda_i = -10$, $\bar{\epsilon} = 0.200$

Figure 7. Control by feedback linearization of the Chua circuit with a discrete-time scheme for various durations τ_u. The period-4 orbit ($5\overline{0}50$) corresponds to the desired dynamics. The rather short transient regime is left in these plots. The Chua circuit is integrated using an Euler scheme with $\delta t = 0.0005$ s, $\tau_s = 10\,\delta t$, and $\tau_d = 2\,\tau_s$. Global gain: $\kappa_g = 1$.

We computed the mean We investigated the effect of the mean control error $\bar{\epsilon}$ for increasing delay τ_d (Figure 6): $\bar{\epsilon}$ progressively increases up to $\tau_d = 12\,\tau_s$, beyond which $\bar{\epsilon}$ jumps by one order of magnitude before ejecting the trajectory to infinity. With respect to the mean duration ($\overline{T} = 2.5$ s) of one oscillation of the Chua circuit, this corresponds to 2.4% of the period of oscillations.

The effect of the duration τ_u between two new u-values is varied for the minimum value $\tau_d = 2\,\tau_s$: by construction, $\tau_u \geqslant \tau_s = 10\,\delta t$ (Figure 8(a)). The mean control error $\bar{\epsilon}$ doubles when $\tau_u = 2\,\tau_s$(Figure 8(b)), and it is already necessary to reduces Λ_i to -10 when $\tau_u = 3\,\tau_s$ to prevent the ejection of the trajectory to infinity, leading to $\bar{\epsilon} = 0.200$. Setting $\tau_u = 4\,\tau_s$ ejects the trajectory to infinity without any possibility to prevent it with an adjusted Λ_i-value. Consequently, when a flat control law based on feedback linearization is used, it is more important to evaluate the u-value at a high-frequency than to reduce the delay τ_d with which derivatives are computed.

The control errors, calculated for the three durations τ_u in Figure 7, confirm (Figure 8) that the control error fluctuates significantly each time the trajectory crosses the switching plane: the fact that the control law u, which is also piecewise, induces this feature.

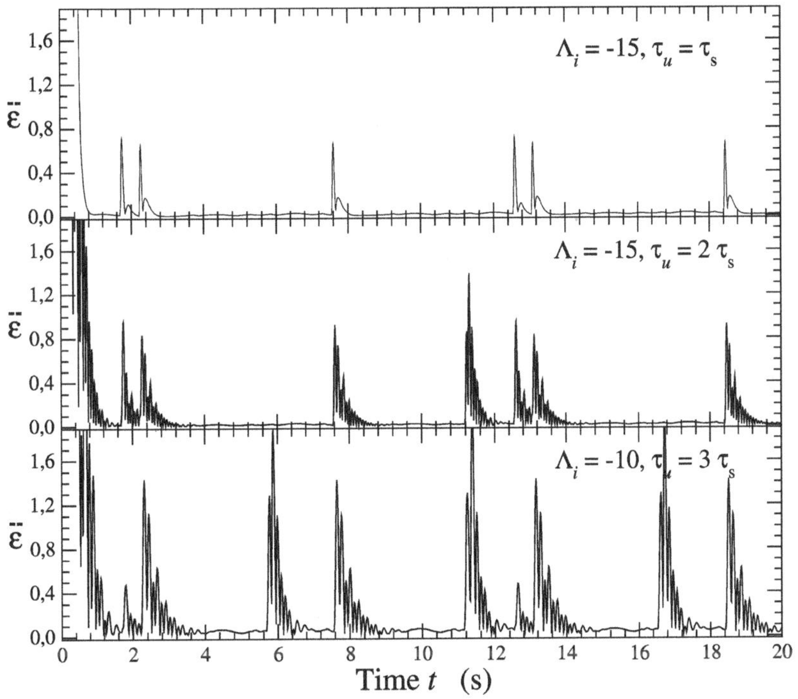

Figure 8. Control error for three different durations τ_u investigated in Figure 7. Mean values of the control error are: $\overline{\epsilon} = 0.035$ for $\tau_u = \tau_s$, $\overline{\epsilon} = 0.078$ for $\tau_u = 2\,\tau_s$, and $\overline{\epsilon} = 0.200$ for $\tau_u = 3\,\tau_s$. Parameters Λ_i and τ_d are reported in each panel. Other parameter values as in Figure 1.

4.3. Comparison with other control laws

In this section, the robustness of the feedback linearization designed (43) in Section 4.1, Equation (43), is compared to three other types of control laws based on widely used theories. The strategy to determine the gain was to plot a bifurcation diagram for selecting the range of gains for which a limit cycle close to the desired orbit is obtained and, thus retaining the gain associated with the smallest mean control error.

When the piecewise linear function $\Psi(x)$ is removed, the Chua circuit is linear and its Jacobian matrix is

$$\mathcal{J}_{\text{lin}} = \begin{bmatrix} -\alpha & \alpha & 0 \\ 1 & -1 & 1 \\ 0 & -\beta & 0 \end{bmatrix}, \tag{54}$$

with a trace $\text{Tr}(\mathcal{J}_{\text{lin}}) = -(\alpha + 1)$, implying that the linear circuit is always dissipative. The source of sensitivity to initial conditions (leading to a positive largest Lyapunov exponent) is thus provided by the piecewise linear function $\Psi(x)$. When the actuator is placed on $\dot{y}$ or $\dot{z}$, the control action reaches the first equation where the switching mechanism acts, after one or two integrators, respectively: then, the control action cannot flatten the switching function as efficiently as when it is directly applied on $\dot{x}$. In agreement with the controllability analysis, the most efficient strategy is thus to place directly the actuator on $\dot{x}$, where the switching function $\Psi(x)$ also enters. Thus, we placed the actuator on $\dot{x}$, used the period-4 orbit as the intended dynamics, and designed three different control laws. In each case, the control error ϵ, was computed according to Equation (52).

The first control law under consideration is a homogeneous control law [9,15] constructed with a sensor placed on z and using the quantity

$$\sigma = k_1 (z - z_\mathrm{d}) + k_2 (\dot{z} - \dot{z}_\mathrm{d}) + (\ddot{z} - \ddot{z}_\mathrm{d}) \tag{55}$$

where $k_1 = \Lambda_1 \Lambda_2$ and $k_2 = -(\Lambda_1 + \Lambda_2)$: our simulations are performed with $\Lambda_i = -1$ $(i = 1, 2)$. The control law

$$u = -\kappa_\mathrm{g}\ \mathrm{sgn}(\mathcal{L}_g \mathcal{L}_f^2 h)\ \mathrm{sgn}(\sigma)\ |\sigma|^\chi\ , \tag{56}$$

where $\chi \in [0, 1]$ is applied on $\dot{x}$: this is also a flat control law. The parameter χ is set according to the confidence one may have in the input: for instance, the better the observability provided by the measured variable, the smaller the χ. This coefficient can also be used for damping the noise effect: the larger the noise perturbations, the larger the χ should be. In the present case, one may be confident about the measured variable (global observability and, in this work, noise free). After parameter sweeping, we found that a value $\chi = 0.3$ provides the smallest control error $\overline{\epsilon}$. There is a large range of gain values ($\kappa_\mathrm{g} > 0.2$) over which the mean control error $\overline{\epsilon} < 0.008$ remains very small (Figure 5(b)). The advantage of this control law with respect to the feedback linearization is its simplicity, in particular, there are fewer parameters to determine. When a bifurcation diagram is computed versus the global gain κ_g (Figure 5(a ,b)), it appears that the feedback linearization and the homogeneous control law are similarly robust against parameter changes. Nevertheless, the smallest mean control error remains larger for the homogenous law than for the feedback linearization.

The second control approach taken into consideration is the passivity-based control law [69]

$$u = -\kappa_\mathrm{g}\,\mathrm{sgn}\,(x - x_\mathrm{d})\ |x - x_\mathrm{d}|^\chi\ , \tag{57}$$

which is based on another approach commonly used in control theory, namely, on the fact, it considers that the control law is applied on $\dot{x}$, where the dynamics are not passive. With a sensor on x and an actuator on $\dot{x}$, the control law cannot be flat. Nevertheless, it has the advantage of only requiring the measurement of variable x, without its derivatives. Nevertheless, the convergence of this law may be not as fast as the previous control laws. It is found that it provides an efficient control to the period-4 orbit for $\kappa_\mathrm{g} \in [0.4088; 0.4200]$ as exhibited by the bifurcation diagram (Figure 5(c)). Nevertheless, the mean control error $\overline{\epsilon} = 0.065$ is about ten times greater than for the two previous flat control laws.

Finally, a third case is tentatively considered, having an input u as designed in Equation (56) but placing the sensor on variable z and the actuator on $\dot{z}$: in Equation (57), we thus replaced x with z. We kept $\chi = 0.3$ and computed a bifurcation diagram as a function of the global gain κ_g (Figure 5(d)): we found a range $\kappa_\mathrm{g} \in [0.636, 0.776]$ for which the controlled trajectory remains still in the neighbourhood of the intended period-4 orbit during a substantial fraction of it, but clearly departs otherwise (Figure 9(c)), leading to a mean error $\overline{\epsilon} = 0.896$, which is more than 10 times greater compared to the error obtained

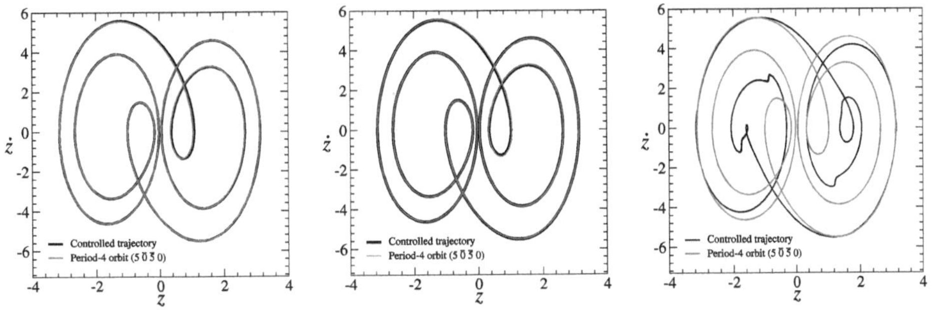

(a) Feedback linearization
Sensor on z, actuator on $\dot{x}$
$\kappa_g = 0.30$: $\bar{\epsilon} = 0.008$

(b) Homogeneous law
Sensor on x, actuator on $\dot{x}$
$\kappa_g = 0.42$: $\bar{\epsilon} = 0.065$

(c) Passivity-based law
Sensor on z, actuator on $\dot{z}$
$\kappa_g = 0.75$: $\bar{\epsilon} = 0.896$

Figure 9. Control of the Chua circuit with a discretized scheme for various sensor and actuator placements. The period-4 orbit ($5\overline{05}0$) is the desired dynamics. The system equations are is integrated using an Euler scheme with $\delta t = 0.0005$ s. These three cases correspond to the three cases considered in Figure 5.

with the previous passivity-based control laws and 100 times larger than the error obtained with flat control laws. Placing correctly the actuator is thus more important than placing the sensor.

From the previous investigations, one can conclude that feedback linearization is optimal for minimizing the error $\bar{\epsilon}$ but, unfortunately, it is very sensitive to the duration τ_u and the delay τ_d as induced for computing deriviatives. First, with a feedback linearization, to prevent breaking the dynamics, we had to set $\Lambda_i = -8$ and to reduce the global gain $\kappa_g = 0.5$. As shown in Figure 5, this allows keeping the error at the level obtained with the non-flat control law (57) for a duration $\tau_u \leqslant 7\,\tau_s$. The conclusion is that the feedback linearization is very efficient when the parameter can be sharply optimized, but its performance degrades very quickly when the processing is not sufficiently fast. The homogeneous control law offers a similar error $\bar{\epsilon}$ and is in fact more robust to the duration τ_u as shown in Figure 10. The control error $\bar{\epsilon}$ increases marginally with the duration τ_u up to

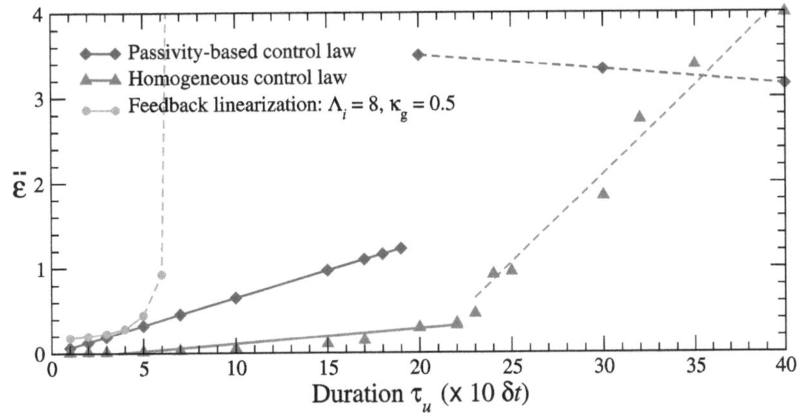

Figure 10. Dependence of the mean control error $\bar{\epsilon}$ on the duration over which the input u is kept constant. Cases of the feedback linearization (green), of the homogeneous control law (red) and of the passivity-based law (blue).

$23\,\tau_s = 0.115$ s, that is, about 20 samples per revolution which is close to the Nyquist criterion $f_t = 8.7 \geqslant f_{\text{Nyquist}} = 7$ Hz. Beyond this threshold τ_u-value, the control law quickly loses its effectiveness, all way up to the point where the mean control error $\bar{\epsilon}$ becomes comparable with the scale of the attractor size. The homogeneous control law, which is a flat control law, therefore behaves quite desirably. In experimental realizations of the Chua circuit, and potentially other chaotic circuits and devices, we therefore recommend to start with the homogeneous control law.

As said, the passivity-based control law, with the sensor on x and the actuator on $\dot{x}$ is not a flat control law. It behaves as the homogeneous control law, that is, with a mean control error $\bar{\epsilon}$ linearly increasing with the duration τ_s, but the slope is more than 3 times greater. The threshold value occurs at $\tau_s = 19\,\tau_s = 0.095$ s ($f_t = 10.5$ Hz): at this τ_u-value, the mean error becomes suddenly of the order of magnitude of the attractor size (Figure 10). The passivity-based control law is, therefore, clearly less efficient than the flat homogeneous control law.

5. Conclusion

Owing to the duality principle, our approach is applicable for every system which has an output providing a global observability of the state space. Clearly, to be not trivial, the output has to be of a dimension lower than the dimension of the system. Ideally, a one-dimensional output is searched for. Among the large set of low-dimensional chaotic systems, many, as the Rössler system and some Sprott systems as reported in Ref. [49], present at least one variable providing a global observability, with the notable exception of the Lorenz-like systems which require measuring two variables for getting a global observability. Such a case should be considered in future work. Moreover, here we limit ourselves to the case where the system structure and parameter values are exactly known. We expect a greater robustness against model mismatch for flat control law, but this should be ascertained in future work. With its piecewise linear function, the Chua circuit is arguably a more demanding case than many other chaotic systems since the flat control law is also based on a switch mechanism. Consequently, our results should be applied to a rather wide class of chaotic systems. The case of hyperchaotic systems would deserve a special attention due to some specific related problems.

We here investigated the reduction in the control efficiency (control error) when a delay – which may be due to the computation of successive derivatives or induced by the sensor and/or actuator. It was observed that a flat control law can be more sensitive than other types of control law (passivity-based or homogeneous control law) to the presence of a delay: nevertheless, the permitted delay still allows to correctly apply a flat control law based on feedback linearization without a significant lack of efficiency.

Three aspects stand central to viable implementation. Firstly, minimizing the number of sensors and actuators. Aside from the obvious remark that both incur complexity and costs related to signal transduction, amplification, and conversion, inherent drawbacks are incurred when excessively instrumenting any entity. In the case of biological systems, taking electrophysiology as an example, large ensembles of external electrodes are notoriously unreliable, and implanted electrodes must be minimized at all costs due to their invasiveness [25,30,77,79]. In this work, we showed that it is indeed possible to work with a single sensor and a single actuator for controlling complex dynamics as the one produced by the

Chua circuit. Future work is motivated to explore how the arguments can be extended to experimental settings, where the target system is incompletely known, measurements have associated uncertainties, and non-ideal behaviours may be significant.

This work provides a compelling example of how the notions of observability and controllability symbolic coefficients, combined with flatness, may guide towards an optimal placement of sensors and actuators. On the one hand, optimal placement allows minimizing the number of sensors and actuators, in this case reaching down to a single sensor and a single actuator. Using such optimal placements leads to a large robustness (the desired states are reached for a large range of gains), and the control error is smaller than otherwise. Thus, for instance, a passivity-based control law which cannot, by construction, be designed with an optimal placement, does not provide a very satisfactory control of the Chua dynamics. Contrary to this, a homogeneous control law, designed with the optimal placements, provides a control of the Chua dynamics with performance close to those obtained with the feedback linearization.

Third, the non-trivial influence of non-instantaneous action, as represented for example by time discretization. For manifold reasons ranging from finite compute and information-transmission time through group delay of signal propagation in filters and even biological conduction times, situations in which the response of an experimentally-realized control system can be approximated as instantaneous represent an exception rather than the norm [28]. We thus showed that when the delay is increased the efficiency of the flat control law is reduced, the control error increasing. In the case of the Chua circuit, we showed that designing a flat control law improves significantly the efficiency of the controller and its robustness against the delay which is necessarily induced by processor-based action. It was also shown that a flat homogeneous control law as in Equation (56), one that is simpler in its algebraic form and less sensitive to the computation of derivatives but also to the quality model, could be preferred over feedback linearization as in Equation (43).

Disclosure statement

No potential conflict of interest was reported by the author(s).

ORCID

Christophe Letellier http://orcid.org/0000-0003-3603-394X

References

[1] E.J. Adam and J.L. Marchetti, Designing and tuning robust feedforward controllers, Comput. Chem. Eng. 28 (9) (2004), pp. 1899–1911.
[2] L.A. Aguirre, Controllability and observability of linear systems: Some noninvariant aspects, IEEE. Trans. Educ. 38 (1) (1995), pp. 33–39.
[3] L.A. Aguirre, L.L. Portes, and C. Letellier, Structural, dynamical and symbolic observability: From dynamical systems to networks, PLoS. ONE. 13 (10) (2018), pp. e0206180.
[4] J. Alvarez-Ramírez, Nonlinear feedback for controlling the Lorenz equation, Phys. Rev. E. 50 (31994), pp. 2339–2342.
[5] L. Angquist, A. Antonopoulos, D. Siemaszko, K. Ilves, M. Vasiladiotis, and H.P. Nee, Open-loop control of modular multilevel converters using estimation of stored energy, IEEE. Trans. Ind. Appl. 47 (6) (2011), pp. 2516–2524.

[6] J.P. Barbot, M. Fliess, and T. Floquet, *An algebraic framework for the design of nonlinear observers with unknown inputs*, in *46th IEEE Conference on Decision and Control*, 2007, pp. 384–389.
[7] J.P. Barbot, A. Levant, M. Livne, and D. Lunz, Discrete differentiators based on sliding modes, Automatica 112 (2020), pp. 108633.
[8] K. Barone and S.N. Singh, Adaptative feedback linearizing control of Chua's circuit, Int. J. Bifurcat. Chaos. 12 (07) (2002), pp. 1599–1604.
[9] S.P. Bhat and D.S. Bernstein, Finite-time stability of continuous autonomous systems, SIAM. J. Control. Optim. 38 (3) (2000), pp. 751–766.
[10] E. Bianco-Martinez, M.S. Baptista, and C. Letellier, Symbolic computations of nonlinear observability, Phys. Rev. E. 91 (6) (2015), pp. 062912.
[11] E. Bilotta and P. Pantano, *A Gallery of Chua Attractors*, World Scientific Publishing, 2008.
[12] S. Boccaletti, C. Grebogi, Y.C. Lai, H. Mancini, and D. Maza, The control of chaos: Theory and applications, Phys. Rep. 329 (3) (2000), pp. 103–197.
[13] M. Bonilla Estrada, M. Figueroa Garcia, M. Malabre, and J.C. Martínez García, Left invertibility and duality for linear systems, Linear. Algebra. Appl. 425 (2-3) (2007), pp. 345–373.
[14] S. Bououden, D. Boutat, G. Zheng, J.P. Barbot, and F. Kratz, A triangular canonical form for a class of 0-flat nonlinear systems, Int. J. Control. 84 (2) (2011), pp. 261–269.
[15] Y. Braidiz, D. Efimov, A. Polyakov, and W. Perruquetti, On robustness of finite-time stability of homogeneous affine nonlinear systems and cascade interconnections, Int. J. Control. 95 (3) (2022), pp. 768–778.
[16] M.A. Brdys and G.J. Kulawski, Dynamic neural controllers for induction motor, IEEE. Trans. Neural. Netw. 10 (2) (1999), pp. 340–355.
[17] R.W. Brockett and C.I. Byrnes, Multivariable Nyquist criteria, root loci, and pole placement: A geometric viewpoint, IEEE. Trans. Automat. Contr. 26 (1) (1981), pp. 271–284.
[18] B. Brogliato and A. Polyakov, *Globally stable implicit Euler time-discretization of a nonlinear single-input sliding-mode control system*, in *54th IEEE Conference on Decision and Control*, IEEE, Osaka, Japan, 2015, pp. 5426–5431.
[19] T.L. Carroll, A simple circuit for demonstrating regular and synchronized chaos, Am. J. Phys. 63 (4) (1995), pp. 377–379.
[20] G. Chen, Controlling Chua's global unfolding circuit family, IEEE. Trans. Circuits. Syst. I. 40 (11) (1993), pp. 829–832.
[21] G. Chen and X. Chen, Controlling Chua's circuit, J. Circuits. Syst. Comput. 03 (01) (1993), pp. 139–149.
[22] L. Chua, The genesis of Chua's circuit, Arch. Electron. Transm. Technol. 46 (1992), pp. 250–257.
[23] L.O. Chua, *CNN: A Paradigm for Complexity*, World Scientific Publishing, 1998.
[24] P. Deregel, Chua's oscillator: A zoo of attractors, J. Circuits. Syst. Comput. 03 (02) (1993), pp. 309–359.
[25] W.L. Ditto, M.L. Spano, V. In, J. Neff, B. Meadows, J.J. Langberg, A. Bolmann, and K. McTeague, Control of human atrial fibrillation, Int. J. Bifurcat. Chaos. 10 (03) (2000), pp. 593–601.
[26] M. Fliess, J.L. Lévine, P. Martin, and P. Rouchon, Flatness and defect of non-linear systems: Introductory theory and examples, Int. J. Control. 61 (6) (1995), pp. 1327–1361.
[27] L. Fortuna, M. Frasca, and M.G. Xibilia, *Chua's Circuit Implementations*, World Scientific Publishing, 2009.
[28] S.A. Frank, *Control Theory Tutorial*, Springer, 2018.
[29] G. Franklin, J. Powell, and A. Emami-Naeini, *Feedback Control of Dynamic Systems*, 8th ed., Pearson, 2015.
[30] A. Garfinkel, J.N. Weiss, W.L. Ditto, and M.L. Spano, Chaos control of cardiac arrhythmias, Trends. Cardiovasc. Med. 5 (2) (1995), pp. 76–80.
[31] J. Gauthier, H. Hammouri, and S. Othman, A simple observer for nonlinear systems applications to bioreactors, IEEE. Trans. Automat. Contr. 37 (6) (1992), pp. 875–880.
[32] R.W. Ghrist, Branched two-manifolds supporting all links, Topology 36 (2) (1997), pp. 423–448.
[33] R.W. Ghrist and P.J. Holmes, An ODE whose solutions contain all knots and links, Int. J. Bifurcat. Chaos. 06 (05) (1996), pp. 779–800.

[34] W. Guo and D. Liu, Adaptive control of chaos in Chua's circuit, Math. Probl. Eng. 2011 (2011), pp. 620946.
[35] T.T. Hartley and F. Mossayebi, Control of Chua's circuit, J. Circuits. Syst. Comput. 03 (01) (1993), pp. 173–194.
[36] R. Hermann and A. Krener, Nonlinear controllability and observability, IEEE. Trans. Automat. Contr.22 (5) (1977), pp. 728–740.
[37] L. Hunt, Renjeng Su, and G. Meyer, Global transformations of nonlinear systems, IEEE. Trans. Automat. Contr. 28 (1) (1983), pp. 24–31.
[38] C.C. Hwang, H.Y. Chow, and Y.K. Wang, A new feedback control of a modified Chua's circuit system, Phys. D. 92 (1-2) (1996), pp. 95–100.
[39] A. Isidori, *Nonlinear Control Systems*, Springer, 1995.
[40] A. Isidori, A. Krener, C. Gori-Giorgi, and S. Monaco, Nonlinear decoupling via feedback: A differential geometric approach, IEEE. Trans. Automat. Contr. 26 (2) (1981), pp. 331–345.
[41] A. Isidori, C.H. Moog, and A.D. Luca, *A sufficient condition for full linearization via dynamic state feedback*, in *25th IEEE Conference on Decision and Control*, 1986, pp. 203–208.
[42] B. Jakubczyk and W. Respondek, On linearization of control systems, Bulletin. Polish. Acad. Math. Sci. Ser. 28 (1980), pp. 517–522.
[43] M. Johansson, *Piecewise Linear Control Systems*, Springer, Berlin - Heidelberg, 2003.
[44] T. Kailath, *Linear Systems*, Information and System Sciences Series, Prentice-Hall, 1980.
[45] R. Kalman, *On the general theory of control systems*, in IFAC Proceedings Vol. 1, 1st International IFAC Congress on Automatic and Remote Control, Moscow, USSR, 1960, pp. 491–502.
[46] C. Letellier, *Chaos in electronic circuits: Chua's contribution (1980–2000)*, in *Chaos, CNN, Memristors and beyond*, A. Adamatzky and G. Chen, eds., World Scientific Publishing, 2013, pp. 211–235.
[47] C. Letellier and L.A. Aguirre, Investigating nonlinear dynamics from time series: The influence of symmetries and the choice of observables, Chaos 12 (3) (2002), pp. 549–558.
[48] C. Letellier and L.A. Aguirre, Graphical interpretation of observability in terms of feedback circuits, Phys. Rev. E. 72 (5) (2005), pp. 056202.
[49] C. Letellier and L.A. Aguirre, Symbolic observability coefficients for univariate and multivariate analysis, Phys. Rev. E. 79 (6) (2009), pp. 066210.
[50] C. Letellier and J.P. Barbot, Flatness for an optimal control of chaotic systems using a minimal numbers of sensors and actuators, Chaos 31 (2021), pp. 103114.
[51] C. Letellier and R. Gilmore, The universal template is a subtemplate of the double-scroll template, J. Phys. A. Math. Theor. 46 (6) (2013), pp. 065102.
[52] C. Letellier, P. Dutertre, and B. Maheu, Unstable periodic orbits and templates of the Rössler system: Toward a systematic topological characterization, Chaos 5 (1) (1995), pp. 271–282.
[53] C. Letellier, J. Maquet, L.L. Sceller, G. Gouesbet, and L.A. Aguirre, On the non-equivalence of observables in phase-space reconstructions from recorded time series, J. Phys. A. Math. Gen. 31 (39) (1998), pp. 7913–7927.
[54] C. Letellier, I. Sendiña-Nadal, and L.A. Aguirre, A nonlinear graph-based theory for dynamical network observability, Phys. Rev. E. 98 (2) (2018), pp. 020303(R).
[55] C. Letellier, I. Sendiña-Nadal, E. Bianco-Martinez, and M.S. Baptista, A symbolic network-based nonlinear theory for dynamical systems observability, Sci. Rep. 8 (1) (2018), pp. 3785.
[56] J. Levine, *Analysis and Control of Nonlinear Systems. A Flatness-Based Approach*, Springer-Verlag, Berlin-Heildeberg, Germany, 2009.
[57] C.T. Lin, Structural controllability, IEEE. Trans. Automat. Contr. 19 (3) (1974), pp. 201–208.
[58] Y.Y. Liu, J.J. Slotine, and A.L. Barabási, Controllability of complex networks, Nature 473 (7346) (2011), pp. 167–173.
[59] C. Lobry, Contrôlabilité des systèmes non linéaires, SIAM. J. Contr. Optim. 8 (4) (1970), pp. 573–605.
[60] R.N. Madan, *Chua's Circuit: A Paradigm for Chaos*, World Scientific Publishing, 1993.
[61] P. Martin, R.M. Murray, and P. Rouchon, Flat systems: Open problems, infinite dimensional extension, symmetries and catalog, Lect. Notes Control Inf. Sci. 264 (2001), pp. 33–57.
[62] S.J. Mason, Feedback theory - some properties of signal flow graphs, Proc. IRE. 41 (9) (1953), pp. 1144–1156.

[63] T. Matsumoto, A chaotic attractor from Chua's circuit, IEEE. Trans. Circuits. Syst. CAS. 31 (12) (1984), pp. 1055–1058.
[64] T. Matsumoto, L.O. Chua, and M. Komuro, The double scroll, IEEE. Trans. Circuits. Syst. CAS. 32 (8) (1985), pp. 797–818.
[65] M.R. Mojallizadeh, B. Brogliato, A. Polyakov, S. Selvarajan, L. Michel, F. Plestan, M. Ghanes, J.P. Barbot, and Y. Aoustin, *Discrete-time differentiators in closed-loop control systems: Experiments on electro-pneumatic system and rotary inverted pendulum*, Research Rep. INRIA Grenoble, 2022. Available at https://hal.inria.fr/hal-03125960.
[66] I.S. Nadal and C. Letellier, Observability analysis and state reconstruction for networks of nonlinear systems, Chaos 32 (8) (2022), pp. 083109.
[67] F. Nicolau, W. Respondek, and J.P. Barbot, Flat inputs: Theory and applications, SIAM. J. Contr. Optim. 58 (6) (2020), pp. 3293–3321.
[68] A. Nuzuri, J. Suykens, and L. Chua, A CNN approach to brain-like chaos-periodicity transitions, Int. J. Bifurcat. Chaos. 08 (11) (1998), pp. 2263–2278.
[69] R. Ortega, Z. Jiang, and D. Hill, *Passivity-based control of nonlinear systems: A tutorial*, in *Proceedings of the 1997 American Control Conference*, Vol. 5, Albuquerque, New Mexico, 1997, pp. 2633–2637.
[70] E. Ott, C. Grebogi, and J.A. Yorke, Controlling chaos, Phys. Rev. Lett. 64 (11) (1990), pp. 1196–1199.
[71] M. Reichhartinger, S. Koch, H. Niederwieser, and S.K. Spurgeon, *The robust exact differentiator toolbox: Improved discrete-time realization*, in *15th International Workshop on Variable Structure Systems*, IEEE, 2018, pp. 1-6.
[72] W. Respondek, *Right and left invertibility of nonlinear control systems*, in *NonLinear Controllability and Optimal Control*, H.J. Sussmann, ed., Routledge, 1990, pp. 133–176.
[73] M. Rosalie and C. Letellier, Systematic template extraction from chaotic attractors: I. Genus-one attractors with an inversion symmetry, J. Phys. A. Math. Theor. 46 (37) (2013), pp. 375101.
[74] T. Saito and K. Mitsubori, Control of chaos from a piecewise linear hysteresis circuit, IEEE. Trans. Circuits. Syst. I. 42 (3) (1995), pp. 168–172.
[75] R. Schenkendorf and M. Mangold, Parameter identification for ordinary and delay differential equations by using flat inputs, Theor. Found. Chem. Eng. 48 (5) (2014), pp. 594–607.
[76] H. Sira-Ramírez and S. Agrawal, *Differentially Flat Systems*, Automation and Control Engineering, CRC Press, 2004.
[77] H. Sohanian Haghighi and A. Markazi, A new description of epileptic seizures based on dynamic analysis of a thalamocortical model, Sci. Rep. 7 (1) (2017), pp. 13615.
[78] H.J. Sussmann and V. Jurdjevic, Controllability of nonlinear systems, J. Differ. Equ. 12 (1) (1972), pp. 95–116.
[79] K. Tsakalis, N. Chakravarthy, and L. Iasemidis, *Control of epileptic seizures: Models of chaotic oscillator networks*, in *Proceedings of the 44th IEEE Conference on Decision and Control*, 2005, pp. 2975–2981.
[80] S. Waldherr and M. Zeitz, Conditions for the existence of a flat input, Int. J. Control. 81 (3) (2008), pp. 439–443.
[81] Z. Yu and J. Zhao*Dynamic behaviors in two novel autonomous circuits with a physical negative resistor characterized by nonsymmetric function*, in *1992 IEEE International Symposium on Circuits and Systems (ISCAS)*, Vol. 6, 1992, pp. 2813–2816.

Appendix A. Special treatment of the observability matrix $\mathcal{O}_{x^3}$

In the matrix

$$\tilde{\mathcal{O}}_{x^3} = \begin{bmatrix} 1 & 0 & 0 \\ \bar{1} & 1 & 0 \\ \bar{1} & \bar{1} & 1 \end{bmatrix}, \tag{A1}$$

we did not consider that the second derivative $\ddot{x}$ is not defined in the singular manifold $\mathcal{M}_s$ related to the switching planes thus defining a singular observability manifold. This can be taken into account by imposing that the third row is all made of non-constant terms, that is, of $\bar{1}$ s. In this case, the first observability matrix is replaced with

$$\tilde{\mathcal{O}}_{x^3} = \begin{bmatrix} 1 & 0 & 0 \\ \bar{1} & 1 & 0 \\ \bar{1} & \bar{1} & \bar{1} \end{bmatrix}. \tag{A2}$$

The symbolic determinant of the matrix $\tilde{\mathcal{O}}_{x^3}$ is thus Det $\tilde{\mathcal{O}}_{x^3} = 1 \otimes 1 \otimes \bar{1}$, that is, made of two constant and one non-constant terms, leading to $N_1 = 2$, $N_{\bar{1}} = 1$ and $N_{\bar{\bar{1}}} = 0$. Using the expression (8) for the symbolic observability coefficients, one gets $\eta_{x^3} = \frac{2}{3} + \frac{1}{9} = 0.78$. The actual observability provided by the variable x is therefore not global as extracted from the graphical analysis, but only local. This was suggested simply from the placement of the switch function on the derivative $\dot{x}$.

Nonlinear dynamics and hyperchaos in a modified memristor-based Chua's circuit and its generalized discrete system

Nadjet Boudjerida, Mohammed -Salah Abdelouahab and René Lozi

ABSTRACT

In this paper, a modified hyperchaotic memristor-based Chua's circuit and its generalized discrete model are reported. First, the dynamics of the continuous system are investigated using different tools, including stability theory, phase portraits, Lyapunov exponents, and bifurcation diagrams, showing that the proposed circuit model possesses a line of equilibrium and exhibits rich dynamics, including the coexistence of regular and chaotic attractors, and mixed-mode oscillations. Lastly, a generalized discrete version of the proposed system is formulated and its dynamics are explored. Using the Jury criterion, we constructed a useful stability test to localize some stable regions in three parameter planes. The presence of an exact periodic solution is numerically highlighted, and the occurrence of chaotic behavior is confirmed using the largest Lyapunov exponent and the 0−1 test.

1. Introduction

In 1971, Leon Chua professor of electronics at the University of California, Berkeley published a paper entitled 'Memristor: the missing circuit element' [15]. Since then, it has aroused the worldwide attention of researchers from different disciplines, such as programmable logic [10], neural networks [36], and electronics [40].

As a result of the nonlinear nature of the memristor element, the application of memristor-based circuits to build continuous chaotic or hyperchaotic systems has gained much attention in recent decades [2,11,29,34,37–39,45], and analog circuits are useful in realizing continuous-time memristors [9,26]. However, digital circuits are increasingly used in practical engineering due to their easier logic programming and higher accuracy. Thus, discrete-time memristor models may become another important tool for the development of memristor-based circuit models [8,30], in particular, a memristor-based Lozi map with hidden hyperchaos was reported in [43].

Because a hyperchaotic system has more than one positive Lyapunov exponent, its dynamic has the characteristic of high security and it is usually used in nonlinear circuits, encryption, secure communications, and electrical circuits [17,27,33,41,46].

For example, replacing Chua's diode in the canonical Chua's circuit with a memristor characterized by a monotone-increasing piecewise linear function has been proposed in [7].

Fractional calculus can be thought of as a generalization of integer calculus and is almost as old as integer differentiation itself. Different definitions exist for fractional derivatives, and it might be the Caputo derivative that is most commonly used in the field of dynamic systems. The Caputo fractional derivative of order $q \in \mathbb{R}^+$ [14] is defined by:

$$D_t^q f(t) = \begin{cases} \frac{1}{\Gamma(m-q)} \int_0^t \frac{f^{(m)}(\tau)}{(t-\tau)^{q-m+1}} \mathrm{d}\tau, & m-1 < q < m, \\ \frac{\mathrm{d}^m}{\mathrm{d}t^m} f(t), & q = m, \end{cases} \tag{1}$$

where $m = \lceil \alpha \rceil$, and Γ is the Γ-function defined by

$$\Gamma(z) = \int_0^{+\infty} \mathrm{e}^{-t} t^{z-1} \mathrm{d}t, \quad \Gamma(z+1) = z\Gamma(z).$$

Discrete fractional-order dynamical systems and their stability were discussed in [4,18,32]. A wide range of phenomena in many scientific and engineering fields can be accurately described by non-integer-order differential or difference equations [1–3,5,16,24,31,42,45]. In [22] a piecewise constant arguments discretization technique (PCADT) for fractional-order differential equations, based on approximating their right-hand side, was reported. Then it has been applied to various fractional differential equations including the Riccati's equations [22], the logistic equation [20], the Chua's system [6] and the prey-predator system [21].

It should be noted that, unlike conventional discrete fractional-order systems, which are non-local [12,28], the resulting discrete fractional-order systems generated using PCADT are local and free from memory effects. This property can enhance its numerical treatment by reducing calculation time. On the other hand, the existence of exact periodic solutions in such systems can be expected, in contrast to conventional ones where exact periodic solutions are not expected [13].

In this paper, a hyperchaotic memristor-based Chua's circuit is proposed, and its corresponding discrete fractional-order system is formulated using the piecewise constant arguments discretization technique. As far as we know, no study has been conducted on discrete fractional-order hyperchaotic memristor-based Chua's circuit. The dynamics of both continuous and discrete systems are explored using different tools, including stability theory, phase portraits, bifurcation diagrams, Lyapunov exponents, and 0−1 test. It has been shown that the new circuit exhibit a rich dynamic including multiple coexisting regular and chaotic attractors. Moreover, the system produces mixed-mode oscillations (MMOs) characterized by the combination of large amplitudes and alternate sub-threshold or small amplitude oscillations [3,31].

The rest of the paper is organized as follows: In Section 2, one introduces a modified memristor-based hyperchaotic model and investigates its dynamics using different tools. Section 3, reveals a generalized discrete version of the proposed model. Finally, in Section 4, a short conclusion is drawn.

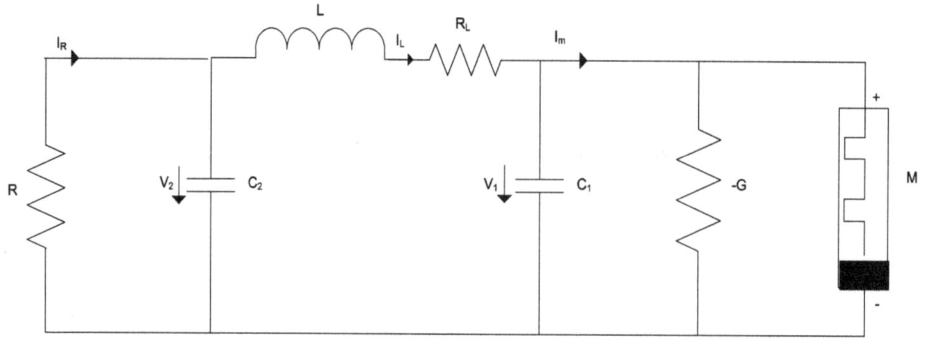

Figure 1. Modified memristor-based Chua's circuit.

2. Modified memristor-based hyperchaotic system and its dynamical behaviors

2.1. Memristor model

A memristor is a nonlinear resistor with a memory effect that may be either flux-controlled or charge-controlled [7]. It is a two-terminal device defined by the relationship

$$f(\varphi, q) = 0.$$

Equations (2) and (3) describe flux-controlled and charge-controlled memristor respectively [15,25]

$$W(\varphi) = \frac{\mathrm{d}q(\varphi)}{\mathrm{d}\varphi}, \quad i = W(\varphi)v, \tag{2}$$

$$M(q) = \frac{\mathrm{d}\varphi(q)}{\mathrm{d}q}, \quad v = M(q)i, \tag{3}$$

where φ and q denote magnetic flux and charge, $W(\varphi)$ and $M(q)$ are called the memductance and memristance respectively.

The flux-controlled memristor considered in this study is described by the piecewise quadratic function $q(\varphi)$.

$$q(\varphi) = -a\varphi + \frac{b}{2}\varphi|\varphi|.$$

Where a and b are both positive parameters and its memductance function is given by

$$W(\varphi) = \frac{\mathrm{d}q(\varphi)}{\mathrm{d}\varphi} = -a + b|\varphi|.$$

In this section, one proposes an alternative memristor-based Chua's circuit by using a negative conductance and a passive flux-controlled memristor described by Equation (2) in parallel, instead of the classical nonlinear diode in the original circuit's and by shifting the position of the inductance which is moved between the two capacitances as shown in Figure 1.

Based on Kirchhoff Laws the suggested circuit is mathematically modelled by the following four-dimensional differential system

$$\begin{cases} \dot{x} = \alpha[z + \xi x - (-a + b|w|)x], \\ \dot{y} = y - z, \\ \dot{z} = -\beta(x - y) - \gamma z, \\ \dot{w} = x, \end{cases} \tag{4}$$

where x, y, z and w are the states and $\alpha, \beta, \gamma, \xi$, a and b are assumed to be positive constant parameters.

2.2. Dissipation

The system (4) is reformulated in its vector notation as

$$\frac{\mathrm{d}X}{\mathrm{d}t} = f(X) = \begin{bmatrix} f_1(x, y, z, w) \\ f_2(x, y, z, w) \\ f_3(x, y, z, w) \\ f_4(x, y, z, w) \end{bmatrix}, \tag{5}$$

where

$$\begin{cases} f_1(x, y, z, w) = \alpha[z(t) + \xi x(t) - (-a + b|w(t)|)x(t)], \\ f_2(x, y, z, w) = y(t) - z(t), \\ f_3(x, y, z, w) = -\beta(x(t) - y(t)) - \gamma z(t), \\ f_4(x, y, z, w) = x(t). \end{cases} \tag{6}$$

One has

$$\mathrm{div} f = \frac{\partial f_1}{\partial x} + \frac{\partial f_2}{\partial y} + \frac{\partial f_3}{\partial z} + \frac{\partial f_4}{\partial w} = \alpha(\xi + a - b|w|) + 1 - \gamma. \tag{7}$$

When $|w| > \frac{\xi + a}{b} - \frac{\gamma - 1}{\alpha b}$, the system (4) is dissipative.

2.3. Equilibrium points and stability

In order to determine the equilibrium points of (4) the next system of equations is solved.

$$\begin{cases} \alpha[z(t) + \xi x(t) - (-a + b|w|)x(t)] = 0, \\ y(t) - z(t) = 0, \\ -\beta(x(t) - y(t)) - \gamma z(t) = 0, \\ x(t) = 0. \end{cases} \tag{8}$$

Thus one obtains the following equilibrium points

$$X_e = \left\{(x, y, z, w);\ x = 0, y = 0, z = 0 \text{ and } w = w_e\right\}. \tag{9}$$

We conclude from X_e that each point on the $w-$ axis is an equilibrium point of (4), and (9) is called the equilibrium set.

The Jacobian matrix at each equilibrium point X_e is

$$J(X_e) = \begin{bmatrix} \alpha(\xi - W(w_e)) & 0 & \alpha & 0 \\ 0 & 1 & -1 & 0 \\ -\beta & \beta & -\gamma & 0 \\ 1 & 0 & 0 & 0 \end{bmatrix} \tag{10}$$

The characteristic polynomial of the Jacobian matrix $J(X_e)$ is given by

$$\begin{aligned} P(\lambda) = \lambda \big[\lambda^3 &+ [\gamma - 1 - \alpha(\xi - W(w_e))]\lambda^2 \\ &+ [(-\gamma + 1)\alpha(\xi - W(w_e)) + (1+\alpha)\beta - \gamma]\lambda \\ &+ \alpha[(\gamma - \beta)(\xi - W(w_e)) - \beta]\big] \end{aligned}$$

For the system parameter values

$$\alpha = 5, \quad \beta = 5, \quad \gamma = 0.1, \quad \xi = 3, \quad a = 1.5, \quad \text{and} \quad W(w_e) = -a + b|w_e|, \tag{11}$$

the characteristic equation becomes

$$\lambda \left[\lambda^3 + (5b\,|w_e| - 23.4)\,\lambda^2 + (50.15 - 4.5b\,|w_e|)\,\lambda + 24.5b\,|w_e| - 135.25 \right] = 0. \tag{12}$$

Clearly, one eigenvalue vanishes for all values of b and w_e. In order to determine the range of b and w_e where system (4) have a three-dimensional stable manifold, one considers the projection of system (4) to the eigenspace corresponding to the three other eigenvalues (regardless of the influence of zero eigenvalue), then the Routh–Hurwitz stability criterion is applied to Equation (13).

$$\lambda^3 + (5b\,|w_e| - 23.4)\,\lambda^2 + (50.15 - 4.5b\,|w_e|)\,\lambda + 24.5b\,|w_e| - 135.25 = 0. \tag{13}$$

Thus the inequalities (14) must be satisfied in order to obtain all the roots of the characteristic equation (13) with negative real parts.

$$\begin{cases} 5b\,|w_e| - 23.4 > 0, \\ 24.5b\,|w_e| - 135.25 > 0, \\ -22.5\,|bw_e|^2 + 331.55b\,|w_e| - 1038.3 > 0, \end{cases} \tag{14}$$

hence

$$5.5204 < b\,|w_e| < 10.221,$$

or

$$\frac{5.5204}{|w_e|} < b < \frac{10.221}{|w_e|}.$$

In contrast, chaos has a greater possibility of occurrence if Equation (12) has one or more roots with positive real parts, that is

$$b|w_e| < 5.5204, \quad \text{or} \quad b|w_e| > 10.221. \tag{15}$$

Figure 2 displays the stable region with yellow color and the unstable region with white color in the (w_e, b) plane separated by bifurcation curves having the equations $b = \frac{5.5204}{|w_e|}$, $b = \frac{10.221}{|w_e|}$ at which the equilibrium point P_e losses or gets its stability.

According to the above results, we deduce that the initial value of the state variable $w(t)$ and the parameter b can affect considerably the dynamical behavior of the system (4).

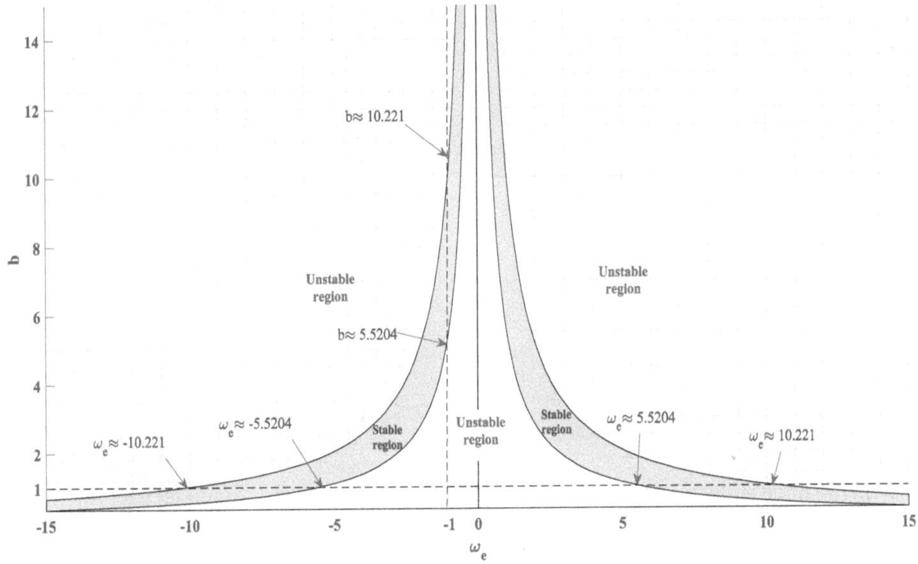

Figure 2. Stable and unstable regions of the projection of system (4) to the eigenspace with non-null eigenvalues in the (w_e, b) plane.

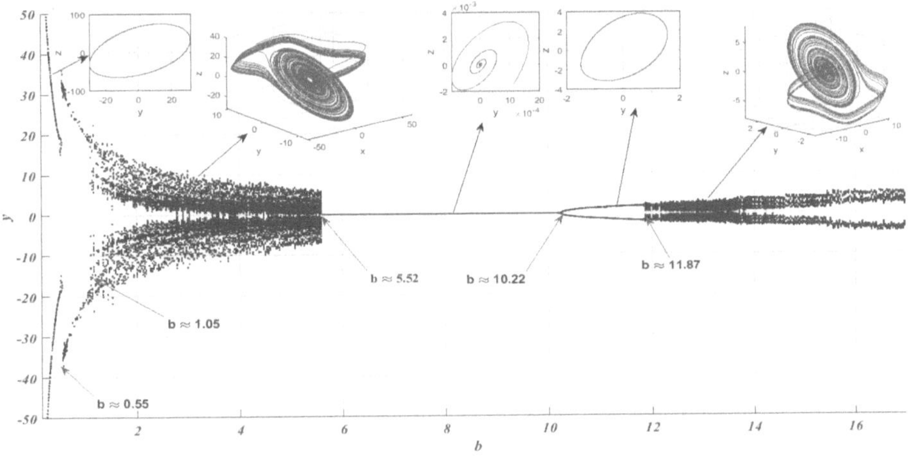

Figure 3. Bifurcation diagram with respect to the parameter b for $w_0 = -1$.

2.4. Bifurcation and Lyapunov exponents

2.4.1. Dynamical behaviors versus the parameter b

In this section, one considers the parameter values $\alpha = 5, \beta = 5, \gamma = 0.1, a = 1.5, \xi = 3$ and letting b vary over a certain interval to discuss the complex dynamics of system (4) with the initial condition $(x_0, y_0, z_0, w_0) = (-0.5, 0.1, 0.01, -1)$. The bifurcation diagram of y and the corresponding Lyapunov exponents spectrum for b varying from 0.2 to 17, with a step size $h = 0.001$ are obtained as illustrated in Figure 3 and Figure 4, respectively, which are in good agreement.

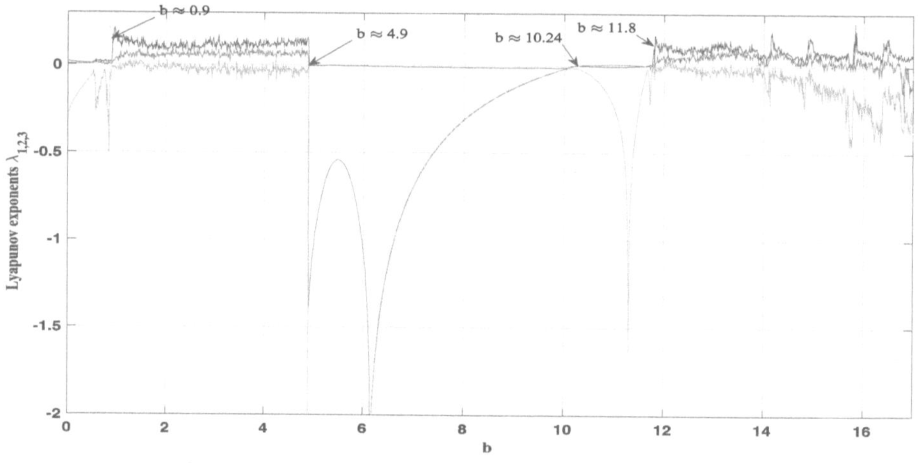

Figure 4. The three largest Lyapunov exponents of system (4) versus the parameter b for $w_0 = -1$.

From these figures it is clear that system (4) displays period 1 orbit for $b \in]0.2, 1.05[\cup]10.22, 11.87[$ and a period 2 orbit for $b \in]10.22, 11.87[$ followed by a period 4 orbit when $b \approx 11.87$ then rapidly it displays hyperchaotic Attractor. For $b \in]1.05, 5.22[\cup]11.88, 17[$ system (4) exhibits chaotic and hyperchaotic behavior.

In particular, for $b = 12$ the Lyapunov exponent are

$$L_1 \approx 0.1625, \quad L_2 \approx 0.0508, \quad L_3 \approx -0.0124, \quad L_4 \approx -33.1206. \tag{16}$$

Since $L_1 + L_2 + L_3 + L_4 = -32.9197 < 0, L_1 > 0, L_2 > 0$, then the system (4) is dissipative and hyperchaotic. The Kaplan–Yorke dimension of its attractor is

$$D_{KY} \approx 3 + \frac{L_1 + L_2 + L_3}{|L_4|} = 3 + \frac{0.1625 + 0.0508 - 0.0124}{33.1206} = 3.0061, \tag{17}$$

which is a fractal dimension.

It is noticeable that for some values of the parameter b amazing mixed-mode bursting oscillations are observed. As an example for $b = 3$, $b = 13$, $b = 14$, the time series is depicted in Figure 5 showing that signals alternate between small amplitude oscillations (S) (around a saddle equilibrium) and large amplitude oscillation (L), (near an unstable periodic orbit, having stable and unstable manifolds), the observed mixed-mode bursting oscillations are of the form (S^3L^1) and (S^4L^1).

2.4.2. Dynamical behaviors versus the parameter w_0

In order to examine the impact of initial condition values on dynamical behavior of the system (4), for the set of parameter values (6), different diagrams are presented to identify chaos.

Considering the initial condition $(x_0, y_0, z_0, w_0) = (-0.5, 0.1, 0.01, w_0)$, the Lyapunov exponents spectrum and the corresponding bifurcation diagram of y, for w_0 varying from -15 to 15 with step 0.01 are obtained as shown in Figure 6 and Figure 7, respectively. From these diagrams, one observes that when the value of initial state w_0 belongs to the following four intervals: $[-15, -11.91]$, $[-5.52, -0.9]$, $[0.9, 5.52]$, $[11.91, 15]$, then system (4) exhibits chaos. Furthermore, the two diagrams indicate a symmetry versus $w_0 = 0$.

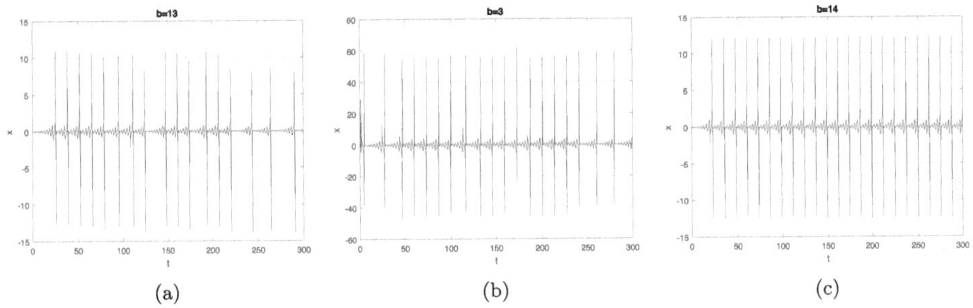

Figure 5. Mixed-mode bursting oscillations obtained in system (4) for some values of b with initial value $w_0 = -1$, (a) $b = 3$, (b) $b = 13$, (c) $b = 14$.

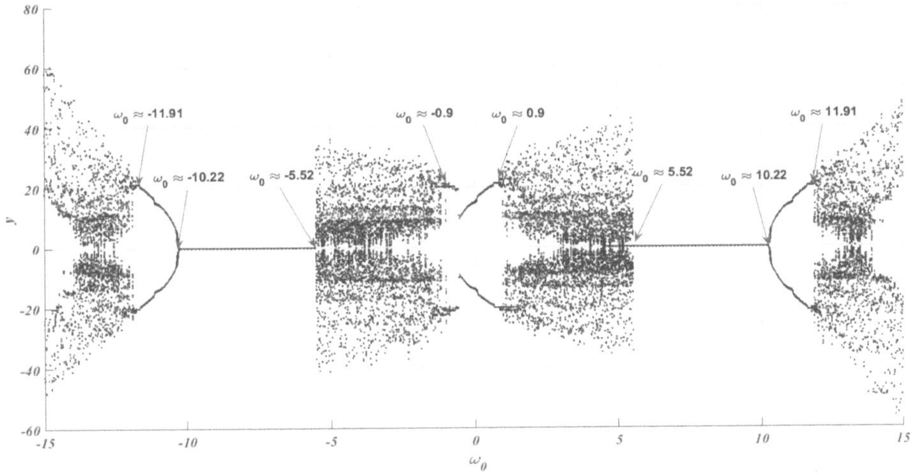

Figure 6. Bifurcation diagram with respect to the fourth coordinate w_0 of initial condition for $b = 1$.

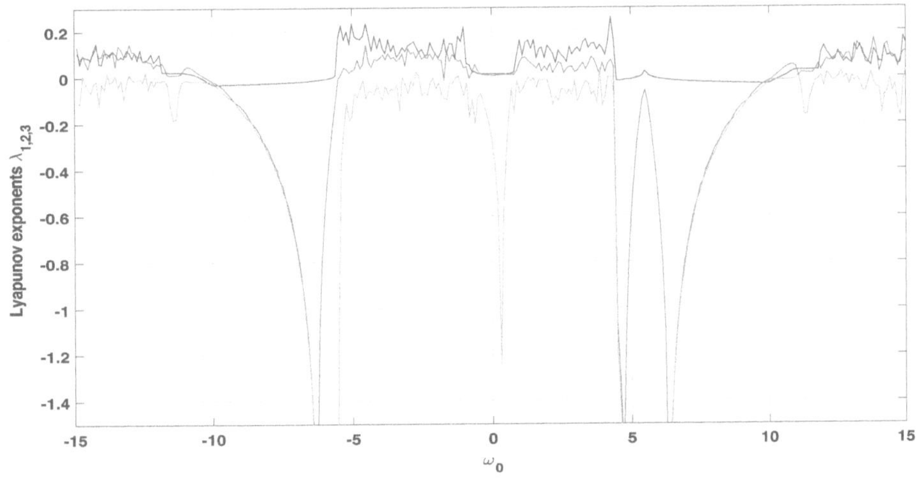

Figure 7. The three largest Lyapunov exponents of system (4) versus the parameter w_0 for $b = 1$.

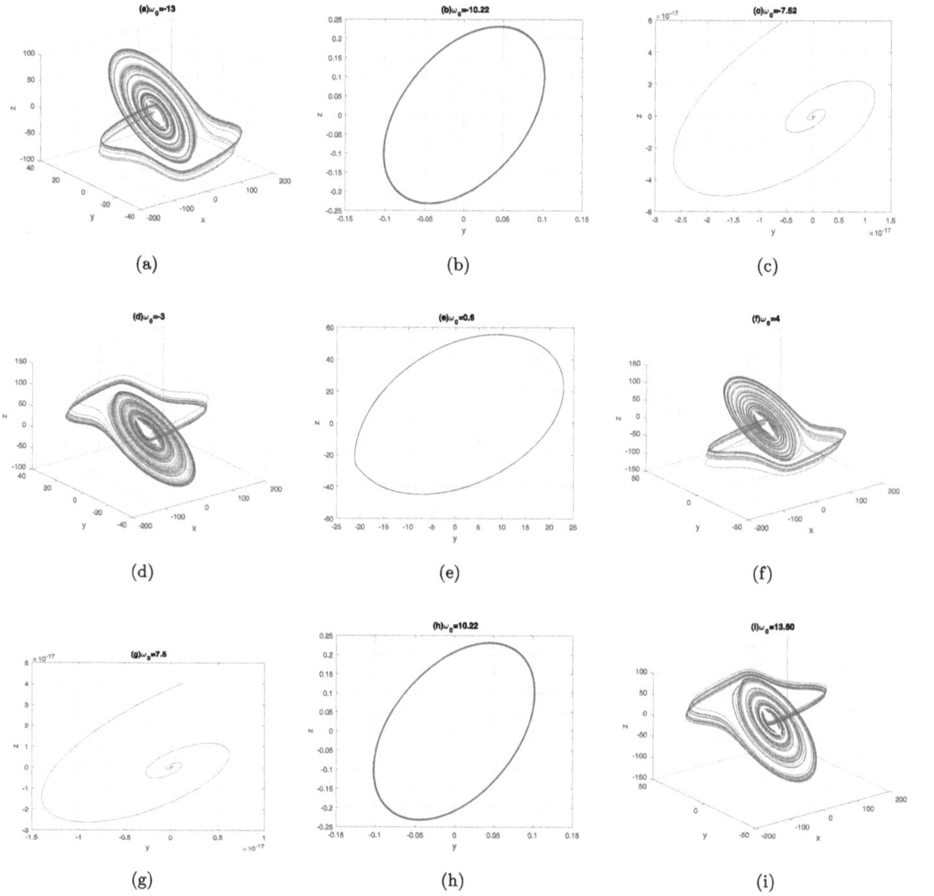

Figure 8. Some attractors for different values of initial condition w_0 : (a) $w_0 = -13$, (b) $w_0 = -10.22$, (c) $w_0 = -7.52$, (d) $w_0 = -3$, (e) $w_0 = 0.6$, (f) $w_0 = 4$, (g) $w_0 = 7.5$, (h) $w_0 = 10.22$, (i) $w_0 = 13.5$.

Particularly, for $w_0 = -3$ the Lyapunov exponents are [44]

$$L_1 = 0.0978, \quad L_2 = 0.0659, \quad L_3 = -0.0338, \quad L_4 = -38.0524. \tag{18}$$

Since $L_1 + L_2 + L_3 + L_4 = -37.9226 < 0$, $L_1 > 0$, $L_2 > 0$, then the system (4) is dissipative and hyperchaotic. The Kaplan–Yorke dimension of its attractor is

$$D_{KY} = 3 + \frac{L_1 + L_2 + L_3}{|L_4|} = 3 + \frac{0.0978 + 0.0659 - 0.0338}{38.0524} = 3.0034, \tag{19}$$

which is a fractal dimension.

Some phase portraits are depicted in Figure 8 for different values of the initial condition w_0. In particular, a period-1 orbits are shown in Figure 8(b,e,h). Moreover, Figure 8(c,g) represents a stable equilibrium point, and Figure 8(a,d,f,i) displays chaotic attractors.

Figure 9 highlights the coexistence phenomena in system (4) where different type of attractors (fixed points, limit cycles, chaotic and hyperchaotic attractors) are depicted for initial conditions: $(-0.5, 0.1, 0.01, -14)$, $(-0.5, 0.1, 0.01, -11)$, $(-0.5, 0.1, 0.01, 0.6)$,

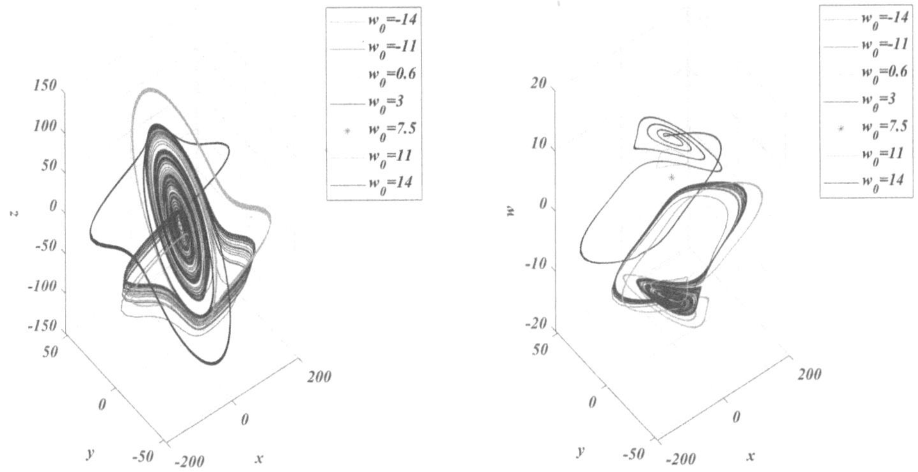

Figure 9. Phase portrait of system (4) showing various coexisting behaviors for different values of w_0: $w_0 = -14, w_0 = -11, w_0 = 0.6, w_0 = 3, w_0 = 7.5, w_0 = 11, w_0 = 14$ with $b = 1$.

$(-0.5, 0.1, 0.01, 3)$, $(-0.5, 0.1, 0.01, 7.5)$, $(-0.5, 0.1, 0.01, 11)$, $(-0.5, 0.1, 0.01, 14)$ with the same parameter value $b = 1$.

3. Discretized fractional-order model and its analysis

The fractional version of the proposed electronic circuit can be constricted by replacing the classical circuit elements with its fractional versions (fractance) [1,2,5], and it can be mathematically represented by the following fractional-order system

$$\begin{cases} D_t^q x = \alpha[z + \xi x - (-a + b|w|)x], \\ D^q y = y - z, \\ D_t^q z = -\beta(x - y) - \gamma z, \\ D_t^q w = x, \end{cases} \tag{20}$$

where D^q is the Caputo fractional derivative of order q defined by (1).

The considered discrete fractional system has not been obtained using a fractional-order discrete operator [12], but it has been derived from the continuous-time fractional-order system (20) in terms of the Caputo derivative via the piecewise constant arguments discretization technique [22], and it is given by

$$\begin{cases} x_{n+1} = x_n + s[\alpha(z_n + \xi x_n - (-a + b|w_n|)x_n)], \\ y_{n+1} = y_n + s[y_n - z_n], \\ z_{n+1} = z_n + s[-\beta(x_n - y_n) - \gamma z_n], \\ w_{n+1} = w_n + s x_n, \end{cases} \tag{21}$$

where $s = \frac{h^q}{\Gamma(1+q)}$.

A special case of this generalized discrete model is the so-called Euler forward discrete model obtained for $q = 1$.

3.1. Existence and stability of fixed points

At the fixed point, we have $x_{n+1} = x_n, y_{n+1} = y_n, z_{n+1} = z_n$ and $w_{n+1} = w_n$. One can then easily deduce that (21) has the w-axis as a line equilibrium (each point in the w-axis is a fixed point of system (21)). Namely, the set of fixed points is given by

$$X_e = \left\{(x, y, z, w);\ x = 0, y = 0, z = 0 \text{ and } w = w_e \in \mathbb{R}\right\}. \tag{22}$$

The Jacobian matrix of system (21) at each fixed point X_e is given by

$$J(X_e) = \begin{bmatrix} 1 + s\alpha(\xi - W(w_e)) & 0 & s\alpha & 0 \\ 0 & 1+s & -s & 0 \\ -\beta s & \beta s & 1 - \gamma s & 0 \\ s & 0 & 0 & 1 \end{bmatrix} \tag{23}$$

The characteristic polynomial of the Jacobian matrix $J(X_e)$ is given by

$$P_4(\lambda) = (\lambda - 1)\,(\lambda^3 + A\lambda^2 + B\lambda + C) = (\lambda - 1)\,P_3(\lambda),$$

where

$$\begin{aligned} A &= -s\alpha(\xi - W(w_e)) - s^2\gamma - 3, \\ B &= 2s\alpha(\xi - W(w_e)) + 2s - 2s\gamma + s^2(\beta - \gamma) + s^2\alpha(\xi - W(w_e)) \\ &\quad + s^2\alpha(\beta - \gamma(\xi - W(w_e))) + 3, \\ C &= -s + s^2\gamma - s\alpha(\xi - W(w_e)) - s^2\beta - s^2\alpha(\xi - W(w_e)) \\ &\quad - s^2\alpha(\beta - \gamma(\xi - W(w_e))) - s^3\alpha(\beta + (\beta - \gamma)(\xi - W(w_e))) - 1. \end{aligned}$$

Clearly, one eigenvalue is $\lambda = 1$ corresponding to the z-axis which is a line equilibrium meaning that X_e is a non-hyperbolic point having z-axis as a central manifold. In order to determine the range of parameters where system (21) has a local three-dimensional stable manifold, we consider the projection of system (21) into the eigenspace corresponding to the roots of $P_3(\lambda)$ (regardless of the influence of the eigenvalue $\lambda = 1$). Let $b_3 = 1 - C^2$, $b_2 = A - BC$, $b_1 = B - AC$, $c_3 = b_3^2 - b_1^2$, $c_2 = b_2 b_3 - b_1 b_2$, and

$$\begin{cases} T_1(p) = B + 1 - |A + C|, \\ T_2(p) = 1 - |C| \\ T_3(p) = \left|1 - C^2\right| - |B - AC|, \end{cases}$$

where p is the vector parameter.

Proposition 3.1: *Let w_e be a real number. The system (21) has a local three-dimensional stable manifold through the fixed point $X_e = (0, 0, 0, w_e)$ if*

$$T(p) = \min_{i=1,2,3} (T_i(p)) > 0. \tag{24}$$

Proof: Suppose that the conditions (24) hold, then $T_i > 0$ for all $i = 1, 2, 3$,

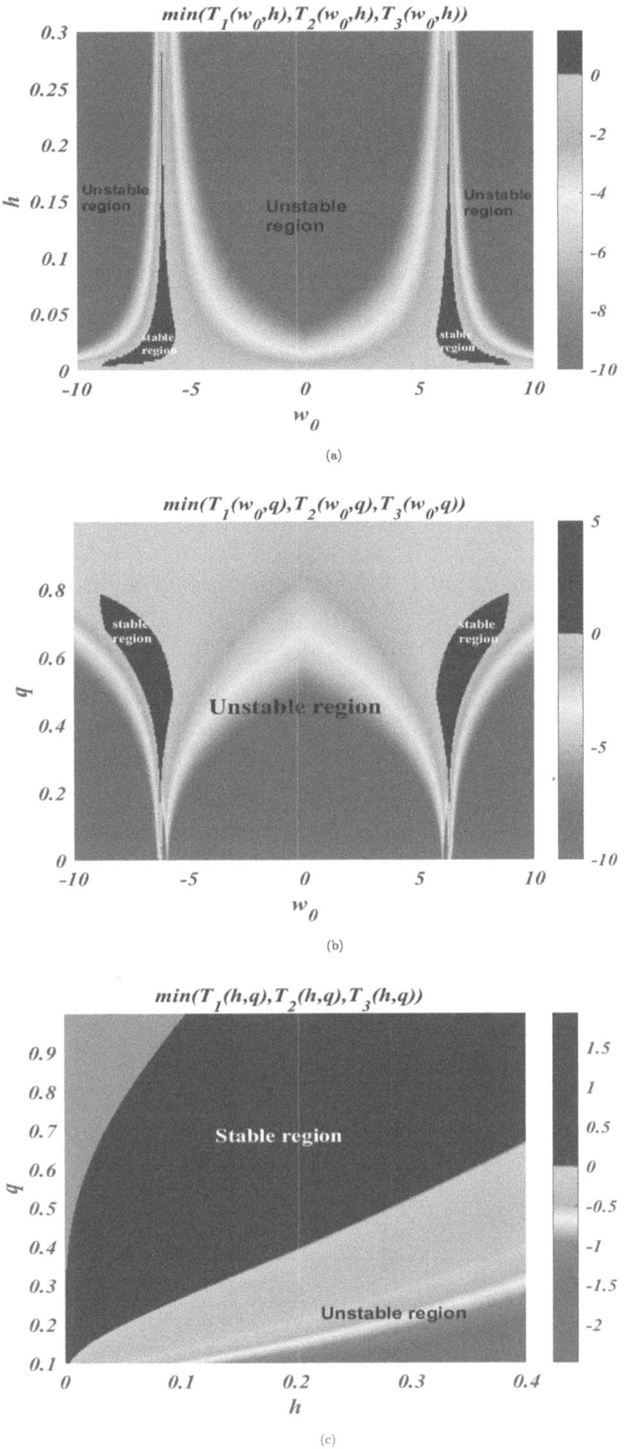

Figure 10. Stability regions in two parameter planes (a) Stable region with dark blue color in the (w_0, h) plane for $q = 0.95$, (b) Stable region with dark blue color in the (w_0, q) plane for $h = 0.01$, (c) Stable region with dark blue color in the (h, q) plane for $w_0 = 6.3$.

$T_1 > 0$ imply that $A+C<B+1$ and $A+C>-B-1$, that is $1-A+B-C>0$ and $1+A+B+C>0$, thus $P_3(1) > 0$, and $(-1)^3P_3(-1) > 0$.

From $T_3 > 0$ one get $|b_0| - |b_2| > 0$,

Where $b_0 = \begin{vmatrix} C & 1 \\ 1 & C \end{vmatrix}$, $b_2 = \begin{vmatrix} C & B \\ 1 & A \end{vmatrix}$.

Then the condition (24) can be written as

$$\begin{cases} P_3(1) > 0, \\ (-1)^3P_3(-1) > 0. \\ |C| < 1, \\ |b_0| > |b_2|. \end{cases}$$

Thus, according to the Jury test [19,35], $P_3(\lambda)$ has all its roots inside the unit disk, meaning that system (21) has a local three-dimensional stable manifold around the fixed point $X_e = (0, 0, 0, w_e)$. ■

Now, let us consider the parameters values as in the continuous part

$$\alpha = 5, \quad \beta = 5, \quad \gamma = 0.1, \quad a = 1.5, \quad b = 1, \quad \text{and} \quad \xi = 3. \tag{25}$$

Based on the stability test (24) one can determine some stability regions in different parameter planes.

Figure 10 illustrates the stability regions with dark blue color in two parameter planes: (w_0, h) in Figure 10(a) for $q = 0.95$, (w_0, q) in Figure 10(b) for $h = 0.01$ and (h, q) in Figure 10(c) for $w_0 = 6.3$.

The boundary of stability region can be viewed as a bifurcation curve where the fixed point X_e losses or gets its stability is determined by the equation $T(p) = 0$.

Unbounded trajectories would be expected in the red region (high possibility of the nonexistence of attractors), while in the cyan region, one can expect the existence of regular or chaotic attractors (high possibility of the existence of attractors).

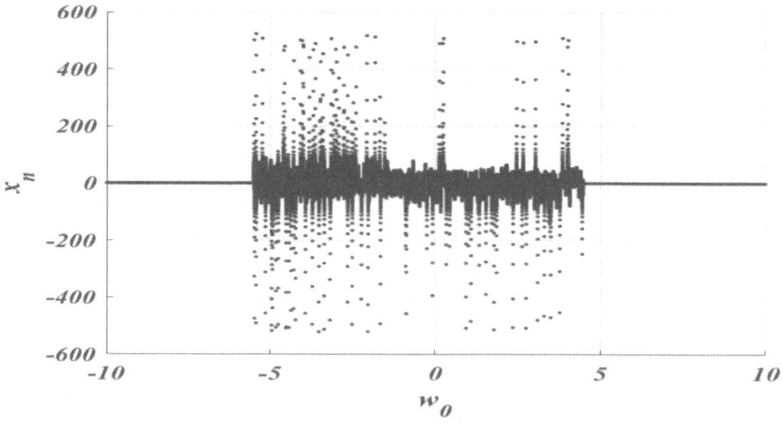

Figure 11. Bifurcation diagram x_n versus $w_0 \in [-10, 10]$ for $h = 0.01$ and $q = 0.95$.

3.2. Bifurcation diagrams and Lyapunov exponents

Numerical simulations of fractional-order discrete systems are performed in this section to validate the theoretical results. One investigates bifurcation diagrams, the largest Lyapunov exponent (calculated using the Lyapunov exponents from the time series algorithm [23]) and k-median (calculated using the 0−1 test algorithm [2]) together with some selected phase portraits to illustrate how the fractional order $0.7 < q < 1$, the step size $0 < h < 0.12$ and the fourth coordinate $-10 < w_0 < 10$ of initial condition affect the system dynamics, where the other parameters values are given by (25).

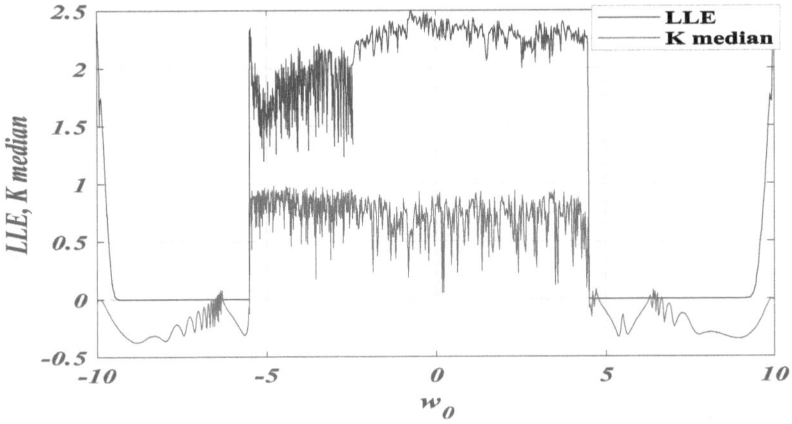

Figure 12. Largest Lyapunov exponent together with the median k versus $w_0 \in [-10, 10]$ for $h = 0.01$ and $q = 0.95$.

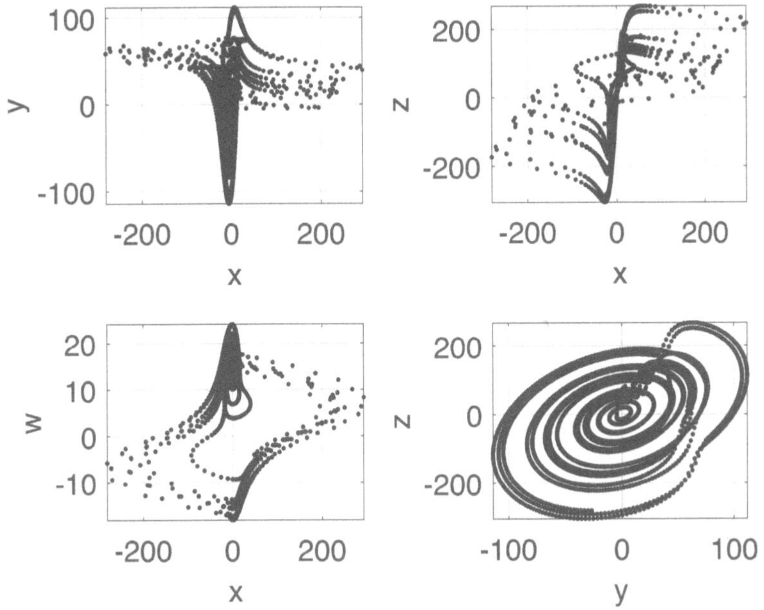

Figure 13. Chaotic attractor for $w_0 = -3$, $h = 0.01$ and $q = 0.95$.

3.2.1. Dynamical behaviors versus the initial coordinate w_0

In this part, one takes $h = 0.01$, $q = 0.95$ and lets w_0 vary over the interval $[-10, 10]$. to discuss the dynamics of system (21) with the initial condition $(x_0, y_0, z_0, w_0) = (-0.5, 0.1, 0.01, -5.2)$.

The bifurcation diagram of w versus w_0 is plotted in Figure 11 and the corresponding largest Lyapunov exponent together with the median k are plotted in Figure 12.

From these diagrams one can conclude that system (21) exhibits a stationary behavior for w_0 in the two intervals $]-10, -5.4[$, $]4.7, 10[$, as illustrated in Figure 14, and it exhibits a chaotic behavior for $w_0 \in]-5.4, 4.7[$, as shown in Figure 13.

Notice that for $w_0 \in]4.7, 5.4[$. the system (21) exhibits a stationary behavior as mentioned above while T is negative as can be seen in Figure 10(a) indicating that the fixed point $(0, 0, 0, w_0)$ is unstable, this is because it loses its stability in favor of another fixed point $(0, 0, 0, w_0 \pm \varepsilon)$ having a sufficiently large basin of attraction for which T will be positive, this is illustrated for example in Figure 14, where the trajectory initiated near the fixed point $(0, 0, 0, 5)$ is attracted by the stable fixed point $(0, 0, 0, 6.043)$.

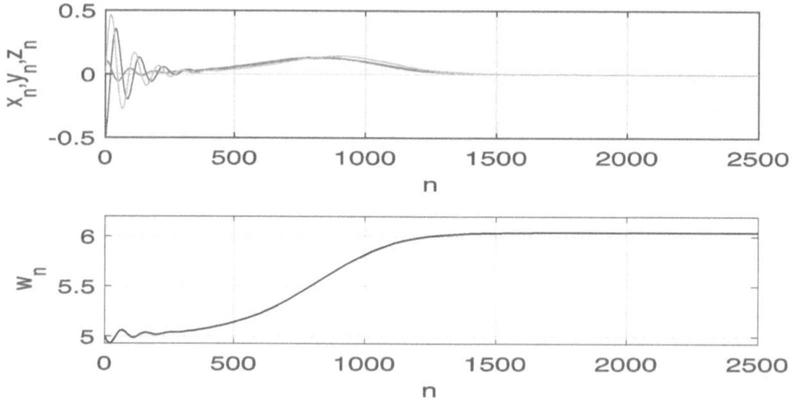

Figure 14. Stable fixed point for $w_0 = 5, h = 0.01$ and $q = 0.95$.

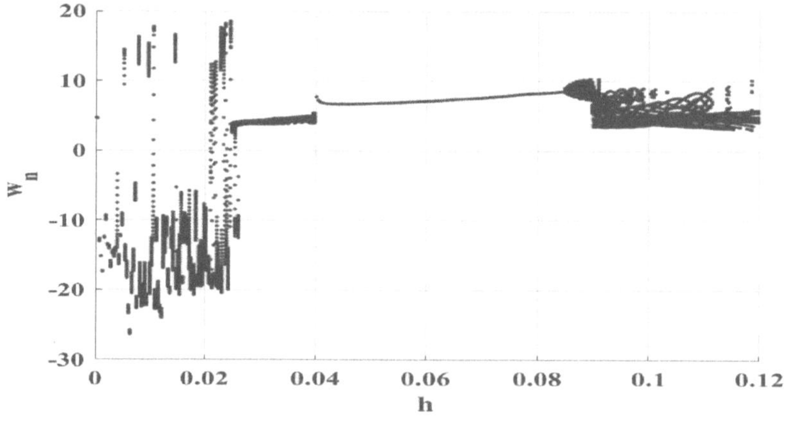

Figure 15. Bifurcation diagram x_n versus $h \in [0, 0.12]$ for $w_0 = 4.7$ and $q = 0.95$.

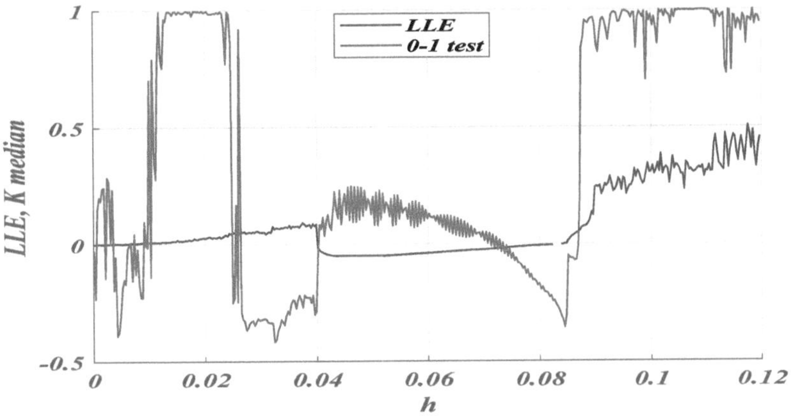

Figure 16. Largest Lyapunov exponent together with the median k versus $h \in [0, 0.12]$ for $w_0 = 4.7$ and $q = 0.95$.

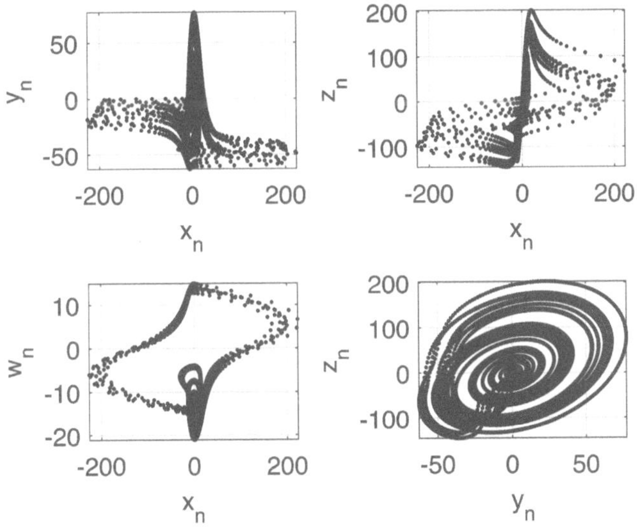

Figure 17. Chaotic attractor for $w_0 = 4.7, q = 0.95$ and $h = 0.007$.

3.2.2. Dynamical behaviors versus the discretization step h

In this part, one takes $w_0 = 4.7$, $q = 0.95$ and lets the discretization step h vary over the interval $[0, 0.12]$. To discuss the dynamics of system (21) with the initial condition $(x_0, y_0, z_0, w_0) = (-0.05, 0.01, 0.01, 4.7)$.

The bifurcation diagram of w versus h is plotted in Figure 15. and the corresponding largest Lyapunov exponent together with the median k are plotted in Figure 16.

From these diagrams, one can conclude that system (21) exhibits a stationary behavior for h belonging to the interval $]0.04, 0.085[$, as illustrated by the time series in Figure 18, and it exhibits a chaotic behavior for $h \in]0, 0.025[$, as shown in Figure 17.

Notice that for $h \in]0.025, 0.04[\cup]0.085, 0.12[$. the trajectories of system (21) are unbounded.

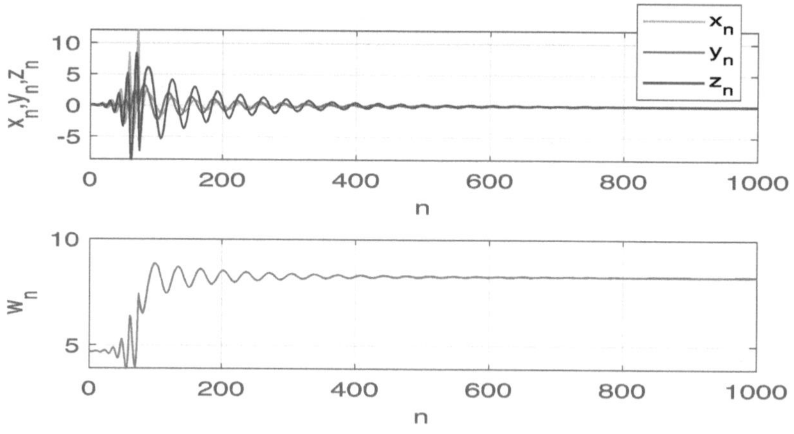

Figure 18. Stable fixed point $X \approx (0, 0, 0, 6.2)$ for $w_0 = 4.7, q = 0.95$ and $h = 0.08$.

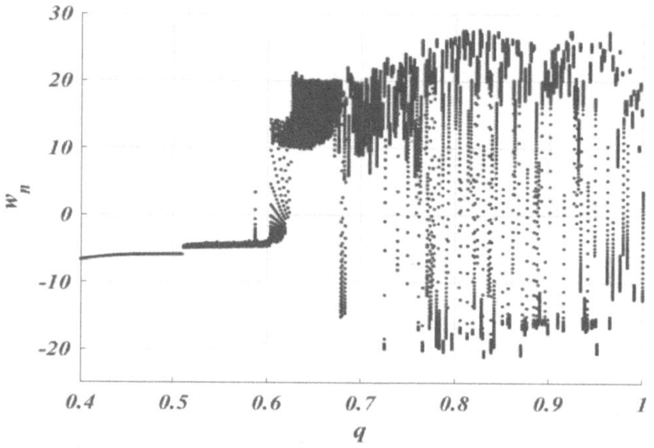

Figure 19. Bifurcation diagram w_n versus the fractional order $q \in [0.4, 1]$ for $w_0 = -5.2$ and $h = 0.005$.

3.2.3. Dynamical behaviors versus the fractional order q

In this part, one takes $w_0 = -5.2$, $h = 0.005$, and lets q vary over the interval $[0.4, 1]$. to discuss the dynamics of system (21) with the initial condition $(x_0, y_0, z_0, w_0) = (-0.5, 0.1, 0.01, -5.2)$.

The bifurcation diagram of w versus q is plotted in Figure 19 and the corresponding largest Lyapunov exponent together with the median k and the stability test value T are plotted in Figure 20.

From these diagrams one can conclude that system (21) exhibits a stationary behavior for q in the interval $]0.4, 0.51[$, as illustrated by the time series in Figure 22, and it exhibits a chaotic behavior for $q \in]0.6, 1[$, as shown in Figure 21.

For $q \in]0.52, 0.59[$. the trajectories of system (21) are unbounded

Notice that T is negative for all $q \in]0.4, 1[$ as can be seen in Figure 20(a) indicating that the fixed point $X_{e0} = (0, 0, 0, -5.2)$ is unstable, while system (21) exhibits a stationary behaviour, because X_{e0} loses its stability in favor of another fixed point $X_e = (0, 0, 0, w)$ having sufficiently large basin of attraction for which T will be positive, as can be observed

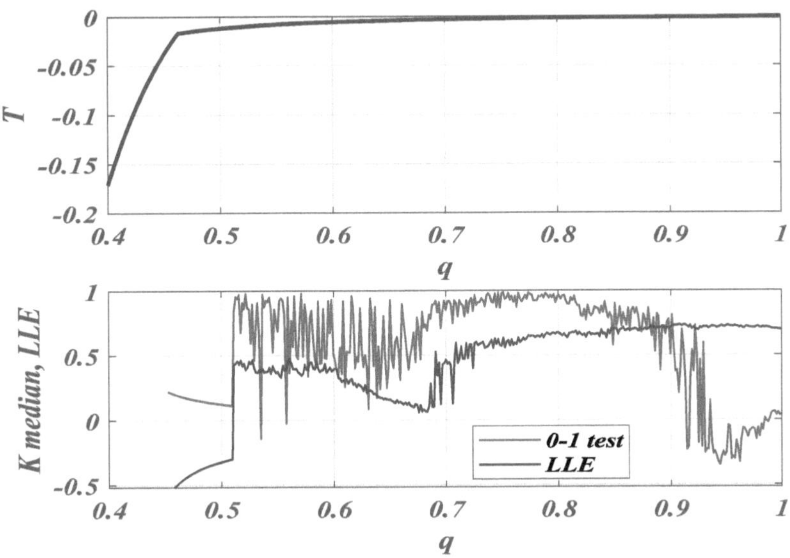

Figure 20. The stability test value T, and the largest Lyapunov exponent together with the median k versus the fractional order $q \in [0.4, 1]$ for $w_0 = -5.2$ and $h = 0.005$.

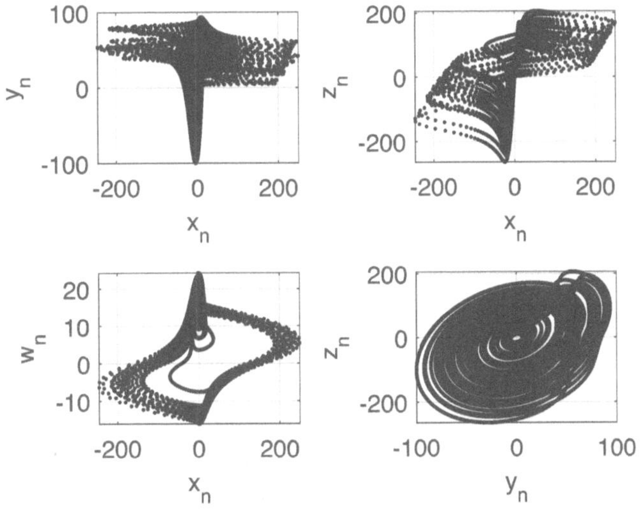

Figure 21. Chaotic attractor for $w_0 = -5.2$, $h = 0.005$ and $q = 0.98$.

in Figure 21, where a trajectory initiated near X_{e0} is attracted by the other fixed point $X_e = (0, 0, 0, -6)$.

Remark 3.1: As mentioned in the introduction the discrete fractional-order systems generated using PCADT are local and free from memory effects, this property allow the possibility of generating exact periodic solutions in such systems, in contrast to those conventional discrete fractional-order systems, where exact periodic solutions are not expected [12], due to non-homogeneous memory effect. To highlight this concept one provides a numerical example for the set of parameters given by (25)

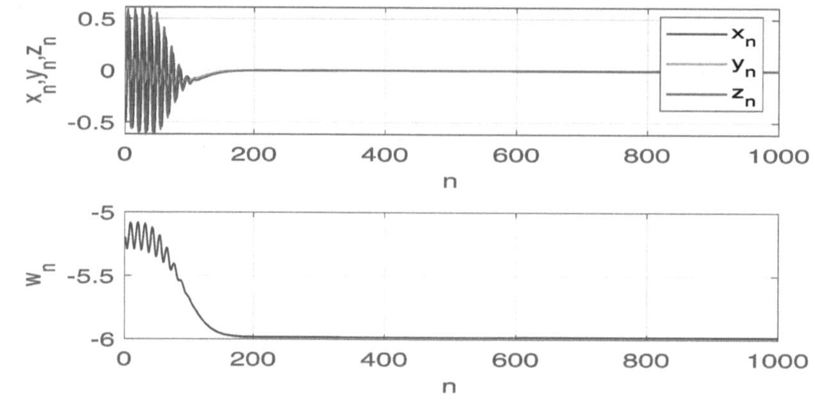

Figure 22. Stable fixed point $X \approx (0, 0, 0, -6)$ for $w_0 = -5.2, h = 0.005$ and $q = 0.45$.

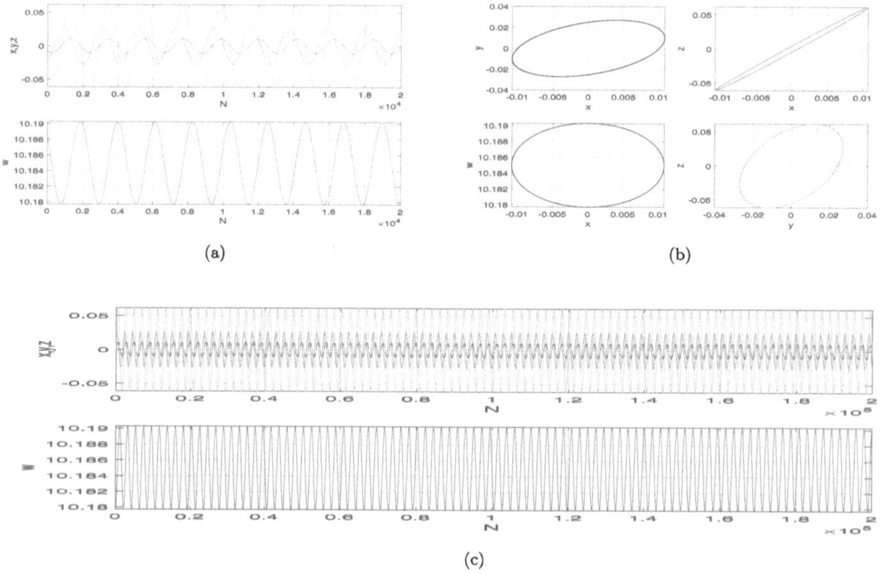

Figure 23. Existence of exact periodic solutions to discrete fractional-order system (21) for the set of parameters given by (25) together with the fractional order $q = 0.95$ and the initial conditions $(x_0, y_0, z_0, w_0) = (0.00021140274268890, -0.024849708951164, -0.0031847591297016, 10.190281422981)$ (a) Time evolution for $t = 20000$ (b) Phase portrait (c) Long time evolution for $t = 200000$.

together with the fractional order $q = 0.95$ and the initial conditions $(x_0, y_0, z_0, w_0) = (0.00021140274268890, -0.024849708951164, -0.0031847591297016, 10.190281422981)$ then one gets an exact periodic solution as shown in Figure 23.

4. Conclusions

An improved hyperchaotic memristor-based Chua's circuit and its generalized discrete model are presented in this paper. Several techniques have been used to analyze the dynamics of continuous and discrete models, including phase portraits, bifurcation diagrams,

Lyapunov exponents, and the 0–1 test. Some stability regions are determined in different parameter planes, based on the Routh–Hurwitz criterion for the continuous model and the Jury test for the discrete one.

According to the obtained results, the proposed circuit model has rich dynamics, including the presence of a line of equilibrium, mixed-mode oscillations, and the coexistence of regular, chaotic and hyperchaotic attractors. Furthermore, the existence of exact periodic solutions to the proposed discrete fractional-order system is numerically confirmed. These properties make it ideal for a wide range of applications, including encryption and secure communications. Therefore, the relevant research is still in its early stages. Our forthcoming articles will discuss the analytical investigation of exact periodic solutions in such fractional-order models and their control and synchronization.

Disclosure statement

No potential conflict of interest was reported by the author(s).

Funding

This work was partially supported by the Algerian General Directorate for Scientific Research and Technological Development (DGRSDT).

References

[1] M-S. Abdelouahab and N. Hamri, *Fractional-order hybrid optical system and its Chaos control synchronization*, Electron. J. Theor. Phys. 11(30) (2014), pp. 49–62.

[2] M-S. Abdelouahab and R. Lozi, *Hopf bifurcation and chaos in simplest fractional-order memristors-based electrical circuit*, Indian J. Ind. Appl. Math. 6(2) (2015), pp. 105–119.

[3] M-S. Abdelouahab and R. Lozi, *Hopf-like bifurcation and mixed mode oscillation in a fractional-order FitzHugh-Nagumo model*, AIP Conf. Proc. 2183(1) (2019), Article ID 100003. https://doi.org/10.1063/1.5136214.

[4] M-S. Abdelouahab, R. Lozi, and L.O. Chua, *Memfractance: a mathematical paradigm for circuit elements with memory*, Internat. J. Bifur. Chaos 24(09) (2014), Article ID 1430023.

[5] R. Abu-Saris and Q. Al-Mdallal, *On the asymptotic stability of linear system of fractional-order difference equations*, Fract. Calc. Appl. Anal. 16 (2013), pp. 613–629.

[6] R.P. Agarwal, A.M.A. El-Sayed, and S.M. Salman, *Fractional-order Chua's system: discretization, bifurcation and chaos*, Adv. Differ. Equ.320 (2013), pp. 1–13. https://doi.org/10.1186/1687-1847-2013-320.

[7] B.C. Bao, J.P. Xu, and Z. Liu, *Initial state dependent dynamical behaviors in memristor based chaotic circuit*, Chin. Phys. Lett. 27(7) (2010), Article ID 070504.

[8] B.C. Bao, H. Li, H. Wu, X. Zhang, and M. Chen, *Hyperchaos in a second-order discrete memristor-based map model*, Electron. Lett. 56 (2020), pp. 769–770.

[9] F. Bayat, M. Prezioso, B. Chakrabarti, H. Nili, I. Kataeva, and D. Strukov, *Implementation of multilayer perceptron network with highly uniform passive memristive crossbar circuits*, Nat. Commun.9 (2018), pp. 1–7.

[10] J. Borghetti, G. Snider, P.J. Kuekes, J.J. Yang, D.R. Stewart, and R.S. Williams, *'Memristive' switches enable 'stateful' logic operations via material implication*, Nature 464 (2010), pp. 873–876.

[11] N. Boudjerida, M-S. Abdeloahab, and R. Lozi, *Modified projective synchronization of fractional-order hyperchaotic memristor-based Chua's circuit*, J. Innov. Appl. Math. Comput. Sci. 2(3) (2022), pp. 69–82.

[12] R. Bououden, M-S. Abdelouahab, F. Jarad, and Z. Hammouch, *A novel fractional piece-wise linear map regular and chaotic dynamics*, Int. J. Gen. Syst. 50(5) (2021), pp. 501–526. https://doi.org/10.1080/03081079.2021.1919102.

[13] S. Bourafa, M-S. Abdelouahab, and R. Lozi, *On periodic solutions of fractional-order differential systems with a fixed length of sliding memory*, J. Innov. Appl. Math. Comput. Sci. 1(1) (2021), pp. 64–78.

[14] D. Cafagna and G. Grassi, *Fractional-order systems without equilibria: the first example of hyper-chaos and its application to synchronization*, Chinese Physics B 24(8) (2015), Article ID 080502. https://iopscience.iop.org/article/10.1088/1674-1056/24/8/080502/meta.

[15] M. Caputo, *Linear models of dissipation whose Q is almost frequency independent-II*, Geophys. J. I.13 (1967), pp. 529–539.

[16] J. Cermak, I. Gyori, and L. Nechvatal, *On explicit stability conditions for a linear fractional difference system*, Fract. Calc. Appl. Anal. 18 (2015), pp. 651–672.

[17] T.I. Chien and T.L. Liao, *Design of secure digital communication systems using chaotic modulation, cryptography and chaotic synchronization*, Chaos Solitons Fractals 24 (2005), pp. 241–255.

[18] L.O. Chua, *Memristor, the missing circuit element*, IEEE Trans. Circuit. Theory 18(5) (1971), pp. 507–519.

[19] J.P. Eckmann, S.O. Kamphorst, D. Ruelle, and S. Ciliberto, *Lyapunov exponents from time series*, Phys. Rev. A. 34(6) (1986), pp. 4971–4979.

[20] S.N. Elaydi, *An Introduction to Difference Equations*, Springer, New York, USA, 1999.

[21] Z.F. El Raheem and S.M. Salman, *On a discretization process of fractional-order logistic differential equation*, J. Egypt. Math. Soc. 22 (2014), pp. 407–412.

[22] A.A. Elsadany and A.E. Matouk, *Dynamical behaviors of fractional-order Lotka Volterra predator-prey model and its discretization*, J. Appl. Math. Comput. 49 (2015), pp. 269–283.

[23] A.M.A. El-Sayed and S.M. Salman, *On a discretization process of fractional order Riccati's differential equation*, J. Fract. Calc. Appl. 4 (2013), pp. 251–259.

[24] L.L. Huang, D. Baleanu, G.C. Wu, and S.D. Zeng, *A new application of the fractional logistic map*, Rom. J. Phys. 61 (2016), pp. 1172–1179.

[25] M. Itoh and L.O. Chua, *Memristor oscillators*, Int. J. Bifurcation Chaos. 18(11) (2008), pp. 3183–3206.

[26] M. Itoh and L.O. Chua, *Dynamics of memristor circuits*, Internat. J. Bifur. Chaos Appl. Sci. Engrg.24 (2014), pp. 1–44.

[27] S. Kaouache and M-S. Abdelouahab, *Modified projective synchronization between integer order and fractional order hyperchaotic systems*, J. Adv. Res. Dyn. Control Syst. 10(5) (2018), pp. 96–104.

[28] A.A. Khennaoui, A. Ouannas, and G. Grassi, *Hidden and Coexisting Attractors in a New Two-Dimensional Fractional Map*, Proceedings of the International Conference on Advanced Intelligent Systems and Informatics 2020. AISI 2020. Adv. Intell. Syst. Comput., Vol. 1261, 2021, pp. 883–889. https://doi.org/10.1007/978-3-030-58669-0_78.

[29] Q. Li, S. Hu, S. Tang, and G. Zeng, *Hyperchaos and horseshoe in a 4−D memristive system with a line of equilibria and its implementation*, Int. J. Circuit Theory. Appl. 42(11) (2013), pp. 1172–1188.

[30] H. Li, Z. Hua, H. Bao, L. Zhu, M. Chen, and B. Bao, *Two-dimensional memristive hyperchaotic maps and application in secure communication*, IEEE Trans. Ind. Electron. 68(10) (2021), pp. 9931–9940.

[31] R. Lozi, M-S. Abdelouahab, and G. Chen, *Mixed-Mode oscillations based on complex Canard explosion in a fractional-order Fitzhug-nagumo Fitzhugh-Nagumo model*, Appl. Math. Nonlinear Sci. 5(2) (2020), pp. 239–256.

[32] D. Mozyrska and M. Wyrwas, *The Z-transform method and delta type fractional difference operators*, Discrete Dyn. Nat. Soc. 2015 (2015), pp. 1–12.

[33] K. Murali and M. Lakshmanan, *Secure communication using a compound signal from generalized chaotic systems*, Phys. Lett. A. 241 (1998), pp. 303–310.

[34] B. Muthuswamy and P.P. Kokate, *Memristor based chaotic circuits*, IETE Tech. Rev. 26(6) (2009), pp. 417–426.
[35] K. Ogata, *Discrete-Time Control Systems*, 2nd ed., Prentice-Hall, Inc, Upper Saddle River, NJ, USA, 1995, p. 185.
[36] Y.V. Pershin and M.D. Ventra, *Experimental demonstration of associative memory with memristive neural networks*, Neural Netw. 23(7) (2010), pp. 881–886.
[37] V.T. Pham, Ch. Volos, and L.V. Gambuzza, *A memristive hyperchaotic system without equilibrium*, Sci. World J. 2014 (2014), pp. 1–9. http://dx.doi.org/10.1155/2014/368986.
[38] O.E. Rössler, *An equation for hyperchaos*, Phys. Lett. A. 71 (1979), pp. 155–157.
[39] S. Rasappan, Y. Li, X. Huang, Y. Song, and J. Lin, *A new fourth-order memristive chaotic system and its generation*, Internat. J. Bifur. Chaos. 25(11) (2015), Article ID 1550151.
[40] S. Shin, K. Kim, and S.M. Kang, *Memristor applications for programmable analog ICs*, IEEE Trans. Nanotechnol. 10(2) (2011), pp. 266–274.
[41] S. Vaidyanathan, C.K. Volos, and V.T. Pham, *Analysis, adaptive control and adaptive synchronization of a nine-term novel 3-D chaotic system with four quadratic nonlinearities and its circuit simulation*, J. Engi. Sci. Technol. Rev. 8(2) (2015), pp. 174–184.
[42] G.C. Wu and D. Baleanu, *Discrete fractional logistic map and its Chaos*, Nonlinear Dyn. 75 (2014), pp. 283–287.
[43] J. Wang, Y. GU, K. Rong, Q. Xu, and X. Zhang, *Memristor-Based lozi map with hidden hyperchaos*, Mathematics 10(3426) (2022), pp. 1–12. https://doi.org/10.3390/math10193426.
[44] A. Wolf, J.B. Swift, H.L. Swinney, and J.A. Vastano, *Determining Lyapunov exponents from a time series*, Physica D 16 (1985), pp. 285–317.
[45] G.C. Wu, M.K. Luo, L.L. Huang, and S. Banerjee, *Short memory fractional differential equations for new memristor and neural network design*, Nonlinear Dyn. 100 (2020), pp. 3611–3623.
[46] G. Ye and J. Zhou, *A block chaotic image encryption scheme based on self-adaptive modelling*, Appl. Soft Comput. 22 (2014), pp. 351–357.

Memristor-based Gauss chaotic maps with hidden/self-exited dynamics

Han Bao, Yang Gu, Jianjun Sun, Xi Zhang and Bocheng Bao

ABSTRACT

Discrete memristors can be used to improve the chaos complexity of existing chaotic maps due to their special nonlinearities of internal states. In this paper, a unified memristor-based Gauss mapping model is presented by coupling discrete memristors with Gauss map and then four two-dimensional (2-D) memristive Gauss maps using four different memristance functions are derived. The memristor-based Gauss mapping model possesses two specific cases of infinitely many line fixed points and no fixed points, resulting in the appearance of hidden dynamics or self-exited dynamics. To address the stability analysis of this kind of hidden dynamics, we use a dimension-reduction conversion method to study the hidden period-doubling bifurcations therein. Afterwards, we simulate and discuss the hidden/self-exited dynamics of the four 2-D memristive Gauss maps using several numerical tools, and perform performance tests and hardware experiments of the generated hidden/self-exited chaotic/hyperchaotic attractors. The results demonstrate that the discrete memristor can make the four 2-D memristive Gauss maps produce complex dynamical behaviours and greatly improve the chaos complexity of the one-dimensional (1-D) Gauss map.

1. Introduction

Nonlinear chaotic systems have been the subject of many researchers in the field of science and engineering. Although chaos was first discovered and studied in continuous dynamical systems, it has also been a subject of interest in discrete dynamical systems. In the past years, numerous discrete chaotic systems have been reported and studied, such as Hénon map [14], Lozi map [27,9], area-preserving Lozi (AP-Lozi) map [24], sine Logistic modulation map [17], multi-cavity hyperchaotic map [36], memristive Logistic (M-Log) map [4], map-based neuronal model [39], memristive Rulkov neuron model [25], hyperchaotic cycloid map [12], Lotka-Volterra predator-prey map [38], fractional-order Lozi map [21], and so on.

In continuous dynamical systems, the algebraic simplicity of the system structure and the topological simplicity of the chaotic attractor are two ways to measure the simplicity of the dynamical system. In three-dimensional (3-D) quadratic chaotic systems, the double-scroll chaotic attractors generated by Lorenz, Chen and Lü systems [2,3,33] are topologically more complex than the single-scroll ones generated by Rössler and most Sprott systems [31,32]. It is always an attractive subject to find lower-dimensional chaotic systems with complex topological attractors and simple algebraic structures [5]. Compared with the continuous dynamical systems, the discrete dynamical systems generally have simpler algebraic structures [8] and higher implementation efficiency in chaos-based applications [23]. Therefore, it is also necessary to seek the lower-dimensional discrete dynamical systems with similar properties to the lower-dimensional chaotic systems [4,8,36]. It has not only in theoretical significance, but also in application value [7,23,29]. To this end, based on a unified memristor-based Gauss mapping model, four 2-D memristive Gauss maps with simple algebraic structures are proposed in this paper, upon which their hidden/self-exited chaotic/hyperchaotic attractors with complex topologies are revealed.

Different from self-excited attractors, hidden attractors are a new class of attractors whose basin of attraction does not overlap with the neighborhoods of unstable fixed points [13,35]. In recent years, the hidden attractors have been reported in multiple discrete dynamical systems, including hidden polynomial maps [37], Vertigo-2 and Vertigo-10 hidden maps [18], hidden quadratic chaotic maps [30], 2-D chaotic map with no fixed points [34], and fractional-order map with no fixed points [22]. However, because there are no any fixed points in these special hidden maps, their stability cannot be analyzed using the stable and unstable manifolds of fixed points. Therefore, the hidden bifurcation mechanism has not been reported in the published literature. To address this issue, a dimension-reduction conversion method is utilized in this paper to study the hidden period-doubling bifurcations in the four 2-D memristive Gauss maps.

Memristor is a special nonlinear element with internal state [1]. Memristors can be used to increase the chaos complexity of existing chaotic maps, and their newly constructed chaotic maps have relatively simple algebraic structures but complex attractor topologies [6]. By introducing discrete memristors into existing chaotic maps, some memristor-based discrete maps with simple algebraic structures have been proposed recently [7,10,23], from which chaotic/hyperchaotic attractors with complex topologies were generated thereby. Due to the high chaos complexity of these memristor-based discrete maps, the generated chaotic/hyperchaotic sequences have been applied to secure communication [23] and generative adversarial nets [7], with excellent application performance. Inspired by these ideas, this paper presents a unified memristor-based Gauss mapping model by coupling discrete memristors with 1-D Gauss map [15] and then derives four 2-D memristive Gauss maps using four different memristance functions. All four 2-D memristive Gauss maps have simple algebraic structures and generate hidden/self-exited chaotic/hyperchaotic attractors with complex topologies.

2. Unified memristor-based Gauss mapping model

This section presents a unified memristor-based Gauss mapping model and then derives four memristive Gauss maps using four different memristance functions.

Table 1. Memristance functions, abbreviations, and mathematical models.

Memristance functions	Abbreviations	Mathematical models
Quadratic memristance	QM	$M(q_n) = q_n^2 - 1$
Absolute value memristance	AM	$M(q_n) = \|q_n\| - 1$
Hyperbolic tangent memristance	HM	$M(q_n) = \tanh(q_n)$
Exponential memristance	EM	$M(q_n) = e^{-q_n} - 1$

2.1. Discrete memristors with different memristances

Several discrete memristors have been proposed using difference modelling theory [6]. For input current $i(t)$, output voltage $v(t)$, and internal state $q(t)$, the corresponding values at the n-th iteration can be denoted as i_n, v_n, and q_n, respectively. Thus, an ideal discrete memristor can be mathematical modelled as

$$\begin{aligned} v_n &= M(q_n)i_n, \\ q_{n+1} &= q_n + i_n, \end{aligned} \tag{1}$$

where q_{n+1} is the value of $q(t)$ at the $(n+1)$-th iteration, and $M(q_n)$ is the value of memristance $M(q)$ at the n-th iteration.

Based on the mathematical models of discrete memristors proposed in [25,6], four representations of discrete memristors are considered in this paper, as listed in Table 1. Note that the discrete memristor with exponential memristance (EM) given in Table 1 is newly proposed in this paper. Therefore, four discrete memristors are obtained, and their mathematical models are very simple.

For the newly proposed discrete memristor with EM in Table 1, we test its fingerprints of the pinched hysteresis loops by applying $i_n = 20\sin(\omega n)$ mA (ω represents the radian frequency). When the initial state q_0 is fixed as $q_0 = 0$ and the radian frequency ω is selected to three different values, the voltage–current relations are simulated and their loci are depicted in Figure 1(a). As can be seen, the discrete memristor with EM displays the radian frequency-dependent pinched hysteresis loops and their lobe areas decrease monotonically as ω increases. Meanwhile, when $\omega = 0.1$ and q_0 is selected to three different values, the loci of the voltage–current relations are depicted in Figure 1(b). As can be found, the discrete memristor with EM exhibits the initial state-dependent pinched hysteresis loops and their lobe areas are related to the initial states. In brief, the results demonstrate that the discrete memristor with EM can satisfy the fingerprints of memristors [1].

2.2. Four 2-D memristive Gauss maps

The 1-D Gauss map is a nonlinear iterative map of the real numbers into a real number interval directly given by the Gaussian function [15], which can be mathematically described as

$$x_{n+1} = G(a, b, x_n) = e^{-ax_n^2} + b, \tag{2}$$

where a and b are two real parameters. Under some parameter settings, the Gauss map can display chaos. However, it has simple system structure and its chaotic behaviours may be easily predicted [26].

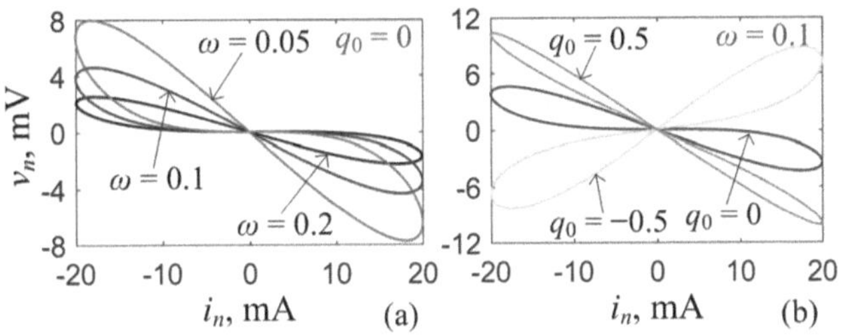

Figure 1. The fingerprints of the discrete memristor with EM by applying $i_n = 20\sin(\omega n)$ mA. (a) Radian frequency-dependent pinched hysteresis loops for fixed $q_0 = 0$ C; (b) initial state-dependent pinched hysteresis loops for fixed $\omega = 0.1$ rad/s.

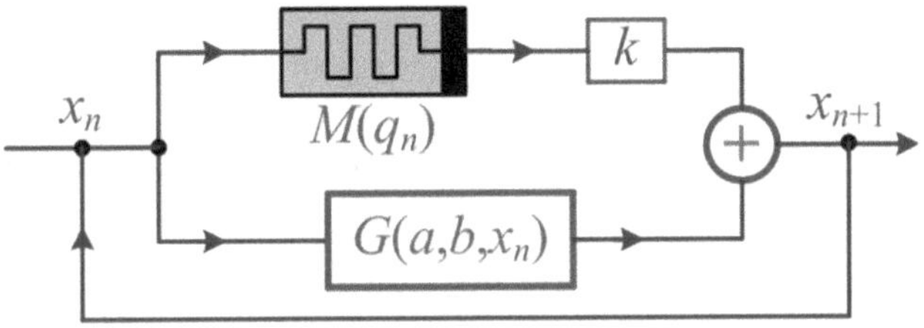

Figure 2. The schematic structure of the unified 2-D memristor-based Gauss mapping model.

To enhance the chaos complexity of the 1-D Gauss map, a unified memristor-based Gauss mapping model can be established by coupling the four discrete memristors given in Table 1 with the 1-D Gauss map. The schematic structure is depicted in Figure 2, in which $G(a, b, x_n)$ represents the 1-D Gauss map with two real parameters a and b, and k represents the coupling coefficient between each discrete memristor and the 1-D Gauss map.

With the schematic structure in Figure 1, the unified 2-D memristor-based Gauss mapping model can be obtained in mathematics as

$$\begin{aligned} x_{n+1} &= e^{-ax_n^2} + b + kM(q_n)x_n, \\ q_{n+1} &= q_n + x_n, \end{aligned} \tag{3}$$

Substituting the four memristance functions given in Table 1 into (3), four 2-D memristive Gauss maps can be thereby yielded and their details are summarized in Table 2, where HD and SE stand for the hidden and self-excited dynamical behaviours respectively. Therefore, the method of constructing memristive Gauss maps is simple and effective.

2.3. *Fixed point and stability*

For the unified 2-D memristor-based Gauss mapping model, the fixed point $S = (X, Q)$ can be solved from the equations

$$\begin{aligned} X &= e^{-aX^2} + b + kM(Q)X, \\ Q &= Q + X, \end{aligned} \tag{4}$$

Obviously, there are two specific cases that need to be discussed.

Table 2. Mathematical models of four memristive Gauss maps with typical settings.

Memristive Gauss maps	Map names	(a, b) (HD/SE)	(x_0, q_0)
$x_{n+1} = e^{-ax_n^2} + b + k(q_n^2 - 1)x_n,$ $q_{n+1} = q_n + x_n.$	QMG map	$(20, -0.4)$ (HD)	$(0.1, 0)$
		$(-0.2, -1)$ (SE)	$(0.1, 0)$
$x_{n+1} = e^{-ax_n^2} + b + k(\lvert q_n\rvert - 1)x_n,$ $q_{n+1} = q_n + x_n.$	AMG map	$(20, -0.4)$ (HD)	$(0.1, 0)$
		$(1.5, -1)$ (SE)	$(0.1, 0)$
$x_{n+1} = e^{-ax_n^2} + b + k\tanh(q_n)x_n,$ $q_{n+1} = q_n + x_n.$	HMG map	$(20, -0.4)$ (HD)	$(0.1, 0)$
$x_{n+1} = e^{-ax_n^2} + b + k(e^{-q_n} - 1)x_n,$ $q_{n+1} = q_n + x_n.$	EMG map	$(20, -0.4)$ (HD)	$(0.1, 0)$

Table 3. Unstable intervals for the QMG and AMG maps when $b = -1$.

Map names	λ_2	Unstable intervals
QMG map	$k(\mu^2 - 1)$	$\lvert\mu\rvert < \sqrt{1 - 1/k}$, $\lvert\mu\rvert > \sqrt{1 + 1/k}$
AMG map	$k(\lvert\mu\rvert - 1)$	$\lvert\mu\rvert < 1 - 1/k$, $\lvert\mu\rvert > 1 + 1/k$

In the first case, one has $b = -1$. The memristor-based Gauss mapping model has infinitely many line fixed points and these line fixed points are described by

$$S = (X,\ Q) = (0,\ \mu), \tag{5}$$

where μ is a constant, representing the initial state of the discrete memristor. The characteristic polynomial equation is obtained by solving the Jacobian matrix at S as

$$P(\lambda) = (\lambda - 1)[\lambda - kM(\mu)], \tag{6}$$

Then the eigenvalues λ_1 and λ_2 are yielded as

$$\lambda_1 = 1,\ \lambda_2 = kM(\mu), \tag{7}$$

Thus, λ_1 is always critically stable because it is on the unit circle, while λ_2 can be stable or unstable, depending on the coupling coefficient k and the initial state μ related to the discrete memristor. In other words, the fixed points for the memristor-based Gauss mapping model can be unstable or critical stable.

However, when $b = -1$, the HMG and EMG maps do no evolve dynamically with parameters or initial states. Therefore, the memristor-based Gauss mapping model can only generate the QMG and AMG maps, but not the HMG and EMG maps. For the QMG and AMG maps, the corresponding unstable intervals associated with the eigenvalue λ_2 are summarized in Table 3. In this particular case, the results indicate that there are multi-stable iterations in these two memristive Gauss maps.

In the second case, one has $b \neq -1$. The memristor-based Gauss mapping model has no any fixed points because there is no real solution in (4). In this specific case, the mapping model is in hidden periodic, chaotic or hyperchaotic iteration.

2.4. Hidden period-doubling bifurcations

Because the memristor-based Gauss mapping model has no any fixed points when $b \neq -1$, we cannot discuss its stability using the stable and unstable manifolds of the fixed points. To address this issue, we use a dimension-reduction conversion method to study the hidden period-doubling bifurcations therein.

The state q in the second equation of (3) can be transformed into the cumulative sum form of the state x. Thus, the memristor-based Gauss mapping model in (3) can be rewritten as

$$x_{n+1} = e^{-ax_n^2} + b + kM\left(q_0 + \sum_{l=0}^{n-1} x_l\right)x_n, \tag{8}$$

where q_0 is the memristor initial state. Note that the memristance function $M(\cdot)$ in (8) is closely dependence on the number n of iterations, so $M(\cdot)$ is a variable parameter rather than a constant parameter.

In the above model (8), k and $M(\cdot)$ are the parameters completely associated with the memristor, and they can be expressed uniformly by a new parameter m, that is, $m = kM\left(q_0 + \sum_{l=0}^{n-1} x_l\right)$. Thus, a dimension-reduction model can be yielded from (8) as

$$x_{n+1} = e^{-ax_n^2} + b + mx_n. \tag{9}$$

Since $M(\cdot)$ is a variable parameter, m is also a variable parameter.

The stability of model (9) can be studied using its fixed point. The fixed point $F = X$ can be obtained by the following transcendental equation

$$X = e^{-aX^2} + b + mX, \tag{10}$$

Therefore, the eigenvalue λ at $F = X$ is given as

$$\lambda = -2aXe^{-aX^2} + m, \tag{11}$$

If $|\lambda| < 1$, the fixed point F is stable; otherwise it is unstable. The result in (11) shows that the parameter m can change the stability of model (9).

As an example, the two real parameters in the 1-D Gauss map are set to $a = 20$ and $b = -0.4$. By solving the transcendental equation in (10), the fixed point trajectory of the dimension-reduction model (9) that changes with the variable parameter m can be obtained, as shown in Figure 3(a), where the blue trajectory represents unstable fixed point (UFP), the red trajectory represents stable fixed point (SFP), and the purple '$*$' represents period-doubling bifurcation point (PBP). When m increases from -0.2 to 2, the fixed point starts with just one unstable point, mutates into one unstable point and one critical point at $m = 0.0935$, and immediately splits into two unstable points and one stable point. When m increases near 1, the one stable point disappears and is transferred into one new unstable point. When m increases further, the unstable point in the middle turns into one stable point at $m = 1.5887$, which leads to the occurrence of the period-doubling bifurcation. Finally, the two unstable points and one stable point merge into one unstable point and one critical point at $m = 1.9065$, and immediately degenerate into one unstable point.

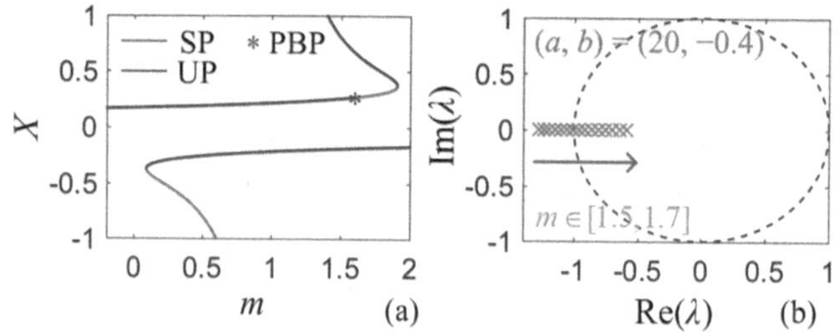

Figure 3. The fixed point trajectory and eigenvalue locus with respect to the memristor-related parameter m. (a) The fixed point trajectory with the increase of m from -0.2 to 2; (b) the eigenvalue locus with the increase of m from 1.5 to 1.7, which goes -1 into the unit circle at $m = 1.5887$, indicating the existence of the period-doubling bifurcation.

To further demonstrate the period-doubling bifurcation in (9), we set the parameter m to increase within $[1.5, 1.7]$ and draw the locus of the eigenvalue λ of (11), as shown in Figure 3(b). As one can see, λ depends on the parameter m when the real parameter a is determined. When m increases from 1.5 to 1.7, λ goes -1 into the unit circle at $m = 1.5887$, indicating the period-doubling bifurcation in the dimension-reduction model. Therefore, the hidden period-doubling bifurcation induced by the discrete memristor can be revealed and interpreted using the dimension-reduction model. However, it should be remarkable that the dimension-reduction model cannot reflect the real dynamical properties and behaviours of the original mapping model, and it is only suitable to approximate the dynamical mechanism of the hidden bifurcation behaviour in the original mapping model.

3. Numerical simulations of hidden/self-exited dynamics

This section simulates and discusses the hidden/self-exited dynamics of the four 2-D memristive Gauss maps using bifurcation plots, phase plane portraits, and basins of attraction.

3.1. Bifurcation plots for the hidden dynamics

When we set the real parameter $b = -0.4$, these four 2-D memristive Gauss maps with no fixed points can show hidden dynamics. To simulate the hidden dynamics, the two-parameter bifurcation plots can be described by measuring the periodicities and Lyapunov exponents (LEs) of the iterative sequence in these memristive Gauss maps.

Using the two-parameter bifurcation plots, the dynamical effects of the discrete memristors on the four 2-D memristive Gauss maps are demonstrated, as shown in Figure 4, where the interception range of the iterative sequence is set to [19900, 20000] and the Wolf's algorithm with the iteration length 2×10^4 is used to compute LEs. The initial states are fixed as $(x_0, q_0) = (0.1, 0)$, and the two variable parameters a and k are employed to construct the parameter plane. In the $a - k$ parameter plane, different colour blocks stand for different dynamical behaviours. The brown block marked by HC stands for hyperchaos with two positive LEs, the red block by CH for chaos with one positive LE, the orange block by QP for quasi-period with zero largest LE, the yellow block by MP for multi-period with

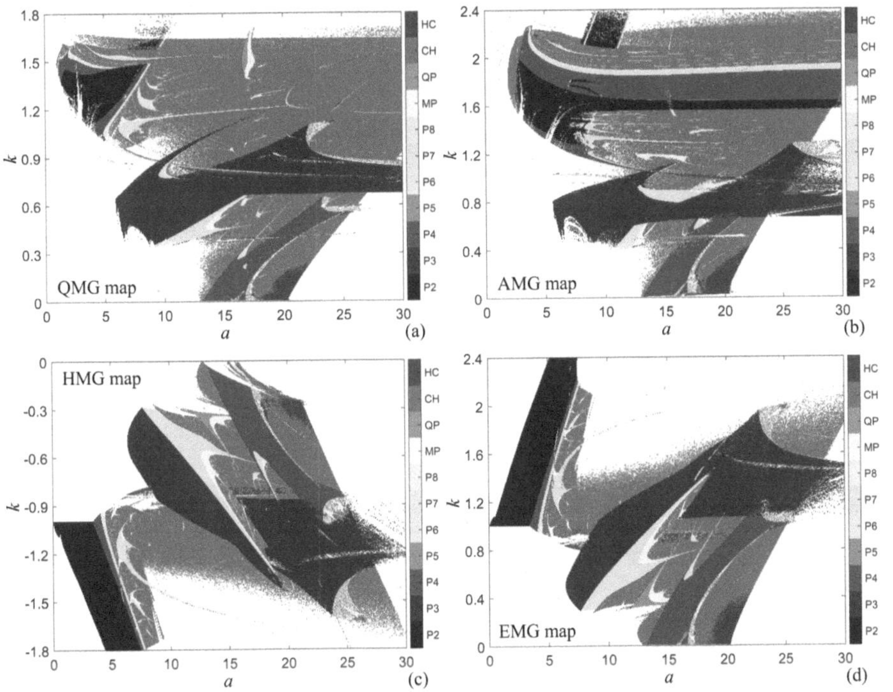

Figure 4. For fixed $b = -0.4$ and $(x_0, q_0) = (0.1, 0)$, the two-parameter bifurcation plots of the four 2-D memristive Gauss maps in the $a - k$ parameter plane: (a) the QMG map with $a \in [27,1]$ and $k \in [0, 1.6]$; (b) the AMG map with $a \in [27,1]$ and $k \in [0.2, 2.2]$; (c) the HMG map with $a \in [27,1]$ and $k \in [-1.4, -0.2]$; (d) the EMG map with $a \in [27,1]$ and $k \in [0.1, 1.7]$.

two negative LEs, and the other colour blocks marked by P2 to P8 for period-2 to period-8 with two negative LEs. Note that the white block stands for unbounded behaviour. Note that the initial states $(x_0, q_0) = (0.1, 0)$ are selected to be the small initial perturbation of these memristive Gauss maps.

The two-parameter bifurcation plots in Figure 4 intuitively exhibit the dynamical effects of the discrete memristors on the four 2-D memristive Gauss maps. In these four images, the distribution regions of chaotic behaviours are the largest, and the distribution regions of period-3 behaviour are the second. Of course, the hyperchaotic and quasi-periodic behaviours can be also observed within these coloured regions. Consequently, the discrete memristors have great influence on the dynamical distributions and dynamical behaviours of the four 2-D memristive Gauss maps.

Following the numerical results in Figure 4, we set the real parameter $a = 20$ and take the coupling coefficient k as the bifurcation parameter. The one-parameter bifurcation plots of bifurcation diagrams (bottom) and LEs (top) are simulated and their numerical results are shown in Figure 5, where the interception range of the iterative sequence and the algorithm for computing LEs are exactly the same as those used in Figure 4. As can be observed, all the four 2-D memristive Gauss maps display the dynamical behaviours of period, chaos, hyperchaos, tangent bifurcation, and chaos crisis. Particularly, both the QMG map in Figure 5(a) and HMG map in Figure 5(c) have the relatively narrow quasi-periodic route to chaos with the variations of the coupling coefficient k [23], whereas both

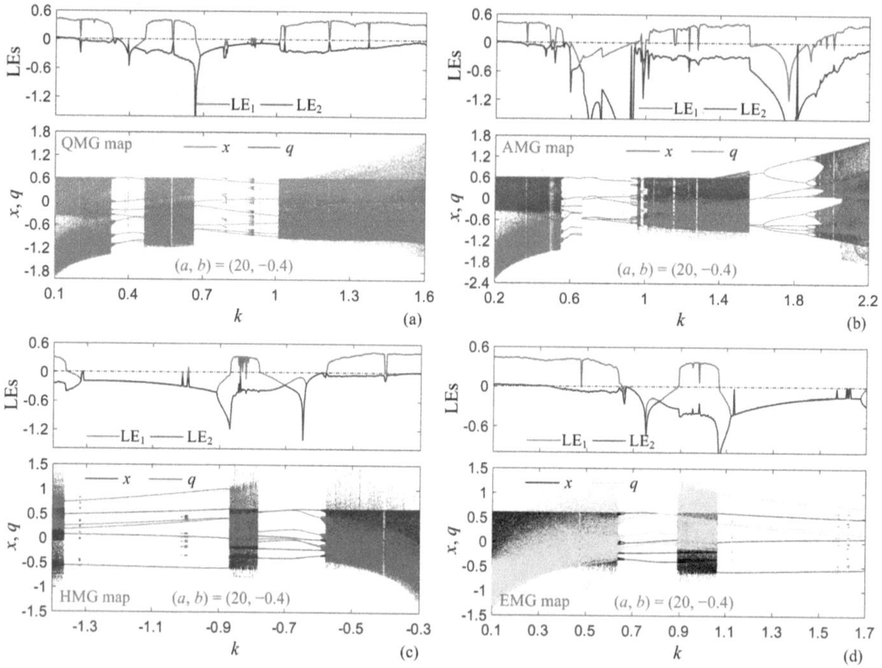

Figure 5. For the four 2-D memristive Gauss maps, the one-parameter bifurcation plots of bifurcation diagrams (bottom) and LEs (top): (a) the QMG map with $k \in [0.1, 1.6]$; (b) the AMG map with $k \in [0.2, 2.2]$; (c) the HMG map with $k \in [-1.4, -0.3]$; (d) the EMG map with $k \in [0.1, 1.7]$.

the AMG map in Figure 5(b) and EMG map in Figure 5(d) have the period-doubling route to chaos with the changes of the coupling coefficient k, in which the AMG map in Figure 5(b) has the relatively wide period-doubling route to chaos. Besides, the local self-similarity with multiple period-doubling bifurcations can be found in Figure 5(a) and (c).

Correspondingly, some parameter settings of (a, k) are selected from the red blocks of the four images in Figure 4. The phase plane portraits of the four 2-D memristive Gauss maps under different parameter settings are plotted and they are shown in Figure 6, where the iteration length is set to 10^5. As can be seen, four different chaotic attractors with complex fractal structures are displayed. For these chaotic attractors, their LEs (LE_1, LE_2) are calculated as (0.3475, −0.1969), (0.3953, −0.1313), (0.4143, 0.0041), and (0.2445, −0.0456), respectively. Consequently, the discrete memristor can greatly enhance the chaos complexity of the 1-D Gauss map, in other words, the four 2-D memristive Gauss maps can generate the chaotic attractors with complex fractal structures.

3.2. Bifurcation plots for the self-excited dynamics

When we set the real parameter $b = -1$, the QMG and AMG maps with infinitely many line fixed points can show self-excited dynamics. To simulate the self-excited dynamics, the two-parameter bifurcation plots can be also described by measuring the periodicities and LEs of the iterative sequence in these two memristive Gauss maps.

Using the numerical method exactly the same as that used in Figure 4, the two-parameter bifurcation plots of the QMG and AMG maps are depicted, as shown in Figure

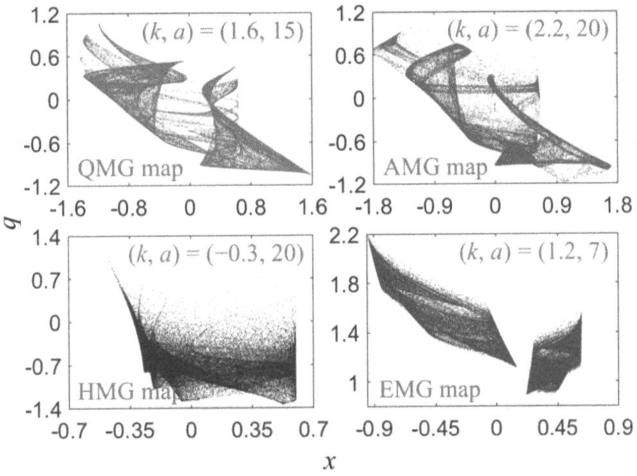

Figure 6. The phase plane portraits of the four 2-D memristive Gauss maps for four parameter settings under $(x_0, q_0) = (0.1, 0)$, where the parameter settings of (k, a) are given in the figures.

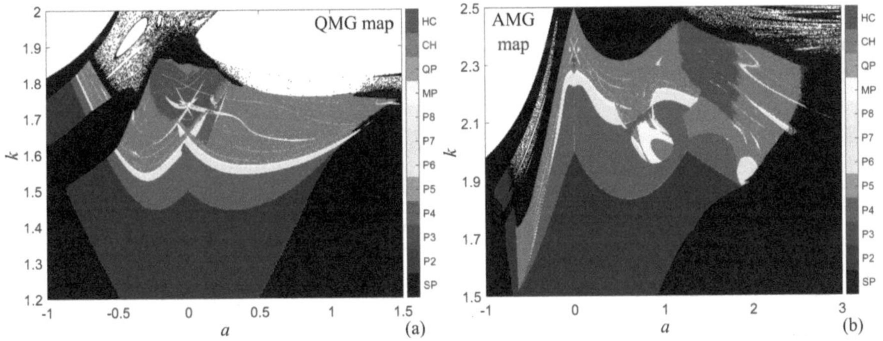

Figure 7. For fixed $b = -1$ and $(x_0, q_0) = (0.1, 0)$, the two-parameter bifurcation plots of the QMG and AMG maps in the a–k parameter plane: (a) the QMG map with $a \in [-1, 1.5]$ and $k \in [1.4, 1.9]$; (b) the AMG map with $a \in [-0.8, 2.8]$ and $k \in [1.5, 2.5]$.

7, where the initial states are fixed as $(x_0, q_0) = (0.1, 0)$, and the two variable parameters a and k are used to construct the parameter plane. In the $a - k$ parameter plane, different colour blocks stand for different dynamical behaviours. Comparing Figure 7 with Figure 4, one can see that the black block representing stable point (SP) is newly added to Figure 7, and the other colour blocks in Figure 7 are the same as those used in Figure 4. Obviously different from the results in Figure 4, the dynamical distributions in Figure 7 have the period-doubling bifurcation structures from period-2 to period-4 to period-8 and finally to chaos. In particular, the dynamical distribution in Figure 7(b) has an interesting border-collision bifurcation structure from period-4 directly to chaos due to the non-smooth nonlinearity of the absolute value memristance function in the AMG map.

According to the numerical results in Figure 7, we set $(a, b) = (-0.2, -1)$ for the QMG map and $(a, b) = (1.5, -1)$ for the AMG map respectively, and take k as the bifurcation parameter. The one-parameter bifurcation plots of bifurcation diagrams (bottom) and LEs

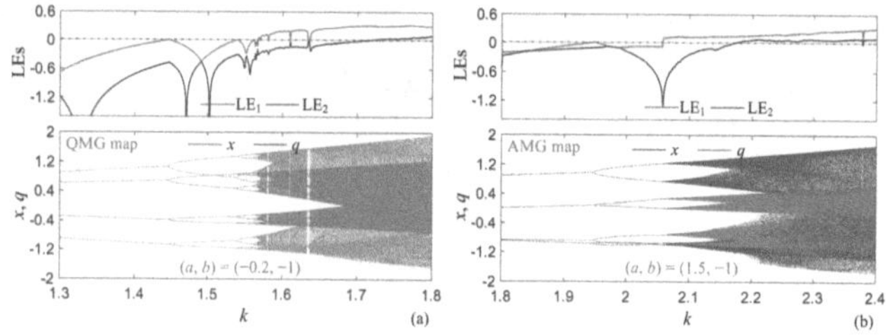

Figure 8. For the QMG and AMG maps, the one-parameter bifurcation plots of bifurcation diagrams (bottom) and LEs (top): (a) the QMG map with $k \in [1.3, 1.8]$; (b) the AMG map with $k \in [1.8, 2.4]$.

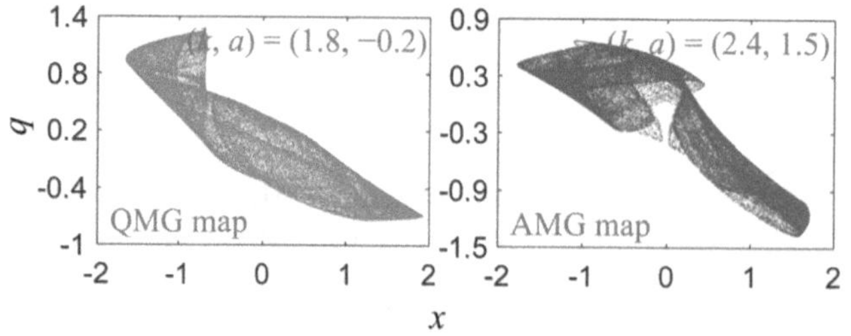

Figure 9. The phase plane portraits of the QMG and AMG maps for two parameter settings under $(x_0, q_0) = (0.1, 0)$, where the parameter settings of (k, a) are given in the figures.

(top) are simulated and their numerical results are shown in Figure 8, where the interception range of the iterative sequence and the algorithm for computing LEs are exactly the same as those used in Figure 4. As can be seen, all the QMG and AMG maps can generate hyperchaos and show dynamical behaviours of period, chaos, hyperchaos, and periodic windows. However, the QMG map in Figure 8(a) has the period-doubling bifurcation route to chaos with the change of k, whereas the AMG map in Figure 8(b) has the border-collision bifurcation route to chaos with the increment of their k and its chaotic behaviour is robust due to fewer periodic windows.

Afterwards, two representative parameter settings of (a, k) are chosen from the hyperchaotic regions in Figure 8. The phase plane portraits of the QMG and AMG maps under two representative parameter settings are depicted and they are shown in Figure 9, where the iteration length is set to 10^5. As can be found, two different hyperchaotic attractors (HCAs) with complex fractal structures are displayed. For these HCAs, their LEs are calculated as $(0.2947, 0.0917)$ and $(0.2897, 0.0880)$ respectively, i.e. these HCAs have two positive LEs.

Meanwhile, for the given real parameters, the infinitely many line fixed points of the QMG and AMG maps can be unstable or critically stable at different initial states of the memristor. This special case will lead to stable point attractors (SPAs) in some initial state regions of these two maps, implying that these two maps will exhibit bi-stability and own two cases of basins of attraction.

To measure the bi-stable behaviours of coexisting attractors, the basins of attraction are employed to classify the initial state regions according to different long-term behaviours of the QMG and AMG maps. For the QMG and AMG maps, their local basins of attraction are drawn by detecting every initial state in the x_0 - q_0 plane and the numerical results are

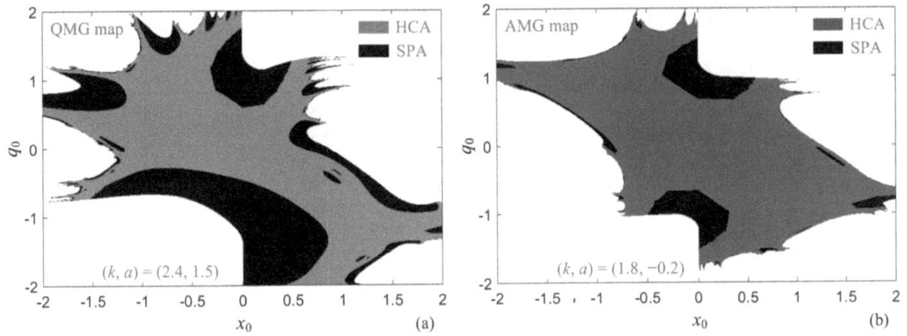

Figure 10. The local basins of attraction in the x_0 - q_0 plane, representing the initial state-dependent coexisting behaviours for the given real parameters: (a) the QMG map with $(k, a) = (2.4, 1.5)$, (b) the AMG map with $(k, a) = (1.8, \text{-}0.2)$.

shown in Figure 10(a) and (b), respectively. The black block represents the region relative to the SPA and the orange/dark-green block represents the region relative to the HCA. Note that the white block represents the unbounded behaviour. Thus, the bi-stability consisting of HCA and SPA is disclosed in the QMG and AMG maps. Besides, it can be observed from Figure 10 that in the small neighborhood of the origin of the initial state plane, the dynamical behaviours of the QMG and AMG maps are robust to the initial states, i.e. the small initial states have no influence on the dynamical behaviours of these maps.

4. Performance test and hardware experiment

This section performs performance tests and hardware experiments of hidden/self-excited chaotic/hyperchaotic attractors generated by the four 2-D memristive Gauss maps.

4.1. Performance test for chaotic/hyperchaotic attractors

The numerically simulated results in Section 3 show that all the four 2-D memristive Gauss maps can generate hidden/self-excited chaotic/hyperchaotic attractors. To quantitatively test the performance of these chaotic/hyperchaotic attractors, we calculate several common performance indicators including the correlation dimension (CorDim), spectral entropy (SpecEn), permutation entropy (PermEn), and Kaplan-Yorke dimension (D_{KY}) [6]. Note that the sequence lengths of these chaotic/hyperchaotic attractors are set as 10^5 for convenience. For the chaotic/hyperchaotic attractors given in Figures 6 and 9, their performance indicators are numerically tested and the test results are listed in Table 4. The test results demonstrate that these chaotic/hyperchaotic attractors all have excellent performance indicators. By contrast, the self-excited hyperchaotic attractor generated by the AMG map has the highest CorDim, the hidden chaotic attractor generated by the HMG map has the highest SpecEn, and the hidden chaotic attractor generated by the AMG map has the highest PermEn. However, all of these chaotic/hyperchaotic attractors have the D_{KY} up to maximum.

It should be noted that under typical parameters $(a, b) = (6.2, -0.5)$, the 1-D Gauss map can generate chaotic sequence, and its performance indicators are CorDim $= 0.9158$,

Table 4. Performance for the chaotic/hyperchaotic attractors of four memristive Gauss maps.

Map names	(k, a, b)	CorDim	SpecEn	PermEn	D_{KY}
QMG map	$(1.6, 15, -0.4)$	1.5001	0.7867	3.5484	2.0000
QMG map	$(1.8, -0.2, -1)$	1.5627	0.8388	3.8947	2.0000
AMG map	$(2.2, 20, -0.4)$	1.5650	0.8828	4.2386	2.0000
AMG map	$(2.4, 1.5, -1)$	1.7292	0.7938	3.5777	2.0000
HMG map	$(-0.3, 20, -0.4)$	1.5703	0.9223	3.5397	2.0000
EMG map	$(1.2, 7, -0.4)$	1.5097	0.9219	2.5857	2.0000

Table 5. Performance comparisons for chaotic attractors of some discrete maps.

Chaotic maps	Parameters	LE_1, LE_2	SpecEn	PermEn	D_{KY}
HMG map	$(-0.3, 20)$	0.4143, 0.0041	0.9223	3.5397	2.0000
Duffing map	$(1.6, -0.5)$	0.0951, −0.7882	0.0854	2.3963	1.1206
Lozi map	$(1.7, 0.5)$	0.4710, −1.1642	0.9124	3.8392	1.4046
Hénon map	$(1.4, 0.3)$	0.4202, −1.6241	0.9315	3.6501	1.2587
NFI_a map	–	0.0620, −0.3240	0.2505	2.2066	1.1913
CF_a map	$(1.2, 2)$	0.1333, −0.1362	0.8419	2.9014	1.9783
NEM_1 map	2	0.1172, −0.2671	0.8223	2.5015	1.4389
AP-Lozi map	$(0.9, -1)$	0.0648, −0.0648	0.6582	3.2145	2.0000
M-Log map	$(0.1, 1.88)$	0.2896, 0.0666	0.7877	3.5772	2.0000

SpecEn = 0.8952, PermEn = 3.0960, and $D_{KY} = 1$, respectively. Therefore, the introduced discrete memristor can indeed enhance chaos complexity of the 1-D Gauss map.

It has been reported that chaotification or modulation method can enhance chaos complexity of some existing discrete maps [17,16]. To make a fair performance comparison, all the chaotification and modulation mapping models are excluded from the comparison object. Taking the proposed HMG map as an example, we compare its performance indicators with those of some existing 2-D chaotic maps under their typical parameters and the test results are listed in Table 5. The existing 2-D chaotic maps include Duffing map [11], Lozi map [9], Hénon map [14], hidden NFI_a map [19], CF_a map with curve fixed points [20], NEM_1 quadratic map [30], area-preserving Lozi (AP-Lozi) map [24], and memristive Logistic (M-Log) map [4]. Thus, only the presented Q-DM map and M-Log map shows hyperchaos with two positive LEs and $D_{KY} = 2$, and other existing 2-D chaotic maps in Table 5 display chaos with only one positive LE and $D_{KY} \leq 2$. Meanwhile, the HMG map can achieve the relatively larger SpecEn and PermEn than most existing 2-D chaotic maps. Overall, the comparative results in Table 5 declare that the proposed HMG map has better chaos complexity.

4.2. Microcontroller-based hardware experiments

A discrete chaotic system is more suitable for hardware implementation in digital than a continuous chaotic system [16]. To demonstrate the simplicity of hardware implementation, a high-performance microcontroller is utilized to develop a digital hardware platform of the four 2-D memristive Gauss maps [6]. The digital hardware platform is built by one 32-bit STM32F407VET6 microcontroller, two 16-bit DAC8563 D/A converters, and some interface level conversion circuits. The powerful microcontroller realizes the four 2-D memristive Gauss maps, the D/A converters provide multi-channel analog voltage

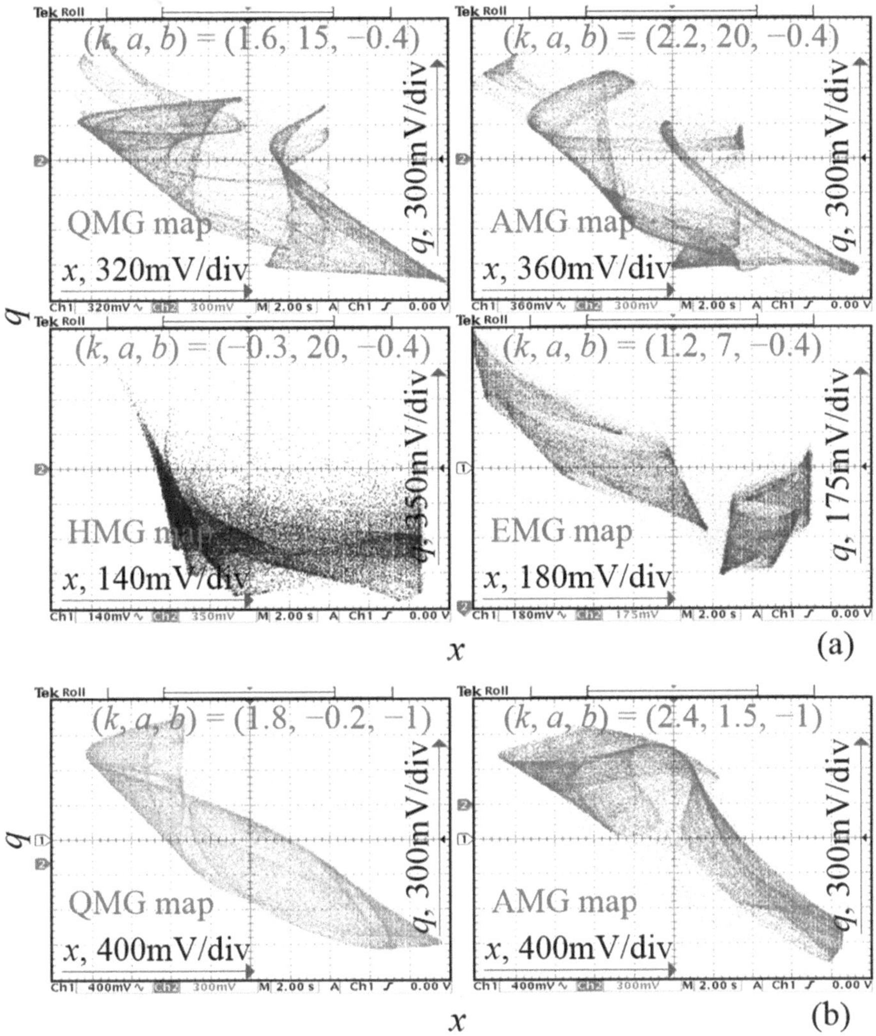

Figure 11. The experimentally acquired phase plane portraits of the four 2-D memristive Gauss maps under different parameter settings and fixed initial states $(x_0, q_0) = (0.1, 0)$, where the parameter settings of (k, a, b) are given in the figures: (a) four hidden chaotic attractors corresponding to Figure 6, (b) two self-excited hyperchaotic attractors corresponding to Figure 9.

sequences, and the interface level conversion circuits output unipolar analog voltages to digital oscilloscope.

Based on the mathematical models of four 2-D memristive Gauss maps listed in Table 2, the software codes are programmed using C language and downloaded to the microcontroller. Six parameter settings given in Figures 6 and 9 are employed and the initial states are set to $(x_0, q_0) = (0.1, 0)$. When switching on the power supply, the generated chaotic/hyperchaotic attractors under six parameter settings are acquired by the digital oscilloscope in the XY mode.

Using the digital hardware platform, the phase plane portraits of the four 2-D memristive Gauss maps under different parameter settings and fixed initial states $(x_0, q_0) = (0.1, 0)$

are captured on the digital oscilloscope and the experimental results are shown in Figure 11. In the figures, Ch1 and Ch2 stand for the x-channel and q-channel respectively, and their values stand for the scales of each division in both channels. To create a better visual effect, the iteration number of the four 2-D memristive Gauss maps is determined as 10^6 and the time scale of the digital oscilloscope is set to 2 s, which ensure that more points are captured per unit of time. The experimental results well verify the numerical ones given in Figures 6 and 9, manifesting the feasibility of the hardware implementation for the four 2-D memristive Gauss maps.

5. Conclusion

This paper presented a unified memristor-based Gauss mapping model and derived four 2-D memristive Gauss maps. Due to the line fixed points or no fixed points, the memristive Gauss maps exhibited hidden dynamics or self-exited dynamics under different parameter settings. Using the dimension-reduction conversion method, stability analysis was performed to explore the hidden period-doubling bifurcations therein. By means of numerical tools, the hidden/self-exited dynamics was revealed and the chaotic/hyperchaotic attractors were generated thereby. The performance tests and hardware experiments declared that the discrete memristor can make the memristive Gauss maps produce complex dynamical behaviours and greatly improve chaos complexity of the 1-D Gauss map. Summarily, the four 2-D memristive Gauss maps have high flexibility in generating hidden/self-exited chaos/hyperchaos with high performance, which are suitable for many applications such as image encryption [28], secure communication [23], and generative adversarial nets [7]. Certainly, the unified memristor-based Gauss mapping model can also generate some new 2-D memristive Gauss maps using new memristance functions, which deserves future study.

Disclosure statement

No potential conflict of interest was reported by the author(s).

Funding

This work was supported by the National Natural Science Foundations of China [grant number 62271088], the Scientific Research Foundation of Jiangsu Provincial Education Department, China [grant number 22KJB510001], and the Postgraduate Research and Practice Innovation Program of Jiangsu Province, China [grant number KYCX22_3044].

References

[1] M.-S. Abdelouahab, R. Lozi and L. Chua, *Memfractance: a mathematical paradigm for circuit elements with memory*, Int. J. Bifurc. Chaos 24 (2014), p. 1430023.

[2] A. Algaba, F. Fernández-Sánchez, M. Merino and A.J. Rodríguez-Luis, *Chen's attractor exists if Lorenz repulsor exists: The Chen system is a special case of the Lorenz system*, Chaos 23 (2013), p. 033108

[3] A. Algaba, F. Fernández-Sánchez, M. Merino and A.J. Rodríguez-Luis, *The Lü system is a particular case of the Lorenz system*, Phys. Lett. A 377 (2013), pp. 2771–2776.

[4] B. Bao, K. Rong, H. Li, K. Li, Z. Hua, and X. Zhang, *Memristor-coupled logistic hyperchaotic map*, IEEE Trans. Circ. Syst. II Exp. Briefs 68 (2021), pp. 2992–2996.

[5] H. Bao, C. Chen, Y. Hu, M. Chen and B. Bao, *2-D piecewise linear neuron model*, IEEE Trans. Circ. Syst. II., Exp. Briefs 68 (2021), pp. 1453–1457.

[6] H. Bao, Z. Hua, H. Li, M. Chen and B. Bao, *Discrete memristor hyperchaotic maps*, IEEE Trans. Circ. Syst. I Reg. Papers 68 (2021), pp. 4534–4544.

[7] H. Bao, Z. Hua, H. Li, M. Chen and B. Bao, *Memristor-based hyperchaotic maps and application in AC-GANs*, IEEE Trans. Ind. Inform. (2021), doi:10.1109/TII.2021.3119387.

[8] H. Bao, Z. Hua, N. Wang, L. Zhu, M. Chen and B. Bao, *Initials-boosted coexisting chaos in a 2-D Sine map and its hardware implementation*, IEEE Trans. Ind. Inform. 17 (2021), pp. 1132–1140.

[9] V. Botella-Soler, J.M. Castelo, J.A. Oteo and J. Ros, *Bifurcations in the Lozi map*, J. Phys. A: Math. Theor. 44 (2011), p. 305101.

[10] Y. Deng and Y. Li, *Nonparametric bifurcation mechanism in 2-D hyperchaotic discrete memristor-based map*, Nonlinear Dyn. 104 (2021), pp. 4601–4614.

[11] R.L. Devaney, *Fractal patterns arising in chaotic dynamical systems*, in *The Science of Fractal Images*, H.O. Peitgen and D. Saupe, eds., Springer, New York, 1988, pp. 137–168.

[12] C. Dong, K. Sun, S. He and H. Wang, *A hyperchaotic cycloid map with attractor topology sensitive to system parameters*, Chaos 31 (2021), p. 083132

[13] D. Dudkowski, S. Jafari, T. Kapitaniak, N.V. Kuznetsov, G.A. Leonov and A. Prasad, *Hidden attractors in dynamical systems*, Phys. Rep. 637 (2016), pp. 1–50.

[14] M. Hénon, *A two-dimensional mapping with a strange attractor*, Commun. Math. Phys. 50 (1976), pp. 69–77.

[15] R.C. Hilborn, *Chaos and Nonlinear Dynamics: An Introduction for Scientists and Engineers*, Oxford Univ. Press, New York, 2004

[16] Z. Hua, Y. Zhou and B. Bao, *Two-dimensional sine chaotification system with hardware implementation*, IEEE Trans. Ind. Informat. 16 (2020), pp. 887–897.

[17] Z. Hua, Y. Zhou, C.M. Pun and C.L.P. Chen, *2D sine logistic modulation map for image encryption*, Inform. Sci. 297 (2015), pp. 80–94.

[18] S. Jafari, V-T. Pham, S.M.R.H. Golpayegani, M. Moghtadaei and S.T. Kingni, *The relationship between chaotic maps and some chaotic systems with hidden attractors*, Int. J. Bifurc. Chaos 26 (2016), p. 1650211.

[19] H. Jiang, Y. Liu, Z. Wei and L. Zhang, *Hidden chaotic attractors in a class of two-dimensional maps*, Nonlinear Dyn. 85 (2016), pp. 2719–2727.

[20] H. Jiang, Y. Liu, Z. Wei and L. Zhang, *A new class of two-dimensional chaotic maps with closed curve fixed points*, Int. J. Bifurc. Chaos 29 (2019), p. 1950094.

[21] A.-A. Khennaoui, A. Ouannas, S. Bendoukha, G. Grassi, R. Lozi and V.-T Pham, *On fractional-order discrete-time systems: Chaos, stabilization and synchronization*, Chaos Solitons Fractals 119 (2019), pp.150–162.

[22] A.A. Khennaoui, A. Ouannas, S. Boulaaras, V.-T. Pham and A.T. Azar, *A fractional map with hidden attractors: Chaos and control*, Eur. Phys. J. Spec. Top. 229 (2020), pp. 1083–1093.

[23] H. Li, Z. Hua, H. Bao, L. Zhu, M. Chen and B. Bao, *Two-dimensional memristive hyperchaotic maps and application in secure communication*, IEEE Trans. Ind. Electron. 68 (2021), pp. 9931–9940.

[24] H. Li, K. Li, M. Chen and B. Bao, *Coexisting infinite orbits in an area-preserving Lozi map*, Entropy 22 (2020), p. 101119.

[25] K. Li, H. Bao, H. Li, J. Ma, Z. Hua and B. Bao, *Memristive Rulkov neuron model with magnetic induction effects*, IEEE Trans. Ind. Inform. 18 (2022), pp. 1726–1736.

[26] M. Liu, S. Zhang, Z. Fan, S. Zheng and W. Sheng, *Exponential H1 synchronization and state estimation for chaotic systems via a unified model*, IEEE Trans. Neural Netw. Learn. Syst. 24 (2013), pp. 1114–1126.

[27] R. Lozi, *Un attracteur étrange du type attracteur de Hénon*, J. Phys. 39 (1978), pp. 9–10.

[28] Y. Ma, C. Li and B. Ou, *Cryptanalysis of an image block encryption algorithm based on chaotic maps*, J. Inf. Secur. Appl. 54 (2020), p. 102566.

[29] H. Natiq, S. Banerjee, M.R.K. Ariffin and M.R.M. Said, *Can hyperchaotic maps with high complexity produce multistability?*, Chaos 29 (2019), p. 011103.

[30] S. Panahi, J. Sprott and S. Jafari, *Two simplest quadratic chaotic maps without equilibrium*, Int. J. Bifurc. Chaos 28 (2018), p. 1850144.
[31] O.E. Rössler, An equation for continuous chaos, Phys Lett A 57 (1976), pp. 397–398.
[32] J.C. Sprott, *Simple chaotic systems and circuits*, Am. J. Phys. 68 (2000), pp. 758–763.
[33] W. Tucker, *The Lorenz attractor exists*, C. R. Acad. Sci. Paris, Ser. I Math. 328 (1999), pp. 1197–1202.
[34] C.F. Wang and Q. Ding, *A new two-dimensional map with hidden attractors*, Entropy 20 (2018), p. 322.
[35] N. Wang, G.S. Zhang, N.V. Kuznetsov and H. Bao., *Hidden attractors and multistability in a modified Chua's circuit*, Commun. Nonlinear Sci. Numer. Simul. 92 (2021), p.105494.
[36] Y. Xiao, K. Sun, M. Yu and X. Xu, *Dynamics of a new multi-cavity hyperchaotic map and its DSP implementation*, Int. J. Bifurc. Chaos 29 (2019), p. 1950194.
[37] X. Zhang and G.R. Chen, *Polynomial maps with hidden complex dynamics*, Disc. Contin. Dyn. Syst. Ser. B. 24 (2019), pp. 2941–2954.
[38] J. Zhao, *Complexity and chaos control in a discrete-time Lotka-Volterra predator-prey system*, J. Differ. Equ. Appl. 26 (2020), pp. 1303–1320.
[39] J. Zhong, L. Zhang and G. Tigan, *Bifurcations of a discrete-time neuron model*, J. Differ. Equ. Appl. 23 (2017), pp. 1508–1528.

Enhanced design and hardware implementation of a chaos-based block cipher for image protection

Mahdi Madani, Safwan El Assad, Fethi Dridi and René Lozi

ABSTRACT
Nowadays chaos-based cryptography systems are widely used to protect sensitive data. A variety of chaos-based cryptosystems with software implementation has been published in the literature. However, due to the rapid growth of connected devices, the development of hardware secure and low-resource cryptographic systems is necessary. In this paper, we enhance the design of one of our software cryptosystems and then we address its FPGA-based implementation. The improvement in the design consists essentially to introduce an Initial Vector (IV) entry to the Chaotic Pseudo Random Number Generator (CPRNG) thus allowing to generate a new dynamic keys for the confusion layer (using a modified 2-D cat map) and diffusion layer (using a logistic map) with each new execution of the system. The CPRNG is formed by XORred a discrete Skew Tent Map (STM) with a discrete Piecewise Linear Chaotic Map (PWLCM). The proposed cryptosystem, which operates in Cipher Block Chaining (CBC) mode, is controlled by a Finite State Machine (FSM) to ensure that both the encryption and decryption are performed by the same reconfigurable architecture. The system is implemented on a Xilinx XC7Z020 ZYNQ FPGA platform. The results obtained in terms of hardware implementation costs (logic resources, throughput, and efficiency) and in terms of security against cryptographic attacks demonstrate the effectiveness of the proposed chaos-based cryptosystem.

1. Introduction

Nowadays, many personal and important informations are transmitted over the different kinds of networks. Therefore, to avoid unauthorized access and to ensure a good resistance against sophisticated cryptographic attacks the community of researchers continues to develop effective and fast techniques to address this problem. In this context, among others, chaos-based algorithms are widely used recently to develop information-security systems [4]. Indeed, due to their sensitivity to initial conditions and control parameters, they are considered as a good pseudo-random number generators [1,3,5,6,9,10,16,25,39]. The first dynamical system considered as chaotic generator was proposed by Lorenz in

1963 [21,22]. Then, many works followed the same principle to develop other chaotic and hyper-chaotic generators, like Rössler, Chen and others [14,18,24,33,35].

The first chaos-based encryption scheme was introduced by Fridrich, J. [12,13]. It became thereafter the basic structure of the majority of chaos-based cryptosystems and has been widely referenced since 1997, such as [44]. The Fridrich encryption scheme is formed by two layers. The confusion layer using the 2-D Baker chaotic map, which is used to calculate the new byte position. The diffusion layer using some arbitrary function of the Gray level. It was chosen as a fixed random permutation, which can be implemented using a simple lookup table. Later, some works have demonstrated the weaknesses of the Fridrich encryption scheme, such as [38] who proved that the Fridrich algorithm could be broken using a chosen ciphertext attack. However, some other works have proposed improvements to overcome these security failures, such as [19], and [11].

In their work, El Assad et al. [11] developed a chaos-based cryptosystem using three adapted chaotic maps. The main property taken into account in the choice of these maps is the satisfaction of the effects of confusion and diffusion which are necessary to obtain a good cryptosystem.

However, their chaotic generator does not contain an IV entry, this means that with the same secret key, the chaotic generator produces the same dynamic keys used in the confusion and diffusion layer.

Unlike [11] who carried out the software study of the cryptosystem, we used Register-Transfer-Level (RTL) abstraction in VHDL to create high-level representations of the proposed architecture using an architectural synthesis technique [26,28,30] to achieve our purpose.

In this study, we improve the design of one of our chaos-based cryptosystem [11] and we perform its hardware implementation. Then, we assess its hardware metric and evaluate its cryptographic robustness. The basic principle consists to divide the plain-image (or any plain-text) to a 1024-pixel (32×32) blocks, then to encrypt each block in Cipher Block Chaining (CBC) mode. The proposed architecture is formed by three main layers. A modified 2-D cat map is used to permute each pixel position in the block (to achieve confusion property) before diffuse it by a logistic map used as diffusion layer. A Chaotic Pseudo Random Number Generator (CPRNG) is used to generate a dynamic keys for the confusion and diffusion layer every round.

The CPRNG is based on a discrete Skew Tent Map (STM) and a discrete Piecewise Linear Chaotic Map (PWLCM) [9,11]. In the hardware design, we develop a Finite State Machine (FSM) that control the synchronization of the chaotic dynamic keys and the processing of the image blocks [27,29]. Then, we propose a global reconfigurable architecture that ensures both the encryption and decryption algorithms in CBC mode.

The proposed architecture is coded using a VHSIC Hardware Description Language (VHDL) and implemented on Field-Programmable Gate Array (FPGA) technology. Its implementation on a Xilinx XC7Z020 ZYNQ FPGA hardware platform achieves a throughput of 767.52 Mbps at an operating frequency of 94.34 Mhz. The robustness of the proposed architecture is tested by evaluating the encrypted images performance: analysing the key and plain-text sensitivities, examining the key space complexity, and investigating the statistical tests (correlation, entropy, chi-square, histogram, hamming distance).

The remainder of this paper is organized as follows. Section 2 describes the general design concept of the proposed cryptosystem. Section 3 presents the FPGA implementation results and the hardware metrics (logic resources, throughput, frequency...). Section 4 investigates cryptanalytic analysis and cryptosystem performance. Finally, Section 5 summarizes the whole paper.

2. Proposed chaos-based cryptosystem

In this section, we present the processing steps of the proposed architecture. We begin by describing the internal modules of the image block cipher shown in Figure 1. Then, we present the additional modules used to form the global image cryptosystem.

2.1. Chaotic pseudo random number generator (CPRNG)

The proposed CPRNG (see Figure 1) is formed by combining the two chaotic maps STM and PWLCM. Each of them is controlled using an *Enable*, *Clock*, and unsynchronized *Reset* signals, a control parameter *P*, an initial value x_i of size $32-bit$. Then, they output a $32-bit$ vector and a validation signal *Done* to indicate the generation of a chaotic value.

The mathematical models of STM and PWLCM are defined by Equations (1) and (2), respectively [11,20].

$$X(n) = \begin{cases} \left\lfloor \dfrac{2^N \times X(n-1)}{Ps} \right\rfloor & \text{if } 0 < X(n-1) < Ps \\ \left\lfloor 2^N \times \dfrac{[2^N - X(n-1)]}{2^N - Ps} \right\rfloor & \text{if } Ps < X(n-1) < 2^N \\ 2^N - 1 & \text{otherwise} \end{cases} \tag{1}$$

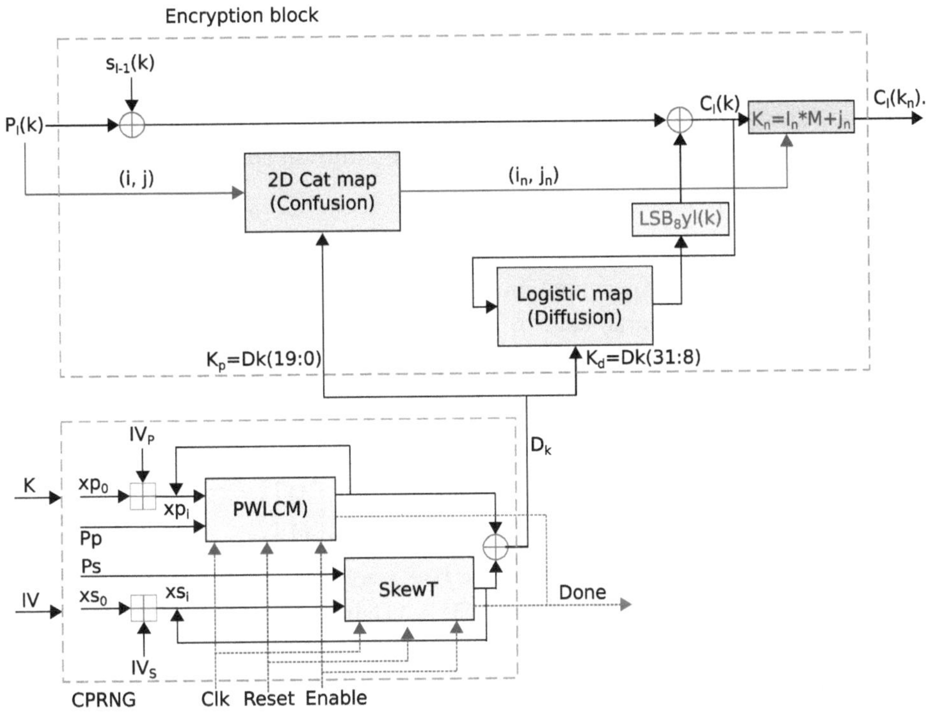

Figure 1. Encryption architecture.

$$X(n) = \begin{cases} \left\lfloor \dfrac{2^N \times X(n-1)}{Pp} \right\rfloor & \text{if } 0 < X(n-1) < Pp \\ \left\lfloor 2^N \times \dfrac{[X(n-1) - Pp]}{2^{N-1} - Pp} \right\rfloor & \text{if } Pp < X(n-1) < 2^{N-1} \\ \left\lfloor 2^N \times \dfrac{[2^N - X(n-1) - Pp]}{2^{N-1} - Pp} \right\rfloor & \text{if } 2^{N-1} < X(n-1) < 2^N - Pp \\ \left\lfloor 2^N \times \dfrac{[2^N - X(n-1)]}{Pp} \right\rfloor & \text{if } 2^N - Pp < X(n-1) < 2^N \\ 2^N - 1 & \text{otherwise.} \end{cases} \tag{2}$$

where Ps and Pp are the control parameters, ranging from 1 to $2^N - 1$ and 1 to $2^{N-1} - 1$, respectively, and $N = 32$ is the finite precision. The inputs of the CPRNG (secret key of size 127$-$*bit* and IV of size 64$-$*bit*) are used to fix the control parameters and to initialize the initial conditions of the chaotic maps, as defined in Equation (3).

$$\begin{array}{ll} Pp <= Key\ (30\ downto\ 0); & Ps <= Key\ (94\ downto\ 63); \\ xp_i <= Key\ (62\ downto\ 31); & xs_i <= Key\ (126\ downto\ 95); \\ IV_p <= IV\ (63\ downto\ 32); & IV_s <= IV\ (31\ downto\ 0); \\ xp_0 <= IV_p \boxplus xp_i; & xs_0 <= IV_s \boxplus xs_i; \end{array} \tag{3}$$

$\boxplus$ is the addition modulo 2^N (for example a and b are 32 bits vectors and s = a+b, the final result is $s = s(30:0) + s(31)$);

The generated outputs of the two chaotic maps are combined using a bitwise *XOR* operation to generate dynamic keys (Dk) to the confusion/diffusion blocks. Note that *Dk* is modified for each block and the *IV* is modified each new execution of the cryptosystem.

2.2. Confusion and diffusion layers

The confusion layer is realized by the modified 2-D cat map (used to achieve the confusion effect). The diffusion layer is realized by the discrete logistic map (to achieve the diffusion effect). Using this structure, the required confusion–diffusion effects are obtained by one encryption round.

2.2.1. Modified 2-D cat map

In addition to the standard permutation cat map tested and analysed by Fridrich [13] and [40], we added two parameters r_i and r_j to overcome the fixed-point problem [19,31].Therefore, each time Kp defined by Equation (4) is changed, the modified 2-D cat map generates a new combination of coordinates (i_n, j_n), as shown in Equation (5).

$$Kp = [u \| v \| r_i \| r_j] \tag{4}$$

$$\begin{bmatrix} i_n \\ j_n \end{bmatrix} = Mod\left(\begin{bmatrix} i + u \times j + Z_1 \\ Z_3 + Z_2 \times j \end{bmatrix}, \begin{bmatrix} M \\ M \end{bmatrix} \right) \tag{5}$$

where M is the root square of the block size (we used $M = 32$), u, v, r_i, and r_j are the system parameters defined in the range of $[0, M-1]$, and Kp is derived from the dynamic key which generated by the CPRNG ($K_p <= Dk(19\ downto\ 0)$).

The implementation of this modified 2-D cat map is carried out by the following optimized process: $Z_1 = r_i + r_j$; $Z_2 = u \times v + 1$; and $Z_3 = v \times i + r_j$. The values of Z_1 and Z_2

are calculated one time per round, but the value of Z_3 is calculated M times per round. This proposed 2D-cat map allows each pixel of coordinates (i, j) to be transferred exactly to one new position of coordinates (i_n, j_n).

2.2.2. Logistic map

The used Logistic map is a non-linear chaotic discrete function that produces random sequences. In the proposed cryptosystem, it is used to achieve and transfer the diffusion effect from one byte to other bytes in the same 1024-bytes block. This structure makes the proposed cryptosystem highly sensitive to the plaintext. The mathematical model of the discrete logistic map is defined by Equation (6).

$$X_{n+1} = \begin{cases} \dfrac{X_n \times (2^N - X_n)}{2^{N-2}} & \text{if } X_n \neq [3 \times 2^{N-2}, 2^N] \\ 2^N - 1 & \text{if } X_n = [3 \times 2^{N-2}, 2^N] \end{cases} \tag{6}$$

where X_{n+1} is the new value calculated from the previous one X_n, N is the output size of the discrete logistic map ($N = 32-bit$ in the proposed cryptosystem). Note that the initial input value to the logistic map is K_d. Then, for the next round, the input is formed by the concatenation of the ciphered pixel ($8-bit$) and K_d ($K_d <= Dk(31\ downto\ 8)$).

2.3. Control unit

To encrypt/decrypt an input image, the proposed architecture works according to a control unit that manages the loading of the image block by block from the first file into RAM block, the loading of the initial block (S_{l-1}) from the second file into another RAM block, the loading of the secret key K and Initial Vector IV from the third file into two vectors of size $127-bit$ and $64-bit$, respectively. This process is ensured using a Finite State Machine (FSM) that swap between the following five states [24,28,30].

Before starting, note that the same FSM is used for the two operating modes (encryption if $mode = 0$ and decryption if $mode = 1$) of the proposed architecture. By convention, we will call $mode_0$ for encryption and $mode_1$ for decryption to simplify the explanation.

- Init : to initialize all the internal signals, load the image file into a block RAM (32×32 pixel), load K ($127-bit$) and IV ($64-bit$) to initialize the CPRNG as presented in Equation (3), load the initial Block S_{l-1} to initialize the temporary block RAM (32×32 pixel). Note that if the last block is not a multiple of 32×32 it will be completed by pixels of value equal to 255.
- Block Load: to load 32×32-pixels (In_Block) from the image block RAM (initialized in the previous step) and check if the last block is achieved. In this case, it indicates the loading of the last block, otherwise it waits for the processing of the current block before loading the next one.
- Pixel Permute: to generate the new position (i_n, j_n) of current pixel in the loaded block from position(i, j). At the end of the permutation, the state of the *Done* signal changes to 1 to indicate the end of this process, and will serve as an interrupt to trigger the next step.
- Pixel Cipher: follow the Permute step to encrypt the permuted pixel in $mode_0$. The received pixel will be combined with the output of the logistic map and the content

of the temporary block RAM at position(k), as defined by Equation (7):

$$c_l(k_n) = p_l(k) \oplus s_{l-1}(k) \oplus LSB_8[f(c_l(k-1))] \quad (7)$$

For k and $k_n = 0$ to $M^2 - 1$; $M = 32$, $\oplus$ is the bitwise XOR operation;
In $mode_1$, the loaded pixel is decrypted as defined by Equation (8):

$$p_l(k) = c_l(k_n) \oplus s_{l-1}(k) \oplus LSB_8[f(c_l(k-1))] \quad (8)$$

For k and $k_n = 0$ to $M^2 - 1$; $M = 32$, $\oplus$ is the bitwise XOR operation;

- Generate: to update the content of the temporary block RAM with the ciphered pixels to be used in the next round of the encryption process. At this state, it indicates also to the Image_Saving block the end of ciphering the current pixel to begin its writing and saving. Then check if the last pixel is encrypted. In this case, it indicates the end of the encryption and writing process, otherwise it waits for the arrival of another pixel. In parallel, the CPRNG is executed to generate a new *Dk* which will be used for the encryption/decryption of the following block by the confusion/diffusion blocks.

The processing steps of the control unit (FSM) are presented in Figure 2.

2.4. Global proposed encryption process

In this section, we present the three modules which have been added to the image encryption block to form the proposed global architecture RTL, as shown in Figure 3. The encryption process is given by Equation (7) (see Figure 1).

2.4.1. Image loading module

As we have already mentioned, the proposed architecture processes blocks of size 1024 pixels. Therefore, this module loads the image to be processed from an external file to an internal block RAM used by our design to save the currently used block (the pixels are

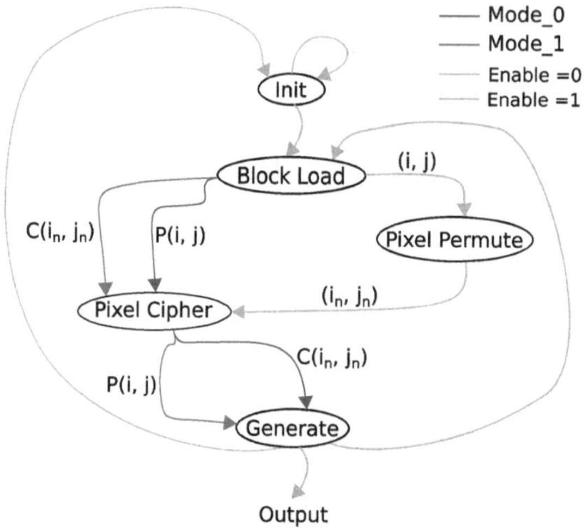

Figure 2. The used control unit (FSM).

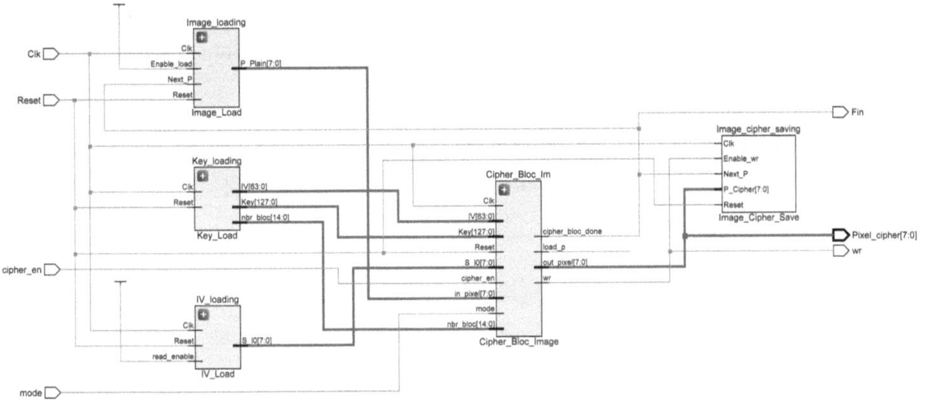

Figure 3. RTL architecture of the proposed cryptosystem.

coded in 8−*bit* binary format). After the generation of the encryption block (state *Generate* of the FSM), the control unit gives the order to load another block of pixels into the block RAM. When the last block is loaded, the control unit indicates to the encryption block that it is the last block to be processed for this image.

2.4.2. Initial vector block module

Like each block cipher in CBC mode [9], we have used an initial vector of the same size as the used block (32×32 pixels in our proposed architecture). This module load the pixels from an external file to an internal block RAM. Then, it will be used to save the encrypted block to be used in the next round.

2.4.3. Secret key and IV load module

This module loads the secret Key and *IV*, initially saved in an external file, into two internal vectors of sizes 127−*bit* and 64−*bit*, respectively.

2.4.4. Image saving module

Unlike the image load module, this module saves the ciphered block each time it is generated (state *Generate* of the FSM). Concretely, the control unit gives the order to save and sets the write bit to 1, and once finished resets it until the arrival of the next block. The ciphered/deciphered pixels are saved line by line in an external file in 8−*bit* binary format. When the last block is ciphered, the control unit indicates to this module that it is the last one to be saved for this image.

2.5. Decryption process

The decryption process is based on the same encryption principle. It begins by loading the ciphered image block by block, the initial vector block, the secret Key and the *IV*. Then, the decryption process starts by deciphering each pixel in the current encrypted block. This process is repeated until the decryption of the last pixel of the last block. The decryption process is given by Equation (8) (see Figure 4).

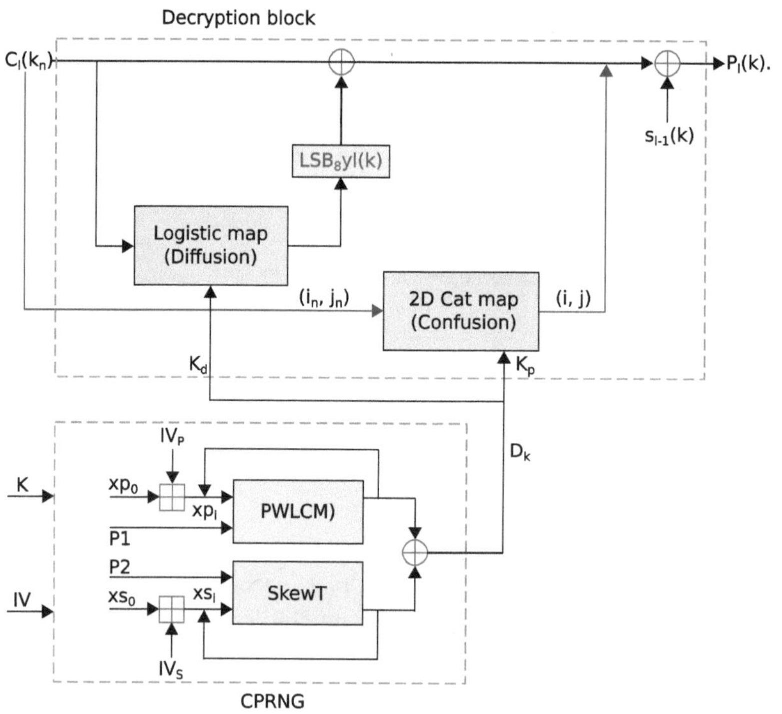

Figure 4. Decryption architecture.

3. Hardware implementation of the proposed cryptosystem

The RTL description of the proposed secure chaos-based cryptosystem has been realized on Xilinx ZYNQ-7000 FPGA prototyping board using VHDL structural description with 32−*bit* fixed-point data formats (signals of the chaotic maps). Then, it has been synthesized, and implemented using the Xilinx Vivado design suite (V.2018.2). Vivado design tools have enabled us to carry out the various known steps from the conception of our design to its implementation on the target FPGA board, as follow: the description, the synthesis, the simulation, the implementation of the design, then programming it on a chip. In Figure 5, we summarize the different steps of the design flow that have been performed under Vivado for the performance evaluation of the proposed architecture [8].

3.1. Conception to implementation steps

Like any Vivado project, we started by creating the hierarchical description formed by the different modules necessary for our architecture described in VHDL. Then, Vivado's tools performed the main analysing steps as follows:

- the synthesis step checks the VHDL description of each module, converts it into a gate-level representation, and creates a netlist.
- the behavioural simulation of the proposed cryptosystem checks its validity and make sure that the results obtained are correct (the decrypted image is the same that the original one). The simulation has been invoked directly by the Xsim simulator integrated

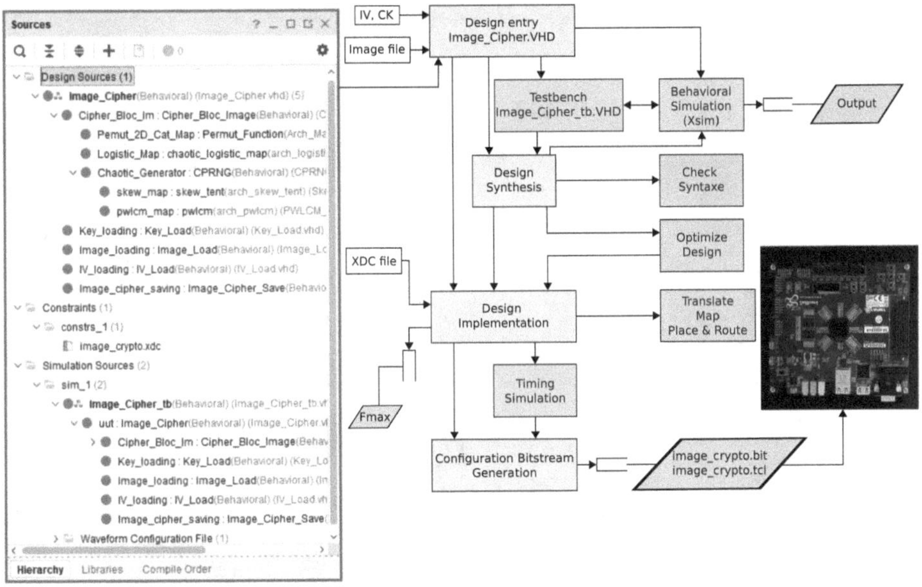

Figure 5. FPGA conception flow (under Vivado) of the proposed cryptosystem.

into the Vivado tools. The obtained results are displayed in the chronogram shown in Figure 6 (Behavioral simulation). At this step, we can assess the statistical performance of the architecture.

- performing a design-by-translation implementation merges the netlists resulting from the design synthesis and the specified constraint file (Xilinx Design Constraint XDC file), then Map adapts the design with the available resources of the FPGA target.
- the Place and Route process places the components and routes them, respecting the constraints specified during the translation to obtain a configuration file. At this step, we get the maximum frequency and the used hardware resources summarized in the implementation reports.
- the post-implementation timing simulation gets the true timing delay information of the design.
- finally, generates a programming file (Bitstream) to program the Xilinx device ZYNQ-7z020-clg484 FPGA.

3.2. *Hardware cost*

In this section, we present the performance analysis of the proposed cryptosystem in terms of occupied logic resources, processing speed, and efficiency.

The synthesis results generated by the Xilinx Synthesis Technology (XST) after place and route are presented in Table 1 which depicts the performance analysis of our implementation, as: the hardware resources in terms of the logic Slice, the LUT Flip-Flops, and LUT logic; the time metrics in terms of the Worst Negative Slack (WNS) defined in Vivado implementation report (the worst negative slack of the clock signal in the intra-clock paths section), maximum frequency and throughput calculated according to Equations (9) and (10), respectively; the efficiency in terms of throughput/slices (see Equation (11)); and

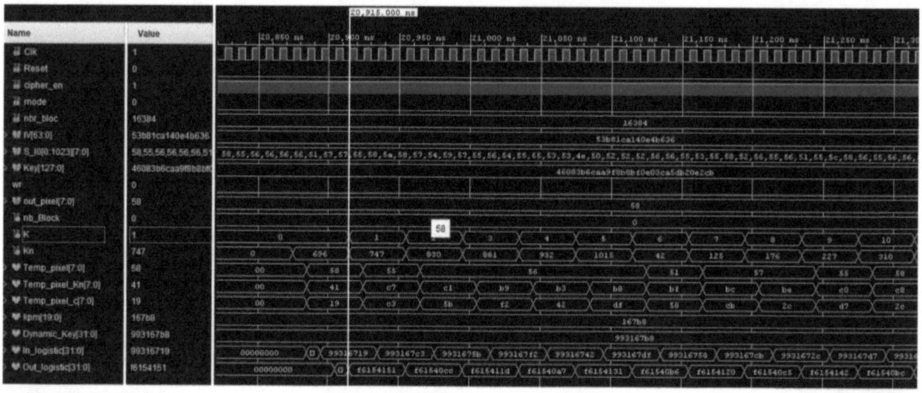

(a)

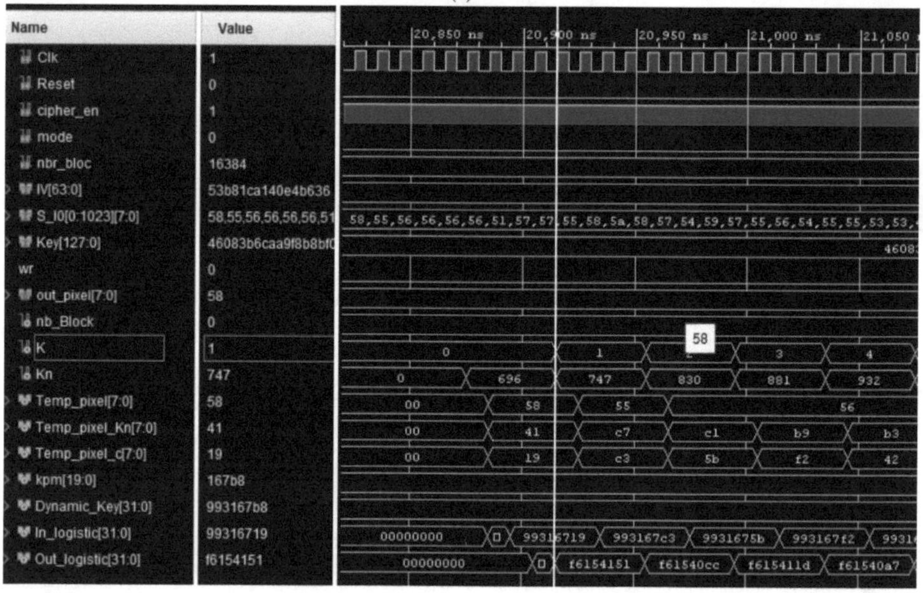

(b)

Figure 6. Behavioral simulation results. (a) Behavioral simulation. (b) Zoomed behavioural simulation.

Table 1. FPGA implementation results of the proposed cryptosystem.

	Parameters	Encryption	Decryption
Board	Family	Zynq-7000	Zynq-7000
	Device	7z020-clg484	7z020-clg484
Hardware resources	Slice	5603/42.13%	7162/53.85%
	LUTs	14188/26.67%	12487/23.47%
	FFs	16796/15.79%	16807/15.80%
Time metrics	WNS (ns)	0.155	0.017
	Maximum Frequency (MHz)	95.74	94.49
	Throughput (Mbps)	767.52	755.92
Efficiency	Efficiency (Mbps/Slices)	0.14	0.11
Consumption	Power (Watts)	0.334	0.223

the power consumption.

$$\text{Max\_}Freq = \frac{1}{T - WNS}\ [\text{MHz}] \tag{9}$$

$$\text{Throughput} = N \times \text{Max\_}Freq \, [\text{Mbps}] \tag{10}$$

$$\text{Efficiency} = \frac{\text{Throughput}}{\text{Slices}} \, [\text{Mbps}/Slices] \tag{11}$$

Note that $T = 10.6$ ns is the target clock period ($F = 1/T = 94.34$ MHz) and WNS is a parameter which is calculated by Vivado tools after the place and route phase of the implementation steps of the design flow. Its value corresponds to the worst slack of all the timing paths for max delay analysis. It can be positive or negative. A negative slack indicates a problem in which the path violates a required setup (or recovery) time. A positive one indicates the satisfactory of the timing requirement in all the paths, like our study.

The performance of our hardware architecture depends on the configured size of the fixed-point data. It results that our digital implementation on a Xilinx ZYNQ-7z020-clg484 device requires only 5603 Slice, 16796 LUTs Flip Flop pairs, and 14188 LUTs Logic. The low power consumption estimated by 0.334 W is suitable for embedded electronic applications. In addition, the precessing speed 767.52 Mbps achieved is suitable for real-time applications.

3.3. Hardware metrics comparison

The comparison of the hardware metrics of several images-cryptosystem implementations is summarized in Table 2. This comparison is difficult to interpret due to the differences in characteristics of the FPGAs tested, in particular for the clock frequency parameter and the difference between the results presented in previous works. However, considering the most important parameters (occupied Slices, clock rate of the FPGA board and the throughput achieved), we can make this comparison. Thus, the proposed cryptosystem presents better hardware metrics (Slices, DSPs, BRAMs) compared to [17,34,43] works. In addition, the timing performances are better than those obtained by Yu et al.[43], Peng et al. [34]. Whereas, the [17] implementation achieved a throughput better than our architecture, but the image size is not the same and the WNS is not defined. By considering those remarks, we can consider that the proposed architecture has a good hardware metrics and timing performance.

Table 2. Hardware metrics usage comparison of several images-cryptosystem implementations.

Parameter	Proposed	[43]	[34]	[17]
Device	ZYNQ-XC7Z020	ZYNQ-XC7Z020	XC7A35TC-PG236	XC6SLX45-CSG324
Slice	5603	–	–	8704
LUTs	14188	15978	10867	–
FFs	16796	21057	1214	–
BRAMs	0	96	48	–
DSPs	9	20	16	–
WNS (ns)	0.155	–	–	–
Max Freq (MHz)	95.74	67.95	–	272.33
Throughput (Mbps)	767.52	173.88	500	4342
Efficiency (Mbps/Slices)	0.14	–	–	0.48
Power (Watts)	0.334	–	0.248	
Image size (pixels)	512×512	256×256	256×256	448×298

4. Performance and security analysis

Below, we determine the performance of the proposed cryptosystem. We start by evaluating its resilience against statistics attacks. Then, we analyse its security and its respect for confusion and diffusion properties. Finally, we finish by evaluating its robustness [2,23].

All the simulations have been implemented in Python 3.7 on a standard computer Intel(R) Core(TM) i7-10710U CPU 1.10 GHz operating under Microsoft Windows 10, 64−*bit*, 16 GB RAM, and 1.6 GHz cpu-speed.

4.1. Key space and confusion property tests

The key space of the proposed system contains 2^{127} different combinations of the secret key, then the cryptosystem can resist to the brute force attack. Note that we consider the IV and s_{l-1} as known parameters.

To consider that a cryptosystem satisfies the confusion theory, it must be very sensitive to small changes in the secret key as defined in Shannon theory [37,42]. This makes the statistical relationship between the secret key, the plain image, and the encrypted image so complex that it is difficult for the attacker to retrieve the secret key even if he obtained many pairs of plain-encrypted images.

To prove this, we carry out the experiments based on eight plain images shown in Figure 7. For each image, 127 different randomly generated secret keys are used to generate 127 encrypted images Ci_j, where $i = 1, 2, \ldots, 8$ represents the indices of images and $j = 1, 2, \ldots, 126$ indicates the indices of secret keys.

4.1.1. Hamming distance test

We calculate the average Hamming Distance (HD) between the plain image and the corresponding encrypted image, as given by Equation (12) over 127 different secret keys.

$$\mathrm{HD}(I, C) = \frac{1}{|nb|} \sum_{k=1}^{nb} (I[k] \oplus C[k]) \tag{12}$$

where nb is the number of bit in an image.

The obtained results presented in Table 3 are very close to the optimal value 50% as defined by the avalanche effect [15] indicating that the probability of bit changes between each ciphered image and its corresponding plain image is 50%.

4.1.2. Histogram and Chi-Square test

The histogram of an image illustrates the distribution of the grey levels of the pixels, which makes it an important characteristic in evaluating the uniformity of the distribution.

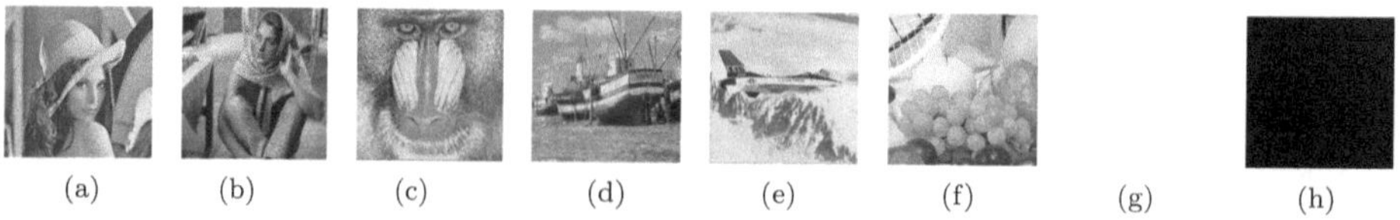

Figure 7. Plain test images. (a) lena (b) barbara (c) baboon (d) boat (e) plane (f) fruits (g) white (h) black.

Table 3. Statistic test results.

Image	Sise	HD(%)	χ^2_{exp}	Entropy: H(P)	Entropy: H(C)
Airplane	512 × 512 × 1	50.01	255.9336	6.7038	7.9993
Lena	512 × 512 × 1	50.05	232.4590	7.4449	7.9994
Barbara	512 × 512 × 1	49.98	266.5684	7.4664	7.9993
Baboon	512 × 512 × 1	50.01	279.6152	7.3582	7.9992
Boat	512 × 512 × 1	50.00	269.0293	7.1914	7.9993
Fruits	512 × 512 × 1	50.01	250.6113	7.4695	7.9993
White	512 × 512 × 1	49.97	229.6016	0.0000	7.9994
Black	512 × 512 × 1	50.02	223.8438	0.0000	7.9994

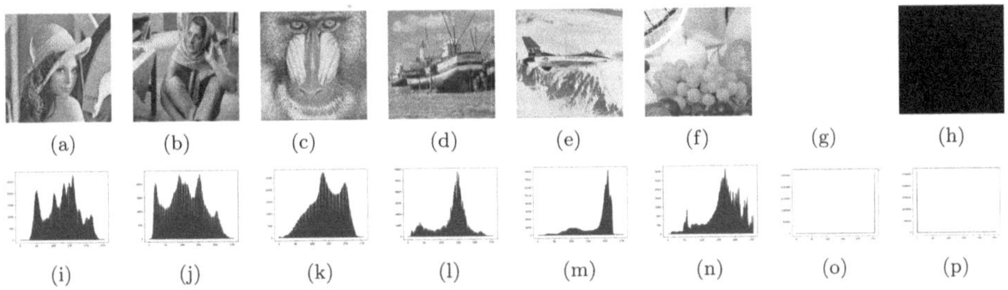

Figure 8. Plain images and their distributions. (a) lena. (b) barbara (c) baboon. (d) boat. (e) plane. (f) fruits. (g) white. (h) black. (i). (j). (k). (l). (m). (n). (o). (p).

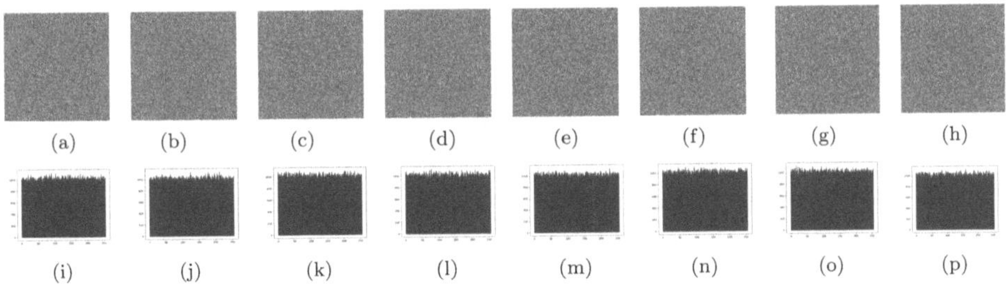

Figure 9. Ciphered images and their distributions. (a) lena (b) barbara (c) baboon (d) boat (e) plane (f) fruits (g) white (h) black (i) (j) (k) (l) (m) (n) (o) (p).

The plain images are shown in the top of Figure 8(a, b, c, d, e, f, g, h) and their corresponding histograms are shown in the bottom of the same figure (Figures i, j, k, l m, n, o, p). Similarly, the ciphered images are shown in the top of Figure 9(a, b, c, d, e, f, g, h) and their corresponding histograms are shown in the bottom of the same figure (Figures i, j, k, l m, n, o, p). From these results, we remarks that the obtained histograms of ciphered images Figure 9(i, j, k, l m, n, o, p) are close to the uniform distribution and are completely different from the plain images distributions Figure 8(i, j, k, l m, n, o, p), which proves the good randomness of the encrypted images.

To confirm statistically the uniformity accurately of the ciphered images, the Chi-Square test [32] is applied using Equation (13).

$$\chi_{exp}{}^2 = \sum_{i=1}^{N_c-1} \frac{(O_i - E_i)^2}{E_i} \tag{13}$$

where $N_c = 2^8 = 256$ is the number of levels, O_i is the calculated occurrence frequency of each grey level, $i \in [0, 255]$ in the histogram of the ciphered image, and E_i is the expected occurrence frequency of the uniform distribution, calculated by $E_i = nb/N_c$, $\alpha = 0.05$, which gives the theoretical value $\chi^2_{th}(255, 0.05) = 293.24$.

The average experimental Chi-Square results χ^2_{exp} over 127 random secret keys for each original test image are presented in Table 3. We remark that the uniformity of all the ciphered images is confirmed because $\chi^2_{exp} < \chi^2_{th}$.

4.1.3. Entropy analysis test

In every image ciphered using a robust cryptosystem, the occurrence probability of each pixel value should be the same or almost the same. To verify this property, the information entropy is used to evaluate uncertainty and

$$H(c) = -\sum_{i=0}^{N_c-1} P(C_i) \times \log_2 P(C_i) \tag{14}$$

randomness properties in the images. The random behaviour of the ciphered image can be quantitatively measured according to Shannon information-entropy formulas given in Equation (14):

Where $H(C)$ is the entropy of the encrypted image, and $P(C_i)$ is the probability of each grey level appearance ($C_i = 0, 1, \ldots, 255$).

Ideally, each level should have equal occurrence probability $Pro(C_i) = 2^{-8}$ to get closer to the maximum value $H(C) = 8$. From the obtained results presented in Table 3, we remark that the average values of the information entropy of the ciphered images are close to the above calculated ideal value. Consequently, from these results, we conclude that the proposed cryptosystem has a high degree level of resilience.

4.1.4. Correlation analysis test

In any plain image, each pixel is highly-correlated with adjacent pixels in a horizontal(H), vertical (V), and diagonal (D) directions. However, secure cryptosystem should break this relation and should produce encrypted images with correlation and redundancy as low as possible (close to zero) between adjacent pixels in the three directions [36]. To evaluate this correlation, 8000 pairs of adjacent pixels have been selected randomly in H, V, and D directions, respectively from the plain image and its corresponding ciphered image. Then, the correlation coefficient ρ_{xy} is calculated according to Equation (15).

$$\rho_{xy} = \frac{\sum_{i=1}^{N_p} [(x_i - \bar{x})(y_i - \bar{y})]}{\sqrt{\sum_{i=1}^{N_p} (x_i - \bar{x})^2} \sqrt{\sum_{i=1}^{N_p} (y_i - \bar{y})^2}} \tag{15}$$

where $Np = 8000$ is the number of the randomly selected pairs of adjacent pixels; x_i, y_i are pixel values; and $\bar{x}$, $\bar{y}$ are their mathematical expectations.

For each plane image of Figure 7, 127 ciphered images have been encrypted using 127 different secret keys, and then the average correlation coefficients have been calculated for the plain and ciphered images. The obtained results are presented in Table 4, where it appears that the correlation coefficients for the plain images are close to 1 in the H,

Table 4. Correlation coefficient results.

Image	size	Plain image H	Plain image V	Plain image D	Ciphered image H	Ciphered image V	Ciphered image D
Airplane	512 × 512 × 1	0.9704	0.9564	0.9331	0.0209	−0.0061	−0.0090
Lena	512 × 512 × 1	0.9674	0.9846	0.9585	−0.0029	−0.0018	−0.0187
Barbara	512 × 512 × 1	0.8658	0.9590	0.8346	−0.0013	0.0171	0.0133
Baboon	512 × 512 × 1	0.8612	0.7523	0.7197	0.0036	0.0065	−0.0035
Boat	512 × 512 × 1	0.9359	0.9724	0.9242	0.0141	−0.0051	0.0058
Fruits	512 × 512 × 1	0.9719	0.9730	0.9504	−0.0063	−0.0148	0.0051
White	512 × 512 × 1	–	–	–	0.0083	0.0108	0.0017
Black	512 × 512 × 1	–	–	–	0.0082	0.0113	0.0016

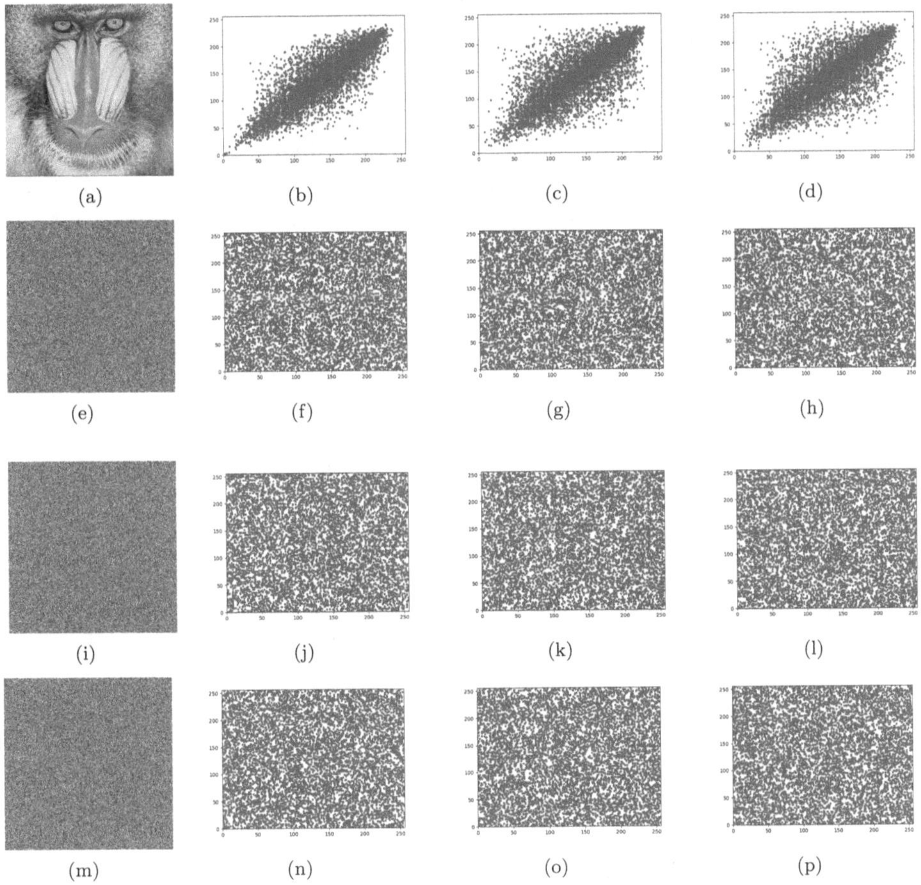

Figure 10. Correlation between adjacent pixels of baboon, white, and black images in three directions. (a) Plain baboon (b) H (c) V (d) D (e) Ciphered baboon (f) H (g) V (h) D (i) Ciphered white (j) H (k) V (l) D (m) Ciphered black (n) H (o) V (p) D.

V, and D directions, which shows that the pixels are highly correlated, whereas for the encrypted images, the correlation coefficients are close to 0 in the three directions, which proves that there is no correlation between the plain and ciphered images. We also shown in Figure 10 the correlation between adjacent pixels of images plain Baboon, ciphered White, and ciphered Black in the horizontal (H), vertical (V), and diagonal (D) directions. This figure proves that the adjacent pixels are highly correlated to each other in the plain images but appear uncorrelated in the ciphered ones. Therefore, we conclude that there is no similarity between plain and encrypted images, proving the very good confusion level achieved by the proposed cryptosystem.

4.2. Key sensitivity and diffusion property

Normally a secure encryption system should be very sensitive to any change in the secret key or in the plain text. The last property, when checked, makes the chosen-plaintext attack (differential attack) infeasible.

To demonstrate that the proposed cryptosystem has these two properties, we carry out the following experiments on the images of Figure 7.

4.2.1. Hamming distance test

First, we evaluate that the sensitivity by ciphering a plain image twice using two keys with only one bit of difference, to obtain two ciphered images C_1 and C_2. Then we calculate the *HD* between C_1 and C_2. Second, we access the plain image sensitivity (the diffusion effect) by measuring the *HD* between two ciphered images C_1 and C_2, which are encrypted from two plain images with only one bit difference.

In the first test, 127 random keys have been used to encrypt a known image. We tested the sensitivity of each bit i of the key, $i = 0, 1, \ldots, 126$. The result is shown in Figure 11. We observe that the average *HD* is close to the optimal value 50%. In the second test, for each image of Figure 7, the pixels at 21 positions have been chosen in turn to change their LSBs, and then after the encryption process, 21 *HD* values have been computed. For each position, the *HDs* of the 8 tested images are shown in Figure 12(a). We observe that when changing a single bit in one of the initial blocks (0–4), the average *HD* is always close to

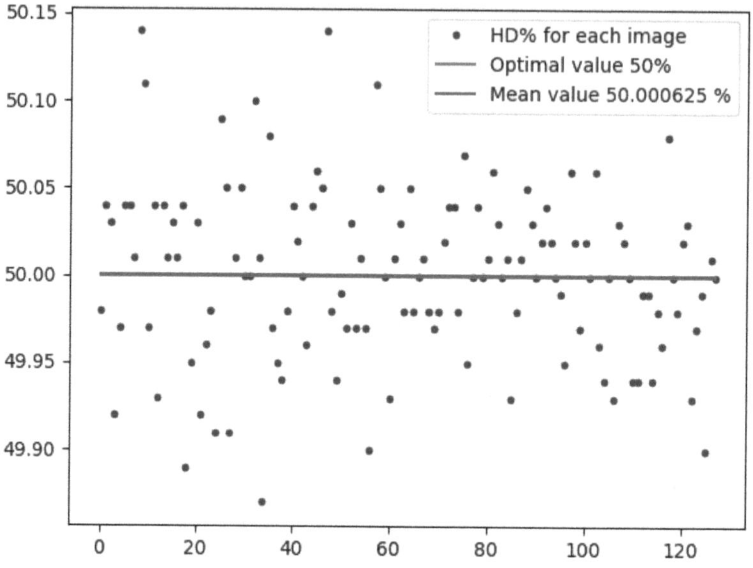

Figure 11. Hamming distance results (secret key sensitivity).

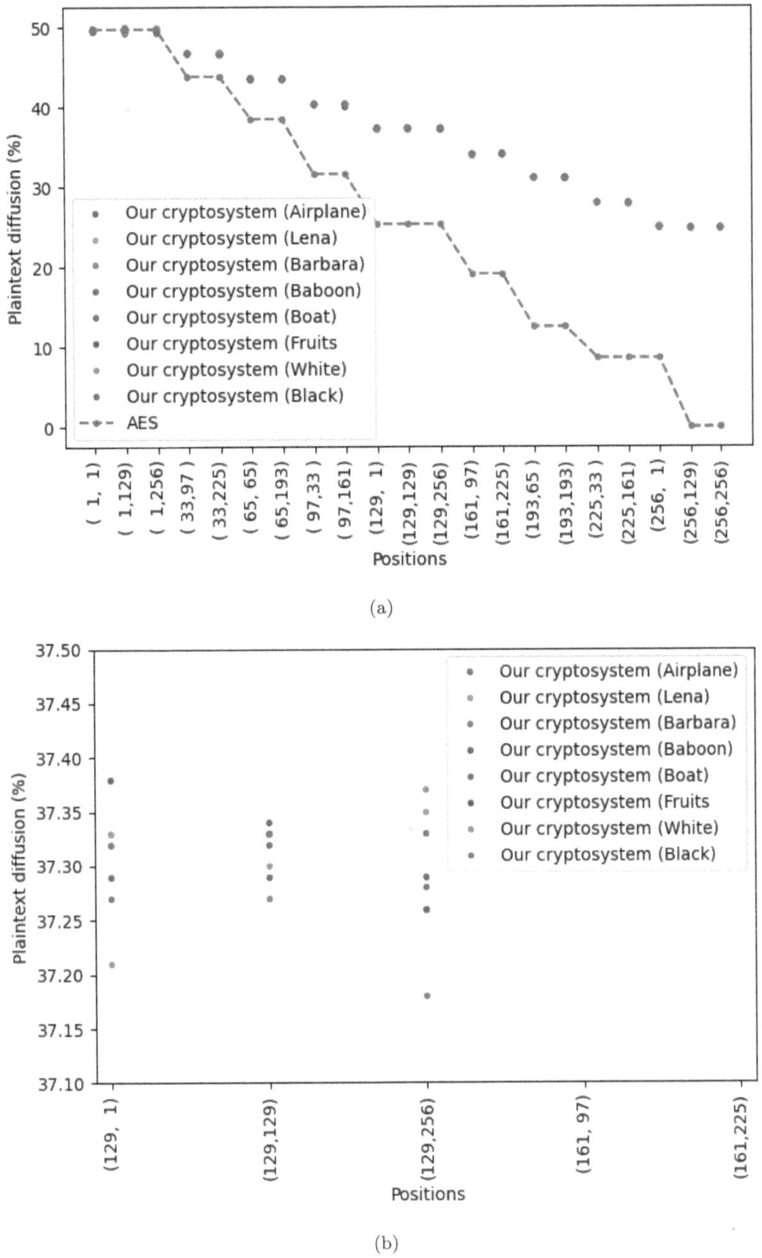

Figure 12. Hamming distance test (plaintext sensitivity).

the optimal value 50%, and then decreases to 25 % when the changes is applied to the last blocks (because of the values too close of the 8 images at each position, we have zoomed in on three positions to distinguish them well, as shown in Figure 12(b)). We remark also that this result is better than the AES sensitivity (red line) in the same CBC mode [7].

In addition, we evaluated the plaintext sensitivity to a single bit changing in the first block for each image of Figure 7, the pixels at positions (0, 0), (0, 511), (0, 1023) have been

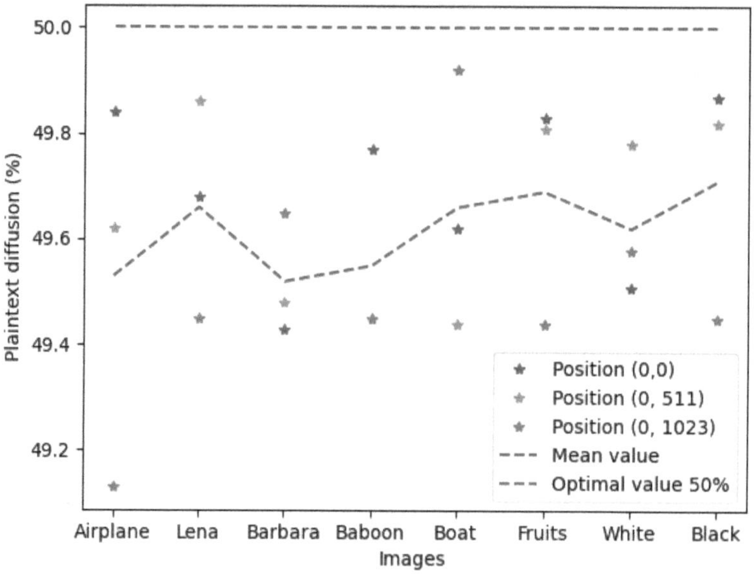

Figure 13. Hamming distance test (first block plaintext sensitivity).

chosen in turn to change their LSBs, and then after the encryption process, 3 *HD* values have been computed. For each position, the *HDs* of the 8 tested images and their average values (red line) are shown in Figure 13. As a result, we remark that the average *HD* is very close to the optimal value 50% indicating that the probability of bit changes between each ciphered image and its corresponding plain image is 50%. Therefore, we conclude that the propped algorithm achieves a good resistance to the key sensitivity and acceptable to the plaintext sensitivity tests.

4.2.2. UACI and NPCR tests

We also tested the sensitivity to the plaintext and the secret key changing (one bit) using the Unified Average Changing Intensity (UACI) and the Number of Pixels Change Rate (NPCR) tests [41], which are commonly used by researchers as defined by Equations (16) and (17), respectively.

$$\text{UACI} = \frac{1}{L \times C \times P} \sum_{l=1}^{L} \sum_{c=1}^{C} \sum_{p=1}^{P} |C_1 - C_2| \times 100\% \tag{16}$$

$$\text{NPCR} = \frac{1}{L \times C \times P \times 255} \sum_{l=1}^{L} \sum_{c=1}^{C} \sum_{p=1}^{P} \left| D(l,c,p) \right| \times 100\% \tag{17}$$

$$D(l,c,p) = \begin{cases} 0, & \text{if } C_1(l,c,p) = C_2(l,c,p) \\ 1, & \text{if } C_1(l,c,p) \neq C_2(l,c,p) \end{cases} \tag{18}$$

where C_1 and C_2 are the ciphered images whose corresponding plain images or secret keys are just one bit different from each other; l, c, p indicate the pixel row, column, and colour plane, respectively; and for a random image, the expected values of NPCR and UACI are 99.6094% and 33.4635%, respectively.

Table 5. NPCR and UACI results.

		Plaintext sensitivity		Secret key sensitivity	
Image	size	NPCR (%)	UACI (%)	NPCR (%)	UACI (%)
Airplane	512 × 512 × 1	98.6390	33.1203	99.5823	33.4343
Lena	512 × 512 × 1	99.0130	33.2805	99.6044	33.5177
Barbara	512 × 512 × 1	98.6169	33.0943	99.6025	33.4808
Baboon	512 × 512 × 1	98.7239	33.1249	99.6143	33.4733
Boat	512 × 512 × 1	98.8659	33.2501	99.6113	33.4388
Fruits	512 × 512 × 1	98.9708	33.2754	99.6304	33.5104
White	512 × 512 × 1	98.7990	33.2586	99.6075	33.4534
Black	512 × 512 × 1	99.0617	33.2269	99.6189	33.4758

To perform these tests, each image of Figure 7 has been encrypted into a ciphered one C_1, then, the LSBs of three pixels ((0,0), (0, 511), (0, 1023)) have been reversed and ciphered to obtain 3 images C_2 after encryption. The secret key sensitivity property has been also tested by changing one bit in the secret key in turn to obtain the ciphered image C_2 for 127 random keys tested.

The average NPCR and UACI results for both the plain image and key sensitivities are presented in Table 5. As we can see from the calculated results, the NPCR and UACI obtained values are very close to the optimal ones. Therefore, we conclude that the proposed cryptosystem is very sensitive to slight modifications of the plain image or the secret key.

5. Conclusion

In this article we have improved the design of one of our software encryption/decryption systems and performed its hardware implementation on a Xilinx XC7Z020 ZYNQ FPGA platform using VHDL description. Then we analysed its hardware metrics and evaluated its robustness against statistical and cryptanalytical attacks. From the results obtained, in one hand, the low area occupied in terms of logic resources, the achieved high throughput and the low energy needed and in the other hand, the resilience against cryptanalytic attacks prove the suitability of the proposed encryption/decryption for secure transmitted, stored sensitive data and IoT. Our future works will be directed towards proposing hardware implementations when using non-volatile FPGA technology, which reduces the possibility of secondary attacks in field applications.

Disclosure statement

No potential conflict of interest was reported by the author(s).

References

[1] A. Akhshani, A. Akhavan, S.C. Lim and Z. Hassan, *An image encryption scheme based on quantum logistic map*, Commun. Nonlin. Sci. Numer. Simul. 17 (2012), pp. 4653–4661.

[2] L. Bassham, A. Rukhin, J. Soto, J. Nechvatal, M. Smid, E. Barker, S. Leigh, M. Levenson, M. Vangel and D. Banks, A Statistical Test Suite for Random and Pseudorandom Number Generators for Cryptographic Applications, Natl Inst Stand Technol, 2010. Sp 800–22 rev.

[3] S. Behnia, A. Akhshani, H. Mahmodi and A. Akhavan, *A novel algorithm for image encryption based on mixture of chaotic maps*, Chaos Solitons and Fractals. 35 (2008), pp. 408–419.

[4] W.D. Chang, *Digital secure communication via chaotic systems*, Dig. Sign. Process. 19 (2009), pp. 693–699.

[5] G. Chen, Y. Mao and C.K. Chui, *A symmetric image encryption scheme based on 3D chaotic cat maps*, Chaos Solitons and Fractals. 21 (2004), pp. 749–761.

[6] F. Chiaraluce, L. Ciccarelli, E. Gambi, P. Pierleoni and M. Reginelli, *A new chaotic algorithm for video encryption*, IEEE Trans. Consum. Electron. 48 (2022), pp. 838–844.

[7] J. Daemen and V. Rijmen, The design of Rijndael: AES-the advanced encryption standard, Information security and cryptography, Springer, 2002.

[8] F. Dridi, S. El Assad, W. El Hadj Youssef, M. Machhout and R. Lozi, *The design and FPGA-based implementation of a stream cipher based on a secure chaotic generator*, Appl. Sci. 11(2) (2021), pp. 621.

[9] S. El Assad, M. Farajallah and C. Vladeanu, *Chaos-based block ciphers: An overview*, 10th IEEE Int. Conf. Communications (COMM) (IEEE), (2014), pp. 1–4.

[10] S. El Assad and H. Noura, *Generator of chaotic sequences and corresponding generating system*, US Patent App., 13/638. (2011), pp. 126.

[11] M. Farajallah, S. El Assad and O. Déforges, *Fast and secure chaos-based cryptosystem for images*, Internat. J. Bifur. Chaos. 26(2) (2016), pp. 1650021.

[12] J. Fridrich, *Image encryption based on chaotic maps*, IEEE Int. Conf. Systems, Man, and Cybernetics, Computational Cybernetics and Simulation, (1997), pp. 1105–1110.

[13] J. Fridrich, *Symmetric ciphers based on two dimensional chaotic maps*, Internat. J. Bifur. Chaos. 8 (1998), pp. 1259–1284.

[14] T.G. Gao, Z.Q. Chen, Z.Z. Yuan and G.R. Chen, *A hyperchaos generated from chens system*, Internat. J. Modern Phys. C. 17(4) (2006), pp. 471–478.

[15] D. Han, L. Min and G. Chen, *A stream encryption scheme with both key and plaintext avalanche effects for designing chaos-based pseudorandom number generator with application to image encryption*, Internat. J. Bifur. Chaos. 26(5) (2016), pp. 1650091.

[16] A. Kassem, H. Al Haj Hassan, Y. Harkouss and R. Assaf, *Efficient neural chaotic generator for image encryption*, Commun. Nonlin. Sci. Numer. Simul. 25 (2014), pp. 266–274.

[17] K. Lata and S. Saini, *Hardware software co-simulation of an AES-128 based data encryption in image processing systems for the internet of things environment*, IEEE International Symposium on Smart Electronic Systems (iSES) (Formerly INiS), Vol. 2020, (2020), pp. 260–264.

[18] Y.X. Li, G. Tang and W.K.S. Chen, *Generating hyperchaos via state feedback control*, Internat. J. Bifur. Chaos. 15(10) (2005), pp. 3367–3375.

[19] S. Lian, J. Sun and Z. Wang, *Security analysis of a chaos-based image encryption algorithm*, Physica. A. 351 (2005), pp. 645–661.

[20] S. Lian, J. Sun and Z. Wang, *A chaotic stream cipher and the usage in video protection*, Chaos Solit. Fract. 34 (2007), pp. 851–859.

[21] E. Lorenz, *Deterministic nonperiodic flow*, J. Atmospheric Sci. 20 (1963), pp. 130–141.

[22] E.N. Lorenz and K. Haman, *The essence of chaos*, Pure Appl. Geophys. 147 (1996), pp. 598–599.

[23] R. Lozi, *Emergence of randomness from chaos*, Int. J. Bifurc. Chaos. 22 (2012), pp. 1250021.

[24] M. Madani, I. Benkhaddra, C. Taougast, S. Chitroub and L. Sieler, *Digital implementation of an improved LTE stream cipher SNOW-3G based on hyperchaotic PRNG*, Secur. Commun. Netw. 2017 (2017), pp. 1–15.

[25] M. Madani, I. Benkhaddra, C. Taougast, S. Chitroub and L. Sieler, *FPGA implementation of an enhanced SNOW-3G stream cipher based on a hyper-chaotic system*, The 4th International Conference on Control, Decision and Information Technologies (CoDIT'17), (2017), pp. 124–128.

[26] M. Madani, S. Chitroub and C. Tanougast, *Two KASUMI components for an optimal implementation of the A5/3 algorithm*, The International Conference on Circuits, System and Simulation (ICCSS 2017), (2017), pp. 124–128.

[27] M. Madani and C. Taougast, *FPGA implementation of an optimized A5/3 encryption algorithm*, Microprocess. Microsyst. 78 (2020), pp. 103212.

[28] M. Madani and C. Taougast, *Combined and robust SNOW-ZUC algorithm based on chaotic system*, The International Conference on Cyber Security and Protection of Digital Services (Cyber Security 2018), Glasgow, Scotland, UK June 11–12, 2018.

[29] M. Madani and C. Taougast, *FPGA implementation of an enhanced chaotic-KASUMI block cipher*, Microprocess. Microsyst. 80 (2021), pp. 103644.

[30] M. Madani and C. Taougast, *Optimized and robust implementation of mobile networks confidentiality and integrity functions*, Computers and Security. 100 (2021), pp. 102093.

[31] Y. Mao, G. Chen and S. Lian, *A novel fast image encryption scheme based on 3D chaotic baker maps*, Internat. J. Bifur. Chaos. 14 (2004), pp. 3613–3624.

[32] S.G. MEINTANIS and Z. HLÁVKA, *Goodness-of-Fit tests for bivariate and multivariate skew-normal distributions*, Scand. J. Statist. 37(4) (2010), pp. 701–714.

[33] S. Nikolov and S. Clodong, *Occurrence of regular, chaotic and hyperchaotic behavior in a family of modified rossler hyperchaotic systems*, Chaos Solitons and Fractals. 22 (2004), pp. 407–431.

[34] X. Peng and Y. Zeng, *Image encryption application in a system for compounding self-excited and hidden attractors*, Chaos Solitons and Fractals. 139 (2020), pp. 110044.

[35] O.E. Rössler, *An equation for hyperchaos*, Phys. Lett. A. 71 (1979), pp. 155–157.

[36] J.L. Rodgers and A.W. Nicewander, *Thirteen ways to look at the correlation coefficient*, Am. Stat.42(1) (1988), pp. 59–66.

[37] C.E. Shannon, *Communication theory of secrecy systems*, Bell. Labs. Tech. J. 28 (1949), pp. 656–715.

[38] E. Solak, C. Cokal, O.T. Yildiz and T. Biyikoglu, *Cryptanalysis of Fridrich's chaotic image encryption*, Internat. J. Bifur. Chaos. 20 (2010), pp. 1405–1413.

[39] X. Wang, D. Luan and X. Bao, *Cryptanalysis of an image encryption algorithm using chebyshev generator*, Dig. Sign. Process. 25 (2013), pp. 244–247.

[40] K.W. Wong, B.S.H. Kwok and W.S. Law, *A fast image encryption scheme based on chaotic standard map*, Phys. Lett. A. 372 (2008), pp. 2645–2652.

[41] Y. Wu, J.P. Noonan and S. Agaian, *NPCR and UACI randomness tests for image encryption*, CYber J. Multidiscip. J. Sci. Technol. Sel. Areas Telecommun. 1 (2011), pp. 31–38.

[42] Y. Wu, Y. Zhou, G. Saveriades, S. Agaian, J.P. Noonan and P. Natarajan, *Local shannon entropy measure with statistical tests for image randomness*, Inf. Sci. 222 (2013), pp. 323–342.

[43] F. Yu, H. Shen, Z. Zhang, Y. Huang, S. Cai and S. Du, *A new multi-scroll Chua's circuit with composite hyperbolic tangent-cubic nonlinearity: complex dynamics, hardware implementation and image encryption application*, Integration. 81 (2021), pp. 71–83.

[44] W. Zhang, K.W. Wong, H. Yu and Z.L. Zhu, *An image encryption scheme using reverse 2-dimensional chaotic map and dependent diffusion*, Commun. Nonlin. Sci. Numer. Simul. 18 (2013), pp. 2066–2080.

On secure communication scheme-based mixed discrete-time hyperchaotic systems

Narjes Khalifa and Mohamed Benrejeb

ABSTRACT

The proposed secure communication system is based on mixed discrete-time chaotic systems in this study. To increase signal transmission security over the public channel, this system is composed of two chaotic systems on the transmitter and receiver sides. Borne and Gentina criterion associated with the Benrejeb arrow form matrix is used for stability study and system description to ensure the synchronization between these chaotic systems. The case of electrocardiogram encryption and decryption is considered an application to illustrate the effectiveness of the proposed chaotic encryption approach. The performance analyses of the proposed secure communication under the Additive White Gaussian Noise (AWGN) channel, brute force attack and some statistical attacks have shown that the proposed encryption scheme exhibited a high level of security, could maintain the confidentiality and integrity of the transmitted information and the original signal can be accurately recovered even in the presence of noise.A comparitive study between different methods is given to show the advantages and the drawbacks of the proposed communication system.

1. Introduction

Chaotic systems have several interesting properties as aperiodicity, irregularity, unpredictability and dependence on the initial conditions. Synchronization of these systems has been of great interest in many fields such as secure communications [11,16–20,23,25–28,37,38,41] since Pecora and Carroll introduced a method for the synchronization of two identical chaotic systems with different initial conditions [34]. In previous years, secure communication-based chaos synchronization has attracted more attention and has been successfully applied to build several secure communication such as the additive chaotic method [12], chaos shift keying [13], chaos modulation [44] and two-channel secure chaotic communication system [43]. Moreover, secure communication-based chaos has shown better performance than other cryptosystems [23,24,39].

On the other hand, chaos synchronization provides an additional layer of security for communication systems. The inherent complexity and sensitivity of chaotic systems make it very difficult for unauthorized parties to predict or to reproduce synchronized behaviour and decrypt transmitted data.

In addition, synchronized chaotic behaviour can be an important source of randomness for generating encryption keys. The unpredictability of chaotic systems ensures that the generated keys are highly secure and resistant to different attacks.

Compared to other cryptographic techniques, chaos synchronization can be implemented with relatively low computational complexity. Chaotic systems can be realized using simple mathematical models or electronic circuits that are computationally efficient and suitable for real-time applications.

A variety of applications such as image encryption, cryptographic key generation and secure communication are utilizing the synchronization of chaotic systems due to their chaotic and dynamic characteristics [31,33,45].

However, there are different ways to attack cryptographic schemes, among them, is brute force attack which is considered, in cryptography, as a strategy that can, in theory, be used against any encrypted data by an attacker. It consists in checking all possible keys until the correct key is found [3]. The key length used in the encryption determines the practical feasibility of performing a brute force attack: longer keys are more difficult to crack than shorter ones. The basic way to prevent a brute force attack is that the designed secure communication system should generate keys of a sufficient length so that the attacker fails to guess the key in a finite time. In fact, among the lot of research was initiated on the choice of the encryption key size, the proposed communication system in [42] is based on an external key combined with a chaotic key and others on the use of a complex encryption key using multiple chaotic systems.

In [35], three chaotic systems (logistic map, Baker map and 4-D hyperchaotic map) were used to ensure video compression and encryption. Moreover, in [40], four one-dimensional discrete chaotic maps have been chosen to compose the encryption key.

In this paper, the designed secure chaotic communication system is based on the use of two discrete chaotic systems on both the transmitter and receiver sides. To make the encryption key more complex, two discrete-time hyperchaotic systems are used for the encryption algorithm implementation in the transmitter one. Then, two observers are implemented on the receiver side to ensure the synchronization respectively with the two chaotic systems on the transmitter side based on the use of the Borne and Gentina criterion for stability study [6,9] associated with the Benrejeb arrow form matrix for the system description [4,5,7,8,14,15,30]. After that, a decryption algorithm is implemented to recover the original signal.

The paper is organized as follows. Section 2 deals with the proposed secure communication-based mixed chaotic systems. In section 3, a discrete chaotic system synchronization-based observer is proposed using the Borne and Gentina stability criterion. In section 4, the obtained results for secure electrocardiogram (ECG) encryption and decryption are presented and discussed. Section 5 presents the performance analysis of the proposed secure communication under the Additive White Gaussian Noise (AWGN) channel, brute force attack and some statistical attacks. In section 6, a comparitive study between different methods is given to show the advantages and drawbacks of the proposed communication system.

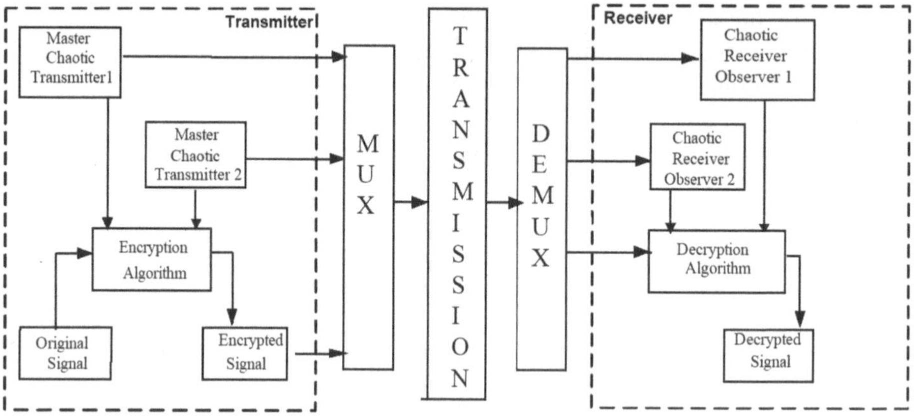

Figure 1. The block diagram of the proposed secure chaotic communication system.

2. The proposed secure communication system based on mixed chaotic systems – basic idea

The basic idea of the proposed secure communication system deals with plain message encryption using two discrete-time hyperchaotic systems on the transmitter side. Then, on the receiver side, two discrete-time observers are designed to fulfil the synchronization process. Thus, the decryption process is performed when the synchronization process is achieved respectively between the two-master and the two-slave hyperchaotic systems.

The proposed communication scheme in Figure 1 has to take into account the following

- the transmitter side is composed of two master hyperchaotic systems,
- the original message to be sent is encrypted using an encryption algorithm which is built in terms of the two encryption keys generated by the two master hyperchaotic systems in the transmitter side. The result is an encrypted message which will be sent through the public channel,
- in the receiver side of the communication system, a discrete-time observer 1 associated with the master chaotic transmitter 1 block is implemented to retrieve the chaotic oscillator signals and get the necessary information for the decryption,
- in the receiver side of the communication system, an associate observer 2 to the master chaotic transmitter 2 block is implemented to retrieve the chaotic oscillator signals and get the necessary information for the decryption,
- once the chaotic signals are retrieved, the decryption algorithm can be used along to get the decrypted message. The details of the main blocks of the communication system are given in the next sections.

In our specific application of ECG encryption, we chose the hyperchaotic 4D Hénon map and 3D Hénon map as the master systems for several reasons. The Hyperchaotic systems exhibit more complex behaviours than chaotic systems, with multiple positive Lyapunov exponents and more state variables. In addition, they are, often, represented by a set of coupled nonlinear differential equations, where the dynamics of each state variable depends on multiple other variables of the system. Furthermore, they are used in applications that require more complex dynamics than chaotic systems can provide, such as secure communications, chaos-based cryptography and control systems.

3. Computational methods for the observers' design-based discrete-time chaotic system synchronization

3.1. Problem formulation

Given two chaotic systems, 4D Hénon map and 3D Hénon map, used as master systems on the transmitter side and their associated observer systems considered as slave systems, the issue of the master-slave synchronization here is to force the response of the slave systems to synchronize with the master ones.

Consider an n-dimensional master chaotic system described by

$$\begin{cases} x_m(k+1) = A(x_m(k)) + f(x_m(k)) \\ y_m(k) = Cx_m(k) \end{cases} \tag{1}$$

and a corresponding associate discrete-time state observer (slave) is given by

$$\begin{cases} x_s(k+1) = A(x_s(k)) + f(x_s(k)) + u(k) \\ y_s(k) = Cx_s(k) \end{cases} \tag{2}$$

where $x_m(k) = [x_{m1}(k)\, x_{m2}(k)\, \ldots\, x_{mn}(k)]^T \in R^n$ and $x_s(k) = [x_{s1}(k)\, x_{s2}(k)\, \ldots\, x_{sn}(k)]^T \in R^n$ are the state vectors of the master and the slave systems, $y_m(k) = [y_{m1}(k)\, y_{m2}(k)\, \ldots\, y_{mn}(k)]^T \in R^n$ and $y_s(k) = [y_{s1}(k)\, y_{s2}(k)\, \ldots\, y_{sn}(k)]^T \in R^n$, respectively, the output vectors of the master and the slave systems, $f(x_m)$ and $f(x_s)$ the non-linear vectors fields, $A = \{a_{ij}\} \in R^{n*n}$ a known constant matrix with a proper dimension, C a constant matrix and $u(k) = [u_1(k)\, u_2(k)\, \ldots\, u_n(k)]^T \in R^n$ the control input vector of the slave system such that

$$u(k) = L(y_m(k) - y_s(k)) \tag{3}$$

with L the observer gain to be determined, $L = \{l_{ij}\} \in R^{n*n}$.

For the error $e(k)$ defined by

$$e(k) = x_m(k) - x_s(k) \tag{4}$$

it comes from the following dynamic error system

$$e(k+1) = (A - LC)e(k) + f(x_m(k)) - f(x_s(k)) \tag{5}$$

If the function $f(.)$ can be factorized as follows

$$f(x_m(k)) - f(x_s(k)) = Q(x_m(k), x_s(k))e(k) \tag{6}$$

with Q an $n \times n$ matrix, then (5) can be rewritten as

$$e(k+1) = A_\alpha((x_m(k), x_s(k))e(k) \tag{7}$$

with

$$A_\alpha((x_m(k), x_s(k)) = A - LC + Q(x_m(k), x_s(k)) \quad (8)$$

The synchronization between the master system (1) and the observer system (2) is achieved if the error system is stable, i.e.

$$\lim_{k \to \infty} |x_m(k) - x_s(k)| = 0 \quad (9)$$

In this paper, the stability study is carried out by the use of the Benrejeb arrow form matrix (See Appendix 2) and the application of the practical Borne and Gentina stability criterion (See Appendix 1).

The proposed synchronization approach investigated the synchronization of the coupled identical 4D Hénon maps and the coupled 3D Hénon maps.

3.2. Application to hyperchaotic 4D Hénon map synchronization

In this sub-section of the paper, we present the synchronization scheme for two identical 4D Hénon map chaotic systems having different initial conditions. The dynamic error system between these two systems and the stability study is considered to achieve synchronization study.

3.2.1. Error system description

Let's consider the following particular fourth-order hyperchaotic Hénon 4D master system described by

$$\begin{cases} x_{m1}(k+1) = 1.76 - x_{m3}^2(k) - 0.1x_{m4}(k) \\ x_{m2}(k+1) = x_{m1}(k) \\ x_{m3}(k+1) = x_{m2}(k) \\ x_{m4}(k+1) = x_{m3}(k) \\ y_m(k) = Cx_m(k) \end{cases} \quad (10)$$

with $x_{mi} \in [-2, 2]$ and the following slave system (11) which the Luenberger system (10) observer

$$\begin{cases} x_{s1}(k+1) = 1.76 - x_{s3}^2(k) - 0.1x_{s4}(k) + u_1(k) \\ x_{s2}(k+1) = x_{s1}(k) + u_2(k) \\ x_{s3}(k+1) = x_{s2}(k) + u_3(k) \\ x_{s4}(k+1) = x_{s3}(k) + u_4(k) \\ y_s(k) = Cx_s(k) \end{cases} \quad (11)$$

with $x_{si} \in [-2, 2]$, $x_m(k) = [\, x_{m1}(k) \; x_{m2}(k) \; x_{m3}(k) \; x_{m4}(k) \,]^T$ is the master state vector, $x_s(k) = [\, x_{s1}(k) \; x_{s2}(k) \; x_{s3}(k) \; x_{s4}(k) \,]^T$ the slave state vector, $y_m(k)$ the master output signal, $y_s(k)$ the slave output signal, C a constant matrix and $u = [\, u_1(k) \; u_2(k) \; u_3(k) \; u_4(k) \,]^T$ the control vector in the form (3).

It comes

$$f(x_m(k)) = \begin{bmatrix} -x_{m3}^2(k) & 0 & 0 & 0 \end{bmatrix}^T \tag{12}$$

$$f(x_s(k)) = \begin{bmatrix} -x_{s3}^2(k) & 0 & 0 & 0 \end{bmatrix}^T \tag{13}$$

$$A = \begin{bmatrix} 0 & 0 & 0 & -0.1 \\ 1 & 0 & 0 & 0 \\ 0 & 1 & 0 & 0 \\ 0 & 0 & 1 & 0 \end{bmatrix} \tag{14}$$

$$C = I_4 \tag{15}$$

$$Q(x_m(k), x_s(k)) = \begin{bmatrix} 0 & 0 & -(x_{m3}(k) + x_{s3}(k)) & 0 \\ 0 & 0 & 0 & 0 \\ 0 & 0 & 0 & 0 \\ 0 & 0 & 0 & 0 \end{bmatrix} \tag{16}$$

Then, the matrix $A_\alpha((x_m(k), x_s(k)) = \{a_{ij}^*\}$ in the form (8) becomes

$$A_a(x_m(k), x_s(k)) = \begin{bmatrix} -l_{11} & -l_{12} & -l_{13} - (x_{m3}(k) + x_{s3}(k)) & -0.1 - l_{14} \\ 1 - l_{21} & -l_{22} & -l_{23} & -l_{24} \\ -l_{31} & 1 - l_{32} & -l_{33} & -l_{34} \\ -l_{41} & -l_{42} & 1 - l_{43} & -l_{44} \end{bmatrix} \tag{17}$$

For l_{23}, l_{24}, l_{32}, l_{34}, l_{42} and l_{43} are chosen such as

$$\begin{cases} 1 - l_{32} = 0 \\ 1 - l_{43} = 0 \\ l_{23} = 0 \\ l_{42} = 0 \\ l_{24} = 0 \\ l_{34} = 0 \end{cases} \tag{18}$$

the matrix $A_\alpha((x_m(k), x_s(k))$ is in Benrejeb arrow form.

3.2.2. *Synchronization study*

The non-linear elements are isolated in one row of the matrice $A_\alpha(x_m(k), x_s(k))$ (17). To test the stability property of the error system (7) with (10)–(18), let's apply the Borne and Gentina stability criterion, for the overvaluing system associated with the vectorial norm $p(z(k)) = [|z_1(k)| \dots |z_n(k)|]^T, z(k) = [z_1(k) \dots z_n(k)]^T$ and the corresponding comparison system

$$z(k+1) = M(A_\alpha((x_m(k), x_s(k)))z(k) \tag{19}$$

with

$$M(A_\alpha((x_m(k), x_s(k))) = \{m_{ij}(.) = |a_{ij}^*| \forall i, j = 1, 2, 3, 4\} \tag{20}$$

All nonlinearities of $(I - M(A_\alpha(.)))$ defined by

$$I - M(A_a((x_m(k), x_s(k))) = \begin{bmatrix} 1-|l_{11}| & -|l_{12}| & -\left|\begin{pmatrix} x_{m3}(k)+ \\ x_{s3}(k) \end{pmatrix} + l_{13}\right| & -|0.1 + l_{14}| \\ -|1-l_{21}| & 1-|l_{22}| & 0 & 0 \\ -|l_{31}| & 0 & 1-|l_{33}| & 0 \\ -|l_{41}| & 0 & 0 & 1-|l_{44}| \end{bmatrix} \tag{21}$$

are isolated in the first row, then the asymptotic stability based on the Borne and Gentina criterion is ensured when the successive principal minors of the matrix $(I - M(A_\alpha(.)))$ are positive [34], i.e.

- $1 - |l_{44}| > 0$,
- $(1 - |l_{33}|)(1 - |l_{44}|) > 0$,
- $(1 - |l_{22}|)(1 - |l_{33})|)(1 - |l_{44}|) > 0$,
- there exists $\varepsilon > 0$, such that

$$\det(I - M(A_a(x_m(k), x_s(k)))) = \left\{ \left[\begin{matrix} (1-|l_{11}|)- \\ \left(\begin{matrix} \dfrac{|l_{12}||1-l_{21}|}{1-|l_{22}|} + \\ \dfrac{|x_{m3}(k) + x_{s3}(k) + |l_{13}|||l_{31}|}{1-|l_{33}|} + \\ \dfrac{|0.1 + |l_{14}|||l_{41}|}{1-|l_{44}|} \end{matrix} \right) \end{matrix} \right] (1-|l_{44}|)(1-|l_{33}|)(1-|l_{22}|) \right. > \varepsilon$$

For, the gains l_{11}, l_{22}, l_{33}, l_{44}, l_{12}, l_{21}, l_{13}, l_{14}, l_{31} and l_{41} chosen such as

$$\begin{cases} l_{11} = 0, l_{22} = 0, l_{33} = 0, l_{44} = 0, \\ l_{12} = 0, l_{21} = 2, \\ l_{13} = 0, l_{31} = 0, \\ l_{14} = 0, l_{41} = 0 \end{cases} \tag{22}$$

the observer gain L can be chosen such as

$$L = \begin{bmatrix} 0 & 0 & 0 & 0 \\ 2 & 0 & 0 & 0 \\ 0 & 1 & 0 & 0 \\ 0 & 0 & 1 & 0 \end{bmatrix} \tag{23}$$

The simulation results of studied 4D coupled hyperchaotic Hénon maps are presented in the next section.

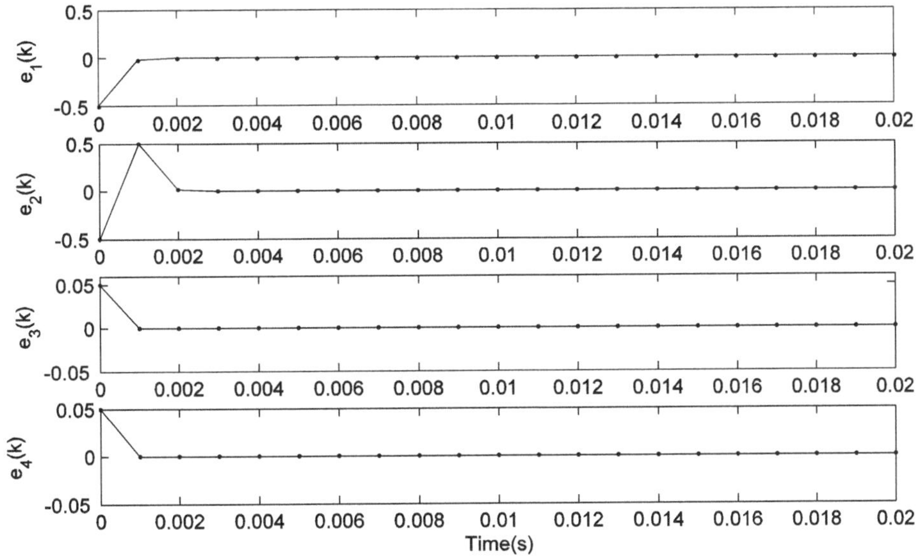

Figure 2. Dynamic errors of the two coupled 4D Hénon maps after the activation of the control law.

3.2.3. *Simulation results*

For the initial conditions of the master system (10): $x_m(0) = (0.1,\ 0.3\ ,0.15,\ 0.25)$ and of the slave corresponding system (11): $x_s(0) = (0.6, 0.8, 0.1, 0.2)$ and for the sampling time $T = 1ms$, Figure 2 shows the finite time convergence of the error system.

The synchronization is achieved between system (10) and system (11) and the error vector reaches exactly zero after three steps, at $t = 2ms$.

3.3. *Application to 3D Hénon map synchronization*

In this subsection, the synchronization of two identical hyperchaotic systems of 3D Hénon having different initial conditions is considered based on the above-proposed synchronization approach.

3.3.1. *Error system description*

The discrete-time master hyperchaotic 3D generalized Hénon given by

$$\begin{cases} x^1_{m1}(k+1) = a - x^1_{m2}(k)^2 - bx^1_{m3} \\ x^1_{m2}(k+1) = x^1_{m1}(k) \\ x^1_{m3}(k+1) = x^1_{m2}(k) \\ y^1_m(k) = Cx^1_m(k) \end{cases} \tag{24}$$

is a hyperchaotic attractor for $a = 1.76$ and $b = 0.1$ and $x^1_{mi}(k) \in [-2, 2]$.

Let the slave system, associated with this master system (24), given by

$$\begin{cases} x_{s1}^1(k+1) = a - x_{s2}^1(k)^2 - bx_{s3}^1 + u_1(k) \\ x_{s2}^1(k+1) = x_{s1}^1(k) + u_2(k) \\ x_{s3}^1(k+1) = x_{s2}^1(k) + x_{s1}^1(k) + u_3(k) \\ y_s^1(k) = Cx_s^1(k) \end{cases} \tag{25}$$

with $x_{si}^1(k) \in [-2, 2]$ and the control law such as

$$\begin{cases} u_1(k) = L_1^1(y_{m1}^1(k) - y_{s1}^1(k)) \\ u_2(k) = L_2^1(y_{m2}^1(k) - y_{s2}^1(k)) \\ u_3(k) = L_3^1(y_{m3}^1(k) - y_{s3}^1(k)) \end{cases} \tag{26}$$

where $x_m^1(k) = [x_{m1}^1(k)\ x_{m2}^1(k)\ x_{m3}^1(k)]^T$ is the state vector of the master system, $x_s^1(k) = [x_{s1}^1(k)\ x_{s2}^1(k)\ x_{s3}^1(k)]^T$ is the state vector of the slave system, $y_m^1(k)$ is the master output signal, $y_s^1(k)$ is the slave output signal and $L^1 = [L_1^1\ L_2^1\ L_3^1]^T$ are the observer gains.

Then, it comes

$$A = \begin{pmatrix} 0 & 0 & -0.1 \\ 1 & 0 & 0 \\ 0 & 1 & 0 \end{pmatrix}, \ C = \begin{bmatrix} c_1 & c_2 & c_3 \end{bmatrix} \tag{27}$$

$$f(x_m^1(k)) = \begin{bmatrix} -x_{m2}^1(k)^2 & 0 & 0 \end{bmatrix}^T \tag{28}$$

$$f(x_s^1(k)) = \begin{bmatrix} -x_{s2}^1(k)^2 & 0 & 0 \end{bmatrix}^T \tag{29}$$

$$Q(x_m^1(k), x_s^1(k)) = \begin{pmatrix} 0 & -(x_{m2}^1(k) + x_{s2}^1(k)) & 0 \\ 0 & 0 & 0 \\ 0 & 0 & 0 \end{pmatrix} \tag{30}$$

and

$$A_a((x_m^1(k), x_s^1(k)) = \begin{pmatrix} -L_1^1c_1 & -L_1^1c_2 - x_{m2}^1(k) - x_{s2}^1(k)) & -L_1^1c_3 \\ 1 - L_2^1c_1 & -L_2^1c_2 & -L_2^1c_3 \\ -L_3^1c_1 & 1 - L_3^1c_2 & -L_3^1c_3 \end{pmatrix} \tag{31}$$

The characteristic matrix $A_a((x_m^1(k), x_s^1(k))$ is in the Benrejeb arrow form matrix [21,24,39,43], if L_2^1, L_3^1, c_2 and c_3 are chosen such that

$$\begin{cases} L_2^1c_3 = 0 \\ 1 - L_3^1c_2 = 0 \end{cases} \tag{32}$$

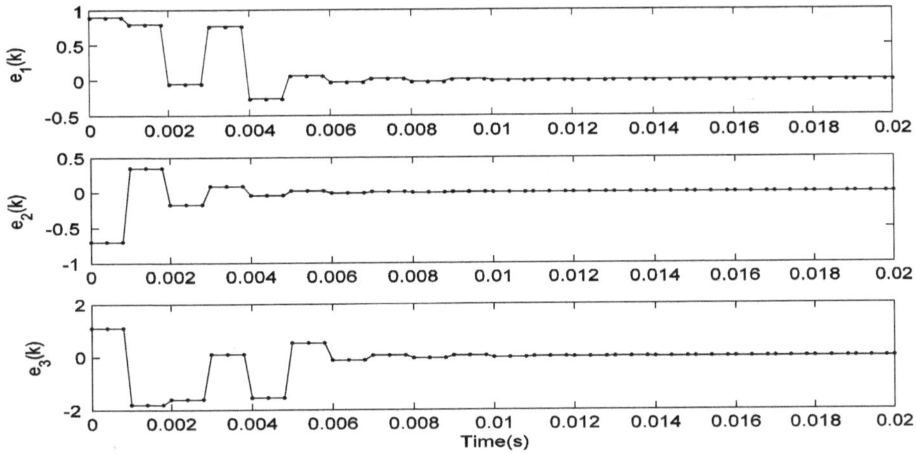

Figure 3. Dynamic errors of the two coupled 3D Hénon maps after the activation of the control law.

3.3.2. Synchronization study

For: $L_2^1 = 1, L_3^1 = 2,\ c_2 = 0.5,\ c_3 = 0$, it comes to the error system (7) with (8), (25)–(32) such that

$$I - M(A_a((x_m^1(k), x_s^1(k)) = \begin{pmatrix} 1 - |L_1^1 c_1| & -|L_1^1 c_2 + x_{m2}(k) + x_{s2}(k)| & -|0.1 - L_1^1 c_3| \\ -|1 - L_2^1 c_1| & 1 - |L_2^1 c_2| & 0 \\ -|L_3^1 c_1| & 0 & 1 - |L_3^1 c_3| \end{pmatrix} \quad (33)$$

By the application of the practical Borne and Gentina stability criterion, (33) is asymptotically stable, if the following conditions are satisfied

- $1 - |L_3^1 c_3| > 0$,
- $1 - |L_2^1 c_2| > 0$,
- $$\begin{aligned} \det(I - M(A_a(x_{m2}(k), x_{s2}(k)))) &= (1 - |L_1^1 c_1|)(1 - |L_2^1 c_2|)(1 - |L_3^1 c_3|) \\ &\quad - |L_1^1 c_2 + x_{m2}(k) + x_{s2}(k)||1 - L_2^1 c_1|(1 - |L_3^1 c_3|) \\ &\quad - |0.1 - L_1^1 c_3||L_3^1 c_1|(1 - |L_2^1 c_2|) > 0. \end{aligned}$$

For: $L_1^1 = 0$ and $c_1 = 1$, it comes $L = \begin{bmatrix} 0 & 1 & 2 \end{bmatrix}^T$ and $C = \begin{bmatrix} 1 & -0.5 & 0 \end{bmatrix}$. The simulation results of studied 3D coupled hyperchaotic Hénon maps are presented in the next section.

3.3.3. Simulation results

The synchronization between the two considered hyperchaotic 3D Hénon maps is achieved, as shown in Figure 3, in a short time 0.01s, for the following initial values of master and slave systems, respectively, $(x_{m1}^1(0), x_{m2}^1(0), x_{m3}^1(0)) = (0.5, 0.3, 0.09)$ and $(x_{s1}^1(0), x_{s2}^1(0), x_{s3}^1(0)) = (-0.4, 1, -1)$ and for the sampling time $T = 1$ms.

4. Results and discussion

Based on the above results of the considered discrete-time chaotic synchronization, the encryption and decryption-proposed approaches are presented. As an application, ECG encryption and decryption results are illustrated and performance analysis is given to show the effectiveness of the proposed secure communication under additive white gaussian noise and brute force attack.

4.1. Encryption and decryption proposed approaches

The proposed encryption algorithm for ECG signals is based on an exponential function, which is commonly used to secure sensitive data, including those involved in online transactions and computer communication. In contrast to XOR encryption, which is relatively weak and is readily cracked by an attacker with access to the ciphertext and knowledge of the key, the exponential encryption scheme is robust. It can be helpful for simple data obfuscation or for providing more protection to a more powerful encryption technique, but it is not appropriate for protecting sensitive information.

The proposed encryption approach encrypts an original signal $m(k)$ with an encryption function $en(k)$ based on the use of two discrete-time chaotic systems. The first 4D Hénon chaotic system generates the first encryption key $K_{m1}(k)$ and the second 3D Hénon chaotic system generates a second encryption key $K_{m2}(k)$.

The encryption function $en(k)$ is such that

$$en(k) = \exp(K_{m2}(k) + m(k)) + K_{m1}(k) \tag{34}$$

with

$$K_{m1}(k) = \sqrt{x_{m1}^2(k) + x_{m2}^2(k) + x_{m3}^2(k) + x_{m4}^2(k)} \tag{35}$$

$$K_{m2}(k) = x_{m1}^1(k) + x_{m2}^1(k) + x_{m3}^1(k) \tag{36}$$

To decrypt the encrypted signal, let $c(k)$ the encrypted signal obtained after its transmission and $d(k)$ the estimated decrypted signal

then

$$d(k) = \log(c(k) - K_{s1}(k)) - K_{s2}(k) \tag{37}$$

with

$$K_{s1}(k) = \sqrt{x_{s1}^2(k) + x_{s2}^2(k) + x_{s3}^2 + x_{s4}^2(k)} \tag{38}$$

$$K_{s2}(k) = x_{s1}^1(k) + x_{s2}^1(k) + x_{s3}^1(k) \tag{39}$$

The proposed encryption and decryption approaches are used to ensure the encryption and decryption of an ECG signal.

4.2. ECG encryption and decryption

Like personal data, biometric information, that is collected to allow identification systems, must be shielded from unwanted intrusion. The electrocardiogram (ECG) contains private health information along with details for personal identification.

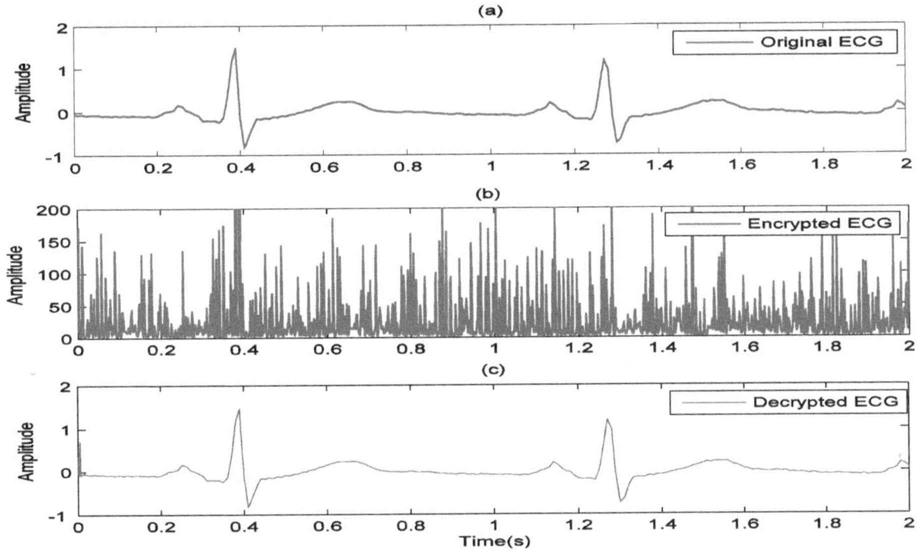

Figure 4. ECG encryption and decryption results without noise: (a) original ECG, (b) encrypted ECG, (c) decrypted ECG.

From the point of view of confidentiality, ECG signals need to be encrypted before transmission through public channels [1,3722,29].

In this work, the design of the chaos-based communication system is proposed to secure ECG signal transmission based on the proposed approach of discrete time chaotic synchronization and encryption and decryption algorithms. An example of an encrypted and decrypted ECG signal is shown in Figure 4; the decrypted ECG is identical to the original one, after 0.01s which consists of the synchronization time between respectively the two discrete-time chaotic systems; therefore, the ECG signal presents a small loss during the transient time.

The encryption and decryption algorithms are implemented using the MATLAB software. It is important to note that the proposed algorithm requires a relatively high computational complexity due to the use of chaotic systems, which can limit its practical implementation on low-power devices or in real-time systems. Therefore, when implementing the algorithm in practice, the required computational resources should be considered, and optimization techniques, such as parallelization or hardware acceleration, may need to be employed to achieve real-time performance. Overall, the proposed algorithm is a promising approach for ECG encryption, but its practical implementation requires careful consideration of the computational resources, chaotic systems synchronization and signal processing techniques to achieve high security and low distortion.

4.3. Key space analysis

An ideal encryption scheme should have a large key space to make brute force attacks infeasible. A brute force attack typically means a brute force search of the key space, testing all possible keys to recover the plain text. It was shown, in recent cryptographic systems, that a key space of size 2^{128} is required to resist brute force attacks.

In this work, the proposed cryptosystem is based on two discrete Hénon 4D and Hénon 3D hyperchaotic systems. Thus, the initial conditions can be used as the encryption key which is presented in 32 bits. The key length is 224 bits. Therefore, the key space is 2^{224}. In fact, the key space of mixed chaotic systems 2^{224} implies that the keyspace has a satisfactory security level against brute-force attacks.

5. Security analysis

5.1. Quality metrics analysis

In this section, the quality metrics used in signal processing such as Mean-Squared Error (MSE), Peak Signal-to-Noise Ratio (PSNR), and Percentage Root mean-square difference (PRD) are computed to measure the difference between original and encrypted ECG signal.

- **Mean Square Error (MSE)** [2] is computed by averaging the squared intensity of the original signal $x(k)$ and the resultant encrypted ECG signal $y(k)$ is defined by

$$\mathrm{MSE} = \frac{1}{N}\sum_{i=1}^{N}(x_i - y_i)^2 \tag{40}$$

  In this case, the MSE should be as high as possible for encryption resistance to attacks.
- **Peak Signal-to-Noise Ratio (PSNR)** [2] is a mathematical measure of signal quality based on the difference between two signals. It calculates the ratio between the maximum possible value of a signal and the power of distorting noise. If these signals are highly similar the PSNR has a large value. In fact, higher values of PSNR indicate the important quality of the used algorithm. In the encryption algorithm, the PSNR measure is an estimate of the quality of the encrypted signal compared with the original signal, it must be low since they are different. The PSNR is defined as

$$\mathrm{PSNR} = 20\log_{10}\left(\frac{\mathrm{MAX}}{\mathrm{MSE}}\right) \tag{41}$$

  with MAX is the maximum original signal value.
- **Percentage Root mean-square Difference (PRD)** [40] is used to measure the distortion of the decrypted signal and to show the robustness of the encrypted signal compared to the original one. The PRD value is defined by

$$\mathrm{PRD} = 100 \times \sqrt{\frac{\sum_{i=1}^{N}(x_i - \tilde{x}_i)^2}{\sum_{i=1}^{N}(\tilde{x}_i)}} \tag{42}$$

Table 1 shows the MSE, the PSNR and the PRD between the original and encrypted ECG signals and between original and decrypted ones.

The results present in one hand large values of MSE and PRD and small values of PSNR between the original and encrypted ECG signal and on the other hand, small values of MSE,

Table 1. Quality metric analysis.

Signal of figure	MSE	PSNR	PRD
4(a) and 4(b)	2.1843×10^3	−28.6826	1.9418×10^4
4(a) and 4(c)	1.3625×10^{-4}	43.3671	4.8499

Table 2. Correlation coefficients.

Signal	Coefficient
4(a) and 4(b)	0.1918
4(a) and 4(c)	0.9988

PRD and large values of PSNR between the original and decrypted ECG signal. Therefore, the proposed encryption algorithm produces an efficient and uniform encrypted signal and a sufficient decrypted one.

5.2. Statistical analysis [42]

5.2.1. Correlation analysis

The correlation is a measure of the intensity of the relation between two or several variables. If the coefficient of correlation is equal to zero, the two signals are totally different. If it's equal to 1, it means that the two signals are equal. The correlation coefficient r is given by

$$r = \frac{\sum_{i=1}^{N} (x_i - \bar{x}_i)(y_i - \bar{y}_i)}{\sqrt{\sum_{i=1}^{N} (x_i - \bar{x}_i)^2}\sqrt{\sum_{i=1}^{N} (y_i - \bar{y}_i)^2}} \tag{43}$$

Table 2 gives the coefficients of correlation between the original and the decrypted ECG signals and between the original and the encrypted ECG signals. The correlation coefficient obtained for encrypted and decrypted ECG signal, as shown in Table 2, is close to 1. Consequently, the obtained decrypted signal is similar to the original one.

5.2.2. ECG histogram characterization

The histogram allows for studying the distribution of a variable. It may be used, in particular, in signal processing. The histogram of the encrypted signal, as shown in Figure 5(b) is uniform, different from the histogram of the original signal, as shown in Figure 5(a). It doesn't give a statistical similarity to the original ECG signal.

5.3. Influence of additive noise on system performances

Let's consider the performance of the proposed secure communication system under the Additive White Gaussian Noise (AWGN) channel. Figure 6 shows that for different chosen

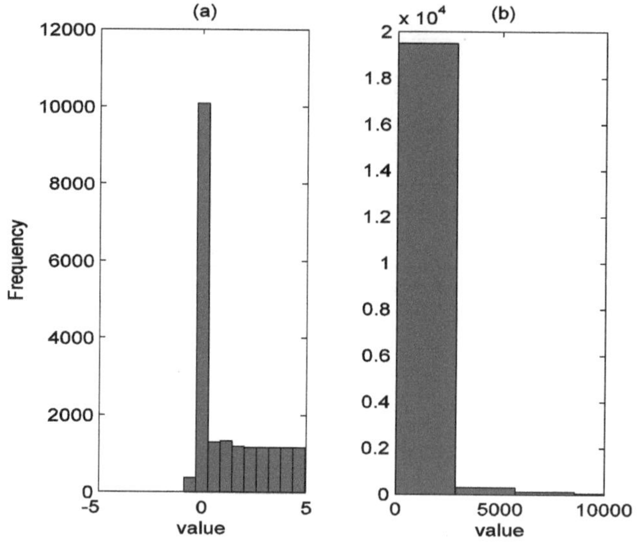

Figure 5. Histograms: (a) Original signal. (b) Encrypted signal.

values of Signal to Noise Ratio (SNR), the recovered ECG signal at the receiver side depends on the noise within the transmission channel.

In fact, the transmitted signal is recovered for SNR = 25db, 35db and 65db, as shown in figure (5b),(5c) and (5d) while, it's corrupted beyond this value 10db, as shown in Figure 5(a).

6. Comparison analysis

In this section, the advantages and disadvantages of the proposed scheme are presented based on comparitive studies with other encryption schemes presented in the literature based on chaos where clinical signals such as ECG are used as plain signals. In Table 3, a comparison is given including encryption characteristics and achieved security analysis. As it is shown in Table 3, the most implemented chaotic encryption method is based on the use of just one chaotic map like the logistic map, Multiscroll map and Lorenz 3D. In contrast with the literature, in this work, the proposed encryption algorithm is based on the use of two hyper-chaotic Hénon maps 3D and 4D to enhance the security level and to increase the key space.

In addition, many research works are based on symmetric encryption where the encryption and the decryption keys are identical and have the same initial conditions.

In fact, the proposed approach based on chaos synchronization is robust against various attacks. Since the chaotic dynamics used for encryption are highly sensitive to initial conditions, even a small change in the initial conditions can lead to a completely different sequence of random numbers. This makes it extremely difficult for an attacker to intercept and decrypt the ciphertext, as any small error or disturbance in the system can result in the loss of synchronization and errors in decryption.

The most disadvantage of the presented work is the loss of small quantity of information due to the transient time between master and slave hyperchaotic systems.

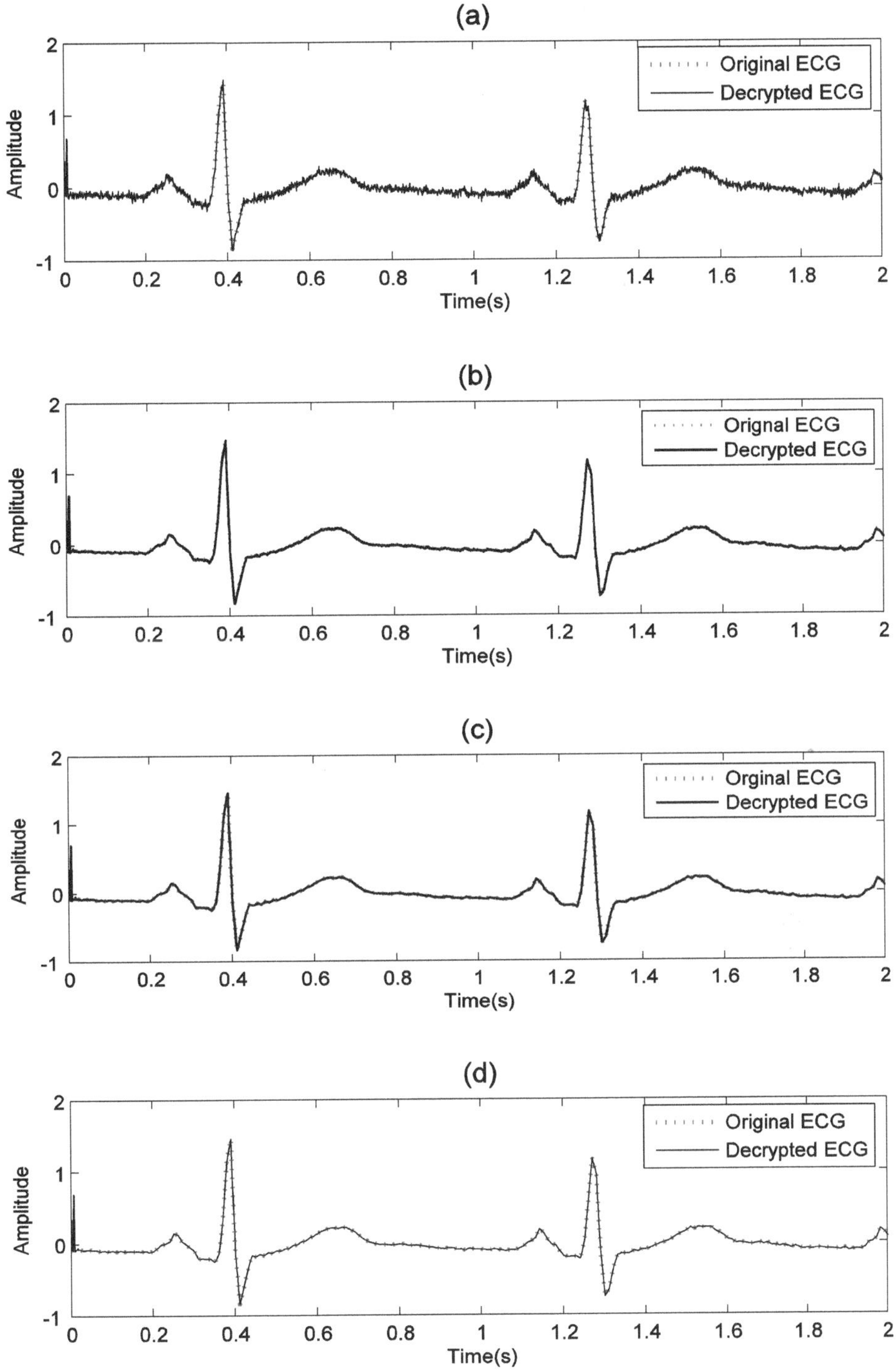

Figure 6. Recovered ECG signal under AWGN channel: (a) SNR = 10 db, (b) SNR = 25 db, (c) SNR = 35 db, (d) SNR = 65 db.

Table 3. Comparison between the proposed method and existing encryption scheme.

Clinical signals	Ref [36]	Ref [32]	Ref [10]	Proposed method
ECG	✓	✓	✓	✓
EEG	–	✓	–	–
BP	–	✓	–	–
Acquisition method				
Live sensing	–	–	Self-developed handheld ECG acquisition device	–
Data Base	Normal Sinus Rhythm Database	PhysioBank ATM	–	PhysioBank ATM
Technique				
Encryption	Based on chaos	Based on chaos	Based chaos synchronization	Based chaos synchronization
Compression	–	–	–	–
Chaos specifications				
Chaotic map	Multi-scroll map	Logistic map	Lorenz 3D	Mixed Hénon 3D and 4D maps
Lyapunov exponent	–	✓	–	–
Security analysis				
Secret keyspace	✓	✓	–	✓
Histogram	–	✓	–	✓
Correlation	–	✓	–	✓
Autocorrelation	–	✓	–	–
Secret key sensitivity	–	✓	–	–
Plain signal sensitivity	–	✓	–	–
Floating frequency	–	✓	–	–
Information entropy	–	✓	–	–
Chosen plain text attack	–	✓	–	–
Power spectral density	–	–	–	–
Mean square error MSE		✓	–	✓
Pick signal-to-noise ratio PSNR	✓	✓	–	✓
Percentage root-mean-square difference (PRD)	✓	–	–	✓
Structural similarity index SSIM	–	✓	–	–
Encryption time	–	✓	–	–
Pseudorandomness analysis				
NIST 800−22 suit	–	✓	–	–
Comparison analysis	✓	✓	–	✓

7. Conclusion

In this paper, a useful design of a secure communication system based on mixed discrete-time chaotic systems and chaotic synchronization is proposed. The used synchronization method, which is based on the Benrejeb arrow form matrix and the Borne and Gentina stability criterion, provides a robust and secure approach to ensure the synchronization between the transmitter and the receiver.

As an application, this scheme is tested, with success, to recover an electrocardiogram signal. Simulation results show the effectiveness of the proposed chaos-based communication system against some statistical attacks and brute force attacks.

Disclosure statement

No potential conflict of interest was reported by the author(s).

References

[1] A.D. Algarni, N.F. Soliman, H.A. Abdallah, and F.E. Abd El-Samie, *Encryption of ECG signals for telemedicine applications*. Multimed Tools Appl. 80(7) (2021), pp. 10679–10703.
[2] Y.A.Y. Al-Najjar and D.C. Soong, *Comparison of image quality assessment: PSNR, HVS, SSIM, UIQI*. Int. J. Sci. Eng. Res. 3(8) (2012), pp. 1–5.
[3] G. Alvarez and S. Li International, *Some basic cryptographic requirements for chaos based cryptosystems*. J. Bifurcation Chaos 16 (2006), pp. 2129–2151.
[4] M. Benrejeb, Sur l'analyse et la synthèse de processus complexes hiérarchisés: application aux systèmes singulièrement perturbés, Université des Sciences et Technologie de Lille 1, Thèse de Doctorat d'Etat es Sciences Physiques, Lille, 14 mai, 1980.
[5] M. Benrejeb, *Stability study of two level hierarchical nonlinear systems*, Plenary Lecture 12th IFAC Symposium on Large Scale Systems: Theory and Applications, Lille, 2010.
[6] M. Benrejeb, P. Borne, and F. Laurent, *Sur une application de la représentation en flèche à l'analyse des processus*. RAIRO Autom. 16(2) (1982), pp. 133–146.
[7] M. Benrejeb and M. Gasmi, *On the use of an arrow form matrix for modeling and stability analysis of singularly perturbed nonlinear* systems. Syst. Anal. Mod. Simul. 40 (2001), pp. 509.
[8] M. Benrejeb and S. Hammami, *New approach of stabilization of nonlinear continuous monovariable processes characterized by an arrow form matrix*, First International Conference on System Engineering Design Applications, SENDA, Monastir, 2008.
[9] P. Borne, *Nonlinear system stability. Vector norm approach*. Syst. Control Encycl. 5 (1987), pp. 3402–3406.
[10] C.-K. Chen, C.-L. Lin, S.-L. Lin, and C.-T. Chiang, *Data encryption and transmission based on personal ECG signals*. Sens. Netw. Data Commun. 4(2) (2015), pp. 1–13.
[11] S. Cheng, J. Sun, and C. Xu, *A color image encryption scheme based on a hybrid cascaded chaotic system*. Int. J. Bifurcat. Chaos 31(9) (2021), pp. 2150125.
[12] K.M. Cuomo, A.V. Oppenheim, and S.H. Strogatz, *Synchronization of Lorenz-based chaotic circuits with applications to communications*. IEEE Trans. Circuits Syst. 40(10) (1993), pp. 626–633.
[13] H. Dedieu, M.P. Kennedy, and M. Hasler, *Chaos shift keying: modulation and demodulation of a chaotic carrier using self-synchronizing Chua's circuits*. IEEE Trans. Circuits Syst. 40 (1993), pp. 634–642.
[14] R.L. Filali, M. Benrejeb, and P. Borne, *On observer-based secure communication design using discrete-time hyperchaotic systems*. Commun. Nonlinear Sci. Numer. Simul. 19 (2014), pp. 1424–1432.
[15] R.L. Filali, S. Hammami, M. Benrejeb, and P. Borne, *Synchronization of discrete-time hyperchaotic maps based on aggregation technique for encryption*, 9th International Multiconference on Systems, Signals and Devices, Chemnitz, 2012.
[16] L.J. Guo and X.Y. Geng, *Chaos communication based on synchronization of discrete-time chaotic systems*. Chin. Phys. 14 (2005), pp. 274–278.
[17] M.F. Hassan and M. Hammuda, *A new approach for constrained chaos synchronization with application to secure data communication*. J. Franklin. Inst. 356(12) (2019), pp. 6697–6723.

[18] H. He, Y. Wu, B. Zhang, D. Zhang, Y. Tian, K. Wang, H. Meng, and M. Hou, *Discrete chaotic synchronization and secure communication design*, International Conference on Natural Computation, Tianjian, 2009, pp. 473–476.

[19] G.Z. Hu, *Global synchronization for coupled Lur'e dynamical networks*. Circ. Sys. Sign. Processing 32 (2013), pp. 2851–2866.

[20] Y.-Y. Huang, Y.-H. Wang, and Y. Zhang, *Shape synchronization of drive response for a class of two-dimensional chaotic systems via continuous controllers*. Nonlinear Dyn. 78(4) (2014), pp. 2331–2340.

[21] N. Kahlifa, R.L. Filali, and M. Benrejeb, *On secure image transmission combining chaotic encryption and watermarking using dead beat synchronization of 4D Henon maps*, 3rd International Conference on Control, Engineering and Information Technologies, Tlemcen, 2015.

[22] G. Kenfack and A. Tiedu, *Secured transmission of ECG signals: Numerical and electronic simulations*. J. Sign. Inform. Proc. 4(2) (2013), pp. 158–169.

[23] N. Khalifa and M. Benrejeb, *On nonidentical discrete-time hyperchaotic systems synchronization: towards secure medical image transmission*, in *Emerging Methodologies and Applications in Modelling, Recent Advances in Chaotic Systems and Synchronization*, Academic Press, 2019, pp. 329–349.

[24] N. Khalifa, R.L. Filali, and M. Benrejeb, *A fast selective image encryption using discrete wavelet transform and chaotic systems synchronization*. Inf. Technol. Control 45 (2016), pp. 235–242.

[25] O.M. Kwon, J.H. Park, and S.M. Lee, *Secure communication based on chaotic synchronization via interval time-varying delay feedback control*. Nonlinear Dyn. 63 (2011), pp. 239–252.

[26] C. Li and G. Chen, *Synchronization in general complex dynamical networks with coupling delays*. Phys. A. 343 (2004), pp. 263–278.

[27] R.G. Li and H.N. Wu, *Secure communication on fractional-order chaotic systems via adaptive sliding mode control with teaching-learning-feedback-based optimization*. Nonlinear Dyn. 95(2) (2019), pp. 1221–1243.

[28] N. Li, Z. Wu, X. Lin, X. Tang, G. Xia, and T. Deng, *Bidirectional digital image secure transmission and recognition based on a long-distance chaos synchronization system with machine learning capability*. Nonlinear Dyn. 104 (2021), pp. 2745–2758.

[29] C.-H. Lin, J.-X. Wu, P.-Y. Chen, C.-M. Li, N.-S. Pai, and C.-L. Kuo, *Symmetric cryptography with a chaotic map and a multilayer machine learning network for physiological signal infosecurity: Case study in electrocardiogram*. IEEE Access 9 (2021), pp. 26451–26467.

[30] Y. Manai and M. Benrejeb, *New condition of stabilization of uncertain continuous Takagi-Sugeno fuzzy system based on fuzzy Lyapunov function*. Int. J. Intell. Sys. Appl. 4(4) (2012), pp. 19–25.

[31] S.-S. Mohanrasu, K. Udhayakumar, T.-M.-C. Priyanka, A. Gowrisankar, S. Banerjee, and R. Rakkiyappan, *Event-triggered impulsive controller design for synchronization of delayed chaotic neural networks and its fractal reconstruction: An application to image encryption*. Appl. Math. Model. 115 (2023), pp. 490–512.

[32] M.A. Murillo Escobar, L. Cardoza-Avendaño, R.M. Lopez-Gutiérrez, and C. Cruz-Hernandez, *A double chaotic layer encryption algorithm for clinical signals in telemedicine*. J. Med. Syst. 4 (2017), pp. 41–59.

[33] G. Narayanan, G. Muhiuddin, M.S. Ali, A.A.Z. Diab, J.F. Al-Amri, and H.I. Abdul-Ghaffar, *Impulsive synchronization control mechanism for fractional-order complex-valued reaction-diffusion systems with sampled-data control: Its application to image encryption*. IEEE Access 10 (2022), pp. 83620–83635.

[34] L.M. Pecora and T.L. Carroll, *Synchronization in chaotic systems*. Int. J. Comput. Cogn. Phy. Rev. Lett. 64 (1990), pp. 821–824.

[35] Q. Qian, Z. Chen, and Z. Yuan, *Video compression and encryption based-on multiple chaotic systems*. Int. J. Innov. Comp. 6 (2010), pp. 315–322.

[36] F. Sufi, F. Han, I. Khalil, and J. Hu, *A chaos based encryption technique to protect ECG packets for time critical telecardiology applications*. Secur. Commun. Netw. 4 (2010), pp. 515–524.

[37] B. Vaseghi, S. Mobayen, S.S. Hashemi, and A. Fekih, *Fast reaching finite time synchronization approach for chaotic systems with application in medical image encryption*. IEEE Access 9 (2021), pp. 25911–25925.

[38] X.-Y. Wang, N. Guan, H. Zhao, S. Wang, and Y. Zhang, *A new image encryption scheme based on coupling map lattices with mixed multi-chaos*. Sci. Rep. 10 (2020), pp. 1–15.
[39] W. Wang, H. Huang, C. Xie, and L. Han, *CBSNTS: A chaotic based security network transmission system*. J. Netw. 9 (2014), pp. 1985–1992.
[40] X. Wang and Q. Yu, *A block encryption algorithm based on dynamic sequences of multiple chaotic systems*. Commun. Nonlinear Sci. Numer. Simul. 14 (2009), pp. 574–581.
[41] B. Wang, X. Zhang, and X. Dong, *Novel secure communication based on chaos synchronization*. IEICE Trans. Fund Electron. Commun. Comp. Sci. E101(7) (2018), pp. 1132–1135.
[42] T. Xiang, K. Wong, and X. Liao, *An improved chaotic cryptosystem with external key*. Commun. Nonlinear Sci. Numer. Simul. 13 (2008), pp. 1879–1887.
[43] T. Yang, *A survey of chaotic secure communication systems*. Int. J. Comput. Cogn. 2 (2004), pp. 81–130.
[44] T. Yang and L.O. Chua, *Secure communication via chaotic parameter modulation*. IEEE Trans. Circ. Sys. 43 (1996), pp. 817–819.
[45] L. Zhou, F. Tan, X. Li, and L. Zhou, *A fixed-time synchronization-based secure communication scheme for two-layer hybrid coupled networks*. Neurocomputing 433 (2021), pp. 131–141.

Appendices

Appendix 1: Borne and Gentina practical stability criterion

Let consider the nonlinear discrete-time system described, in state space, by

$$z(k+1) = A(.)z(k),\ z \in R^n$$

$A = \{a_{ij}\}$ and the overvaluing matrix such as: $M(A(.)) = \{|a_{ij}(.)|\}, \forall i, j = 1, \ldots, n$.

If the nonlinear elements are isolated in only one row of $M(A(.))$, the verification of the Kotelyanski condition enables to conclude to the asymptotic stability of the original system characterized by $A(.)$. Kotelyanski lemma applied to the overvaluing matrix, obtained by the use of the regular vector norm

$$p(z(k)) = [|z_1(k)| \ldots |z_n(k)|]^T z(k) = [|z_1(k)| \ldots |z_n(k)|]^T$$

leads to the following sufficient conditions of asymptotic stability of the original system.

$$(I_n - M(A(.))) \begin{pmatrix} 1 & 2 & \cdots & i \\ 1 & 2 & \cdots & i \end{pmatrix} > 0, \forall i = 1 \ldots n$$

Appendix 2. Benrejeb arrow form matrix

The matrix $A = \{a_{ii}\}$ is on Benrejeb arrow form, if its non null elements are isolated in the diagonal, the last column and the last row

$$A = \begin{pmatrix} a_{11} & 0 & 0 & 0 & a_{1,n} \\ 0 & \ddots & 0 & 0 & \vdots \\ 0 & 0 & \ddots & 0 & \\ 0 & 0 & 0 & a_{n-1,n-1} & a_{n-1,n} \\ a_{n1} & \cdots & \cdots & a_{n-1,n} & a_{nn} \end{pmatrix}$$

If $a_{ii} \neq 0 \forall i = 1, 2, \ldots, n-1$, then it comes

$$\det(A) = \left[a_{nn} - \sum_{i=1}^{n-1} \frac{a_{ni}a_{in}}{a_{ii}} \right] \prod_{i=1}^{n-1} a_{ii}$$

On the use of chaotic dynamics for mobile network design and analysis: towards a trace data generator

Martin Rosalie and Serge Chaumette

ABSTRACT
With the constant increase of the number of autonomous vehicles and connected objects, tools to understand and reproduce their mobility models are required. We focus on chaotic dynamics and review their applications in the design of mobility models. We also provide a review of the nonlinear tools used to characterize mobility models, as it can be found in the literature. Finally, we propose a method to generate traces for a given scenario involving moving people, using tools from the nonlinear analysis domain usually dedicated to topological analysis of chaotic attractors.

1. Introduction

The number of applications that use autonomous devices, for instance robots or unmanned aerial vehicles (UAVs), increases nowadays. In this context, defining and analysing their mobility is particularly important. A mobility model describes the behaviour of an entity considering its capacities, possible moves and speed. The mobility models are described either analytically at the individual level, or by the interactions between the parts of the system (between UAVs, UAVs and planes, UAVs and points to survey, etc.). The resulting behaviours described with these simple rules can induce the emergence of a global intelligent behaviour. Inversely, from the resulting behaviour of such a swarm, these initial simple rules are hard to discover. A similar phenomenon occurs for chaotic dynamics where a chaotic process appears to be random while it arises from a deterministic process. Therefore, the idea is to study the connections between the two concepts.

Literature on *chaotic dynamics* and *nonlinear dynamics* has been developed at the end of the century while the main concepts of this theory come from the early nineties (see Fig. 1 of [1]). Chaos is observed, described and analysed in numerous domains, for instance: electronic circuits [29], chemical reactions [47], laser behaviours [30] or biological models [62]. The first example of the use of chaos (from the Chua system [9]) to design a mobility model has been proposed in 1997 [49]. The authors proposed to use the properties of chaotic dynamics to ensure a good coverage of an area. Chaotic dynamics is defined as follows: the solution of a deterministic process is chaotic if it is sensitive to initial conditions, aperiodic and globally time invariant. These properties induce that this chaotic solution will be unpredictable when considering a long term behaviour. Nowadays, several tools improve the understanding of a chaotic behaviour, chaotic mechanism and bifurcation diagrams and some authors use them to provide mobility models. Because ' little is known of the potential relationship between swarm and chaotic systems' [21], the main goal of this paper is to explore the uses of chaotic dynamics in the domain of mobility models. This can be done either by designing chaotic deterministic models or by using tools coming from nonlinear analysis.

In this paper, we first present a review of mobile networks including chaotic dynamics using chaotic maps or ordinary differential equations (ODEs). We then review the tools used to analyse these systems. In Section 3, we propose a method to generate traces using nonlinear analysis tools. This method is supported by numerical simulations using the Lorenz system and permits to reproduce congestion and distribution patterns. We finally present the conclusion along with our future work in Section 4.

2. Mobility models and chaotic dynamics

In this section, we present models from the literature including chaotic dynamics and the tools used to analyse them. In these models, chaos is obtained from well-known discrete or continuous systems. The chaotic variables of these systems are used to design mobility models. Sections 2.1 and 2.2 detail the two main approaches used to generate chaotic dynamics for mobility models using: a chaotic attractor from a discrete system or a chaotic attractor solution of a continuous system. Section 2.3 is dedicated to the tools from nonlinear analysis and their applications. The last section (Section 2.4) details our previous contributions in this domain in which we proposed to use periodic orbits of chaotic dynamics to define mobility models of UAVs in order to enhance the coverage of an area.

2.1. Models using chaotic maps

Chaotic maps are nonlinear recurrent relations. The most used in the literature is the *logistic map* defined by $x_{n+1} = \alpha x_n(1 - x_n)$. Introduced by Verhulst [58] it represents the growth of population x at each step where α is the growth rate. This map contains only one nonlinear term and can exhibit both periodic and chaotic dynamics. When varying α from 2 to 2.4, its bifurcation diagram results in a period doubling cascade: a classical route to chaos found in several systems. For details on the logistic map, the reader is referred to [5]. Charrier et al. [7] propose to use the logistic map to reproduce the behaviour of flocks. In the original paper [39], Reynolds introduced *boids* to reproduce the flocking behaviour with three rules (collision avoidance, velocity matching and flock centring) that generates a force vector for each agent in the swarm. In the Charrier et al. model the synchronization between agents is performed by the environment: the control parameter of the logistic map is updated depending on the neighbourhood of each agent. This model does not use the standard rules of flocking but reproduces their behaviour. The authors use a *bifurcation diagram* to emphasize the convergence of the system and to show the global dynamics of their agents and the transition to chaotic dynamics.

A chaotic map (the standard map [24]) can also be used to produce chaotic motion for a robot [28]. From such a two-dimensional map, a planning is assigned to the robot using two coordinate points obtained from the chaotic map. Then a robot visits these points in their order of appearance. As the purpose of the robot is to cover a square surface, the authors evaluate their system using coverage rate. Curia et al. [11] proposed to use another map, the Hénon map, to move a mobile robot with unpredictable trajectories. Their mobility model combines a guiding line that the robot follows with a chaotic motion obtained from the map. While the robot follows the guideline, the chaotic motion controls the evolution of the robot around this line. Here again, the authors use a bifurcation diagram to underline the chaotic properties of their system. Even if their system is made of six equations (including the Hénon map), they prove that it has a chaotic solution when $a = 1.4$ as it is the case for the Hénon map. The additional equations dedicated to the movement around the line do not influence the chaotic dynamics.

2.2. Models using ODE systems

In this section, we present models that use a set of ODEs as a source of chaotic dynamics. We are now considering chaotic continuous solutions instead of discrete solution from a map. However, there is a way to discretely represent these continuous chaotic dynamics. These solutions are embedded in the *phase space*, for instance the Lorenz attractor [26] is a famous attractor in a three-dimensional phase space. Introduced by Poincaré [37]; the *Poincaré section* is a transversal surface of the flow that provides a discrete description of the chaotic dynamics obtained from continuous systems. The original idea is to consider only the discrete points when the flow crosses a Poincaré section instead of the whole trajectory in the phase space. The discrete sequence of points contains a synthesis of the dynamical properties of the attractors. Most of the references presented below use the chaotic dynamics from the chaotic attractor using a Poincaré section.

There are many examples of such systems where nonlinearity induces chaotic behaviour. Among the most studied one: Lorenz, Rössler, Chua systems; there are several articles about robots, the mobility models of which use a set of ODEs. For instance, Nakamura and Sekiguchi [34] use the Arnold equations to model the behaviour of a mobile robot with chaotic motion. They prove that the coverage of their chaotic robot is better than the coverage of a robot using a random walk. Further to this work, Bae et al. [3] proposed a 'chaotic UAV' with the Chua system, Arnold equations and Van Der Pol equations. They also introduce an obstacle avoidance method without decreasing the performance of the model in terms of coverage.

Fallahi and Leung [16] proposed a cooperative set of four mobile robots synchronized using Chen [8] and Lorenz [26] systems. They used one of the variables of these systems to define the movements of their robots. In their system, one of the robots is the master, and the others are synchronized with it. This system is efficient compared to unsynchronized robots or random walks in terms of coverage rate and travelled distances. Similarly, Mukhopadhyay and Leung [33] present synchronized robots with chaotic path planners. They add a *symbolic dynamic* description of their robots. The symbolic dynamic is a nonlinear analysis tool used to describe chaotic dynamics. Its purpose is to label a discrete trajectory obtained from a Poincaré section. Thus, it gives a symbol according to the dynamical aspect of the solution depending on the topological period of the system. At the end, the solution is no longer a variable but a sequence of symbols indicating dynamical aspects of the system studied. The authors use this sequence of symbols to evaluate the synchronization rate between their robots. We also would like to mention the work of Bezzo et al. [4] where synchronization of chaos is used to detect changes in the topology of a mobile robotic network. The authors study the motion of mobile agents through an unknown environment with obstacles (see also [54,55] for details on this method).

Volos and co-workers [59] proposed another use of chaos which consists in designing a path planning generator for autonomous mobile robots. From the double scroll chaotic circuit (Chua system [9]), they obtained a chaotic true random bit generator. They use it to define the path planning of the robots, i.e. the list and the order of the points the robots have to visit. The efficient coverage rate and the unpredictability of the robot trajectories are the main characteristics of this system. This system is similar to the system using chaotic map [28]. Comparing their coverage rate in terms of the number of planned points, the system designed by Volos and co-workers [59] is ten times better. From the same authors, similar results are obtained when the Arnold map is used to generate waypoints for path planning [12]. Finally, we would like to point out a recent work done by Pimentel-Romero and co-workers [36] using Poincaré sections of chaotic attractors as threshold to generate random numbers. They conclude that these particular random number generators (RNGs) using chaotic dynamics are efficient for generating random paths for autonomous mobile robots.

Another way to include chaos to support mobile robot mobility is presented by Rosyid et al. [48]. Their method does not use any well-known ordinary differential equations (ODE) system to drive the robots. Robots communicate using sound that all can hear. Each robot moves in a direction depending on the 'total amount' of sound received. This is modelled by an ODE that details how this system is synchronized because their relative positions influence the sound emitted and received. The authors use the coverage rate to compare their system to the previously presented methods for robots driven with Lorenz or Arnold equations to prove that their model has better performance. They also compute the *largest Lyapunov exponent* (LLE) to ensure that chaos occurs in their system. This value is a measure of the separation rate of two infinitely initially closed trajectories [35,60]. The LLE refers to the predictability of a system and is commonly used as an indicator of a chaotic behaviour. This is a metric approach that does not permit to distinguish chaotic solutions that have distinct structures in the state space. The reader is referred to [6,17,31] for details about chaotic mechanisms (e.g. folding mechanism or tearing mechanism). The 0–1 test can also be used as an indicator to distinguish chaotic dynamics from periodic one (the reader is referred to [19] for details). We also mention that diffusion coefficient can be computed for first return maps to measure the difference between the dynamics when a parameter is varied [20,23].

We recently proposed mobility models [44–46] using chaotic behaviour based on the Rössler system using Poincaré section and periodic orbits. In the next section, we will present nonlinear tools used to analyse mobility models. Then, in Section 2.4, we present our mobility models and the nonlinear tools used to build and analyse them.

2.3. Nonlinear analysis tools

We gave above examples of mobility models built from chaotic dynamics. Some of them have been analysed with tools coming from the domain of nonlinear analysis. In this section, we present studies carried out on mobility models using these nonlinear tools to understand and describe their behaviour.

In 2014, Timme and Casadiego wrote an article entitled 'Revealing networks from dynamics' [56]. This paper gives an overview of approaches considering collective nonlinear network dynamics, but few details are given about tools from the domain of nonlinear analysis that could be used when chaotic systems are identified. These tools are well introduced by Qu et al. [38] in their paper about emergence in swarming systems. The authors detail the emergence phenomenon and the following tools: Lyapunov exponents, attractor, recurrence plot, Poincaré section. They also extract *periodic orbits* from a Poincaré section. The 'periodic' term of periodic orbits refers to the state space and not to the time space (topological period). Periodic orbits are time invariant while the system evolves in a chaotic state (from initial condition, the solution evolves and successively visits the unstable periodic orbits). From a chaotic time series and with a Poincaré section, orbits can be extracted, and this acquisition is a preliminary step of the *topological characterization* [17]. For dissipative systems, the purpose of this method is to obtain the structure of the chaotic mechanism from a topological invariant (the linking number) computed between periodic orbits (the reader is referred to [18] for details).

Hazan et al. [22] opt for this approach because the aim of their work is to classify the behaviour of robots using periodic orbits. The authors explain that their method is not based on a metric (for instance the LLE) because it is 'highly sensitive to perturbations such as noise contamination' [22]. They use the topological characterization tool after building an embedding (they reconstruct a phase space from one variable) of the behaviour from the x-axis motion of the robot. They build a Poincaré section in order to extract periodic orbits and then describe the behaviour of the robots using the linking numbers between the orbits.

Das and co-authors [13,14] propose to use Lyapunov exponents to distinguish transitions in a multi-agent swarm system. The system is designed to solve an optimization problem where the agents have to reach a particular point. Computing Lyapunov exponents enables the authors to find the range of parameters where chaos occurs and where the system is no longer periodic. The authors propose an application of their system: each robot is an automatic fire extinguisher, and they have to reach a burning place. ' In any swarming dynamics, emergence of chaos is a very important situation to be dealt with' [13] and this is illustrated by the work of Wu and co-workers [61]. In the latter article, the authors use Lyapunov exponents to analyse their swarming system and concluded that the chaos in swarm model becomes weaker while the emergence becomes stronger. The Lyapunov exponents are also used to analyse a swarm model of self propelled particles (SPP) by Shiraishi and Aizawa [52,53]. This tool permits to understand the relations between the behaviour of the system and the number of agents: ' the Lyapunov exponents reflect the biological sensitivity hidden behind the motion of swarm [52].

Chaos in neuronal network is also studied with Lyapunov exponents [15] or more recently, using reconstructed attractors with time-delay coordinates and with a Poincaré section [25]. These latest tools are robust to well define the chaotic mechanism because the first return map to the Poincaré section with unimodal structure indicates that there is a stretching and folding mechanism: it is a signature of the classical 'horseshoe' mechanism (also known as *folding mechanism*). This type of chaotic mechanism is also present in the work of Sato and co-workers [50,51] as illustrated in Fig. 6 of [51]. They work on a multiagent system using reinforcement learning that is modelled with coupled differential equations. This is applied to game theory: Matching Pennies and Rock-Scissors-Paper games. The stretching and folding mechanisms describe the effect of mutual adaptation and memory loss with non-transitive structure for their system. This leads to Hamiltonian chaos if there is no memory loss and to a dissipative system where there is memory loss. The dissipative system exhibits limit cycles, intermittency and deterministic chaos. To study their system, they employ Lyapunov exponents, Poincaré section, bifurcation diagram and extract periodic orbits. These tools are also used to study languages and learning mechanisms where chaotic dynamics appears [32].

2.4. Mobility models based on periodic orbits

We recently proposed mobility models using chaotic behaviour based on the Rössler system [44–46]. These mobility models permit to enhance the coverage of an area compared to random mobility models. We used the *first return map* from a Poincaré section of a chaotic attractor solution of the Rössler system and considered the periodic orbits to build efficient mobility models in terms of coverage rate. The Rössler system [47] is given by the equations

$$\begin{cases} \dot{x} &= -y - z, \\ \dot{y} &= x + ay, \\ \dot{z} &= b + z(x - c), \end{cases} \tag{1}$$

and its Poincaré section is defined as follows:

$$P = \{(y_n, z_n) \mid x_n = 0, \dot{x}_n > 0\}. \tag{2}$$

These tools are used to obtain the topological structure of the Rössler attractor [42] (Figure 1) where $\rho_n \in [0; 1]$ is the normalized value of y_n in the Poincaré section (Figure 2). We introduce a new concept to provide trajectories for UAVs: the dynamics of the first return map enable us to obtain a local direction. The first return map is a step-by-step process used to update the direction of the UAVs based on a three symbols dynamic (L for left, R for right and A for ahead). The periodic orbits of an attractor are considered as its skeleton because they structure the dynamics of the system. From the first return map, we extract the periodic orbits of attractors to obtain these recurrent points often visited with the same order. Thus, our UAVs can follow these specific patterns that allow them to explore a wide area. We obtained straight lines and wide turns with respectively period one orbit (AAAA ... for ahead, ahead, ahead, ...) and period two orbit (ARARAR ... for ahead, right, ahead, ...) (Figure 2). The period one orbit leads to a straight forward line to enable exploration while the period two orbit enables the UAV to make large right turns to change its direction. The reader is referred to [46] for details about the periodic orbits of attractors for mobility models of UAVs. The increase in performance provided by these mobility models using first return maps indicates that they deserve further investigation.

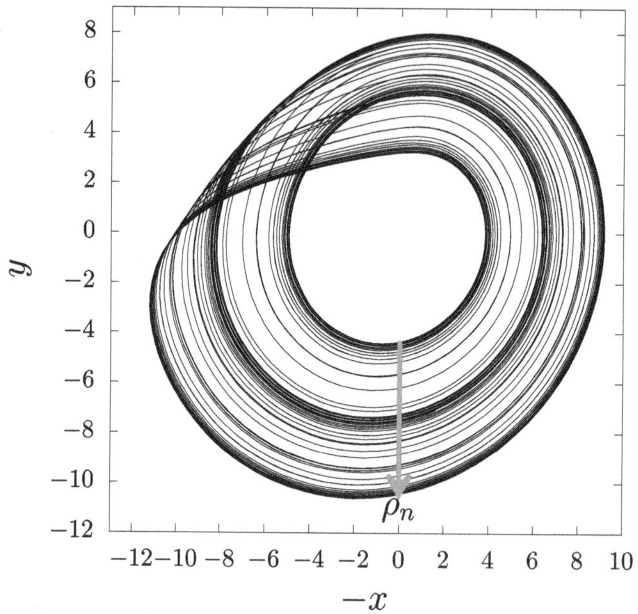

Figure 1. Chaotic attractor solution of the Rössler system (1) (values of parameters $a = 0.1775$, $b = 0.215$ and $c = 5.995$) with the Poincaré section (2) represented by an arrow.

In the next section, we propose a method to generate traces for mobile agents using this specific tool from nonlinear analysis. We would like to answer the following question: from a global point of view, can first return maps be useful to produce and/or analyse traces of UAVs?

3. Traces generation from multi-components Poincaré section

In this section, we first present the concept of *partial first return map* as a tool to describe data with an unknown part. Conversely to the classical first return map, this partial first return map allows input and output which is a mandatory property to provide data traces. The main purpose of this article is to analyse and generate traces of agents in an open environment which means that the agents can be added or removed in the model. Then, Section 3.2 provides a scenario where a partial first return map can describe the behaviour of agents. In Section 3.3, we present a proof of concept supported by numerical experiments using a theoretical dynamical system: the Lorenz system [26].

3.1. Concept

The purpose of this section is to propose a methodology to generate traces of mobile entities named *agents* (robots, persons, UAVs, …) using the most accurate tools related to the chaotic behaviour. We consider that the agents move in a well-defined area (the environment). The agents can enter, move and exit this area after a while. We suppose that the behaviour of the agents is mainly induced by constraints of the environment that force them to follow certain paths. As a consequence, we can make an analogy between the

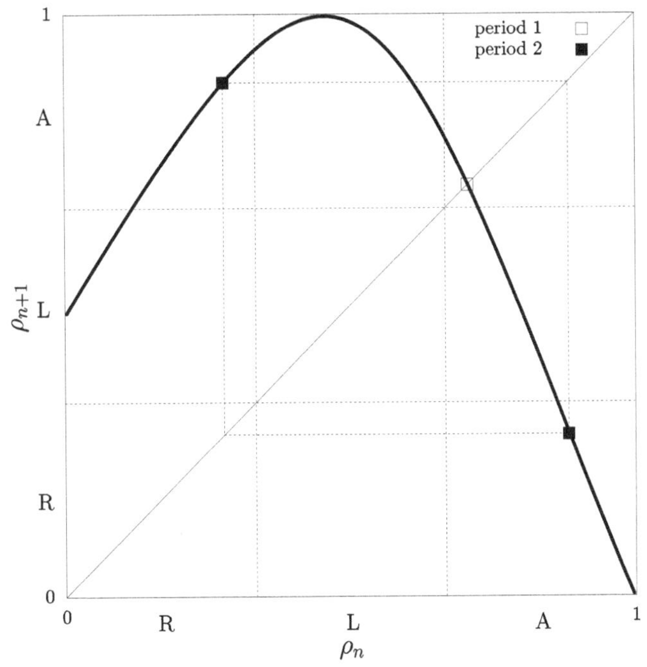

Figure 2. First return map to the Poincaré section of the Rössler attractor (Figure 1). This map is partitioned in three parts that give the UAV directions: L (left), A (ahead) and R (right). Orbits of period 1 and 2 illustrate patterns (AAAAA...) and (ARARA...), respectively straight lines and large turns. These patterns are efficient to cover an area with UAVs [46].

environment and the phase space of a deterministic dynamical system. We do consider that these paths are similar to unstable periodic orbits of an attractor in a phase space that agents might follow.

The classical method applied in this case is to reconstruct the whole phase space from one measured variable (for instance one coordinate of the position). Contrary to Hazan et al. [22], we do not consider that the traces contain the whole dynamical properties, but only a portion of it because of the capability of the considered robot to enter and exit the area. They use the standard way to reconstruct the phase space. For the same reason, we are not able to use even better global modelling methodology for time series data [27].

Based on our experiment in mobility models design [44–46], we can say that an approach using the orbits of an attractor as guidelines is very efficient in terms of coverage of an area. This efficiency is due to the patterns followed by the UAV during the exploration process. Such an approach including patterns repetition can be applied to the generation of traces of mobile agents. Figure 2 is the first return map of a Poincaré section of a Rössler attractor used in several research projects [44–46]. This first return map is an unimodal map made of an increasing branch and a decreasing branch: this illustrates the folding mechanism ('horseshoe' mechanism). However, this first return map details the whole chaotic dynamics of a bounded and globally time invariant process of a given chaotic attractor. Thus, the periodic points of Figure 2 describe the entire periodic orbits and considering the analogy with the mobility model, this prevents entrance or exit of agents. As we are aiming to obtain a trace data generator, this global perspective is a drawback to overcome, because the agents cannot be considered as permanently evolving in a dedicated well-defined environment. Therefore, we propose a new method using partial data from a chaotic attractor.

For attractors bounded by high genus torus, the Poincaré section is made of several components that can be used to properly describe chaotic dynamics using symbolic dynamics [42]. For instance, the multispiral chaotic attractor (Figure 3) introduced by Aziz-Alaoui [2] is bounded by a genus–5 torus. To describe the dynamics in a discrete way, the Poincaré section has to detail the transitions between the spirals. Consequently, the Poincaré section will be made of several *components*. These components are chosen accordingly to the bounding torus theory [57] using the fixed points. For the multispiral attractor (Figure 3) this theory indicates that four components are required to build the Poincaré section. The Poincaré section is no longer one plane but a set of planes. The flow crosses these planes and from a continuous flow we obtain discrete values. We used a concatenation of these values to build one variable representing the whole Poincaré section: ρ_n. Each component is represented by a range of values in ρ_n: component c_i for $\rho_n \in [i-1, i]$ (see Figure 3). Thus, ρ_n is a variable between 0 and 4 that describes the Poincaré section. The four components synthesized in ρ_n describe the entire dynamic of the system.

In a first return map to this Poincaré section, orbits can be extracted, and a symbolic dynamic can be assigned as it has been done for the Rössler attractor (see Figure 2 that details periodic points associated to periodic orbits). Figure 4 shows a first return map of a Poincaré section made of four components for the multispiral attractor (Figure 3) by plotting ρ_{n+1} versus ρ_n. This first return map contains both horseshoe mechanisms and tearing mechanisms. This map is a discrete description of the flow and permits to obtain information concerning possible transitions between components of the Poincaré section. The reader is referred to [42] for details on the methodology used to build a Poincaré section with several components using the properties of fixed points and the bounding torus theory. One of the way to build a bounding torus is to place the hole where the fixed point of

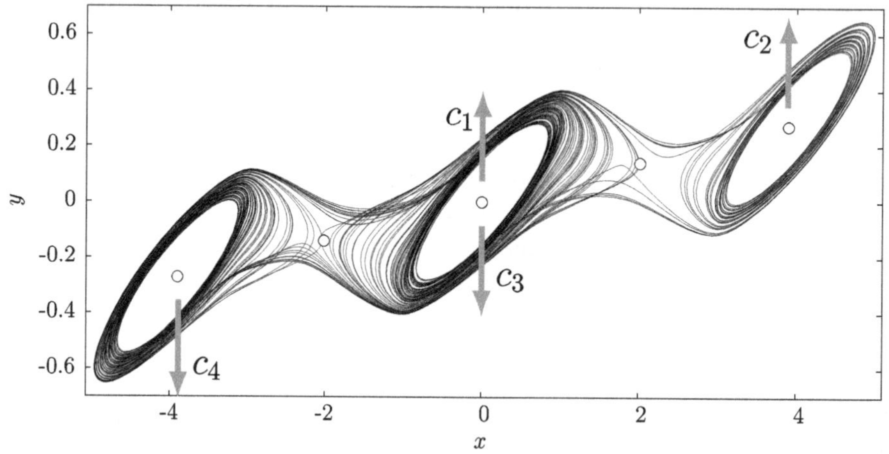

Figure 3. The multispiral attractor defined in [2] with the fixed points and the four components of the Poincaré section: c_1, c_2, c_3 and c_4 [42]. This attractor is bounded by a genus–5 torus (the five aligned holes of the bounding torus are the fixed points and indicated with a dot in a circle ⊙).

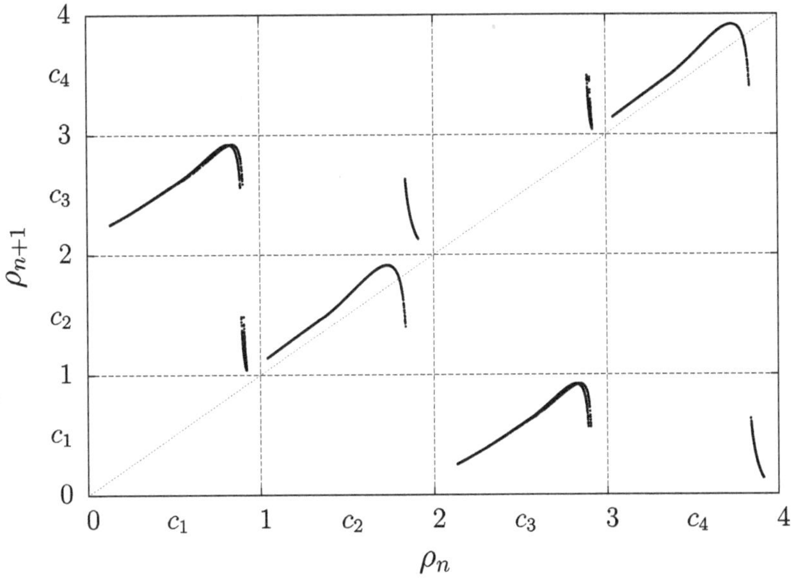

Figure 4. First return map of a Poincaré section describing a multispiral attractor (Figure 3) using a Poincaré section made of four components [42].

the attractor are. For the multispiral attractor there are details in [42] and for the Rössler system, there are details in [40] including parameters variation ensuring the robustness of the method.

In this multispiral attractor there are transitions from a component to the same component (e.g. c_2 to c_2) and transitions to another component (e.g. c_2 to c_3). The first return map (Figure 4) details the chaotic mechanism occurring to perform these transitions and describes the whole dynamics of the system. The novelty of our approach lies on the use of only a part of the data to introduce chaotic mechanisms as a model for traces generation. We develop a new tool to handle this unknown part of the data: the *partial first return map*. A partial return map is an incomplete map with at least one component with incoming flow, the *initial components*, and at least one component with outgoing flow, the *final components*. The flow is split in such a way that the partial first return map describes the transitions between the initial components and the final components using *transitional components*. With this repartition of components, we propose to follow a particular ordering to build the partial first return map: the initial components, the transitional components and the final components.

Our new concept is to consider only a part of the trajectory in the environment. Thus, a *partial first return map* is only an uncompleted first return map with transitions between a subset of components of a Poincaré section. This permits to represent experimental data without taking into account the rest of the trajectories. Consequently, this missing or unknown data can be considered as input and output for traces model indicating where agents enter or leave the environment. In the next section we first present a scenario for our methodology and in Section 3.3 we detail the method to build partial first return map.

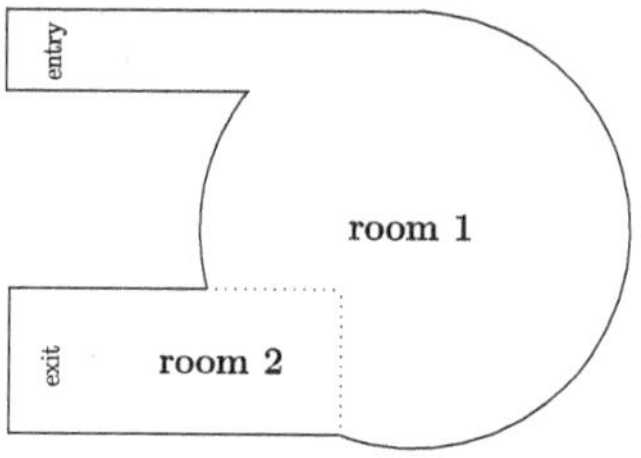

Figure 5. Scenario using an exhibition centre with two rooms. The shape is intentionally similar to the considered phase space (Figure 6) to underline the asset of our method but this is not mandatory: only transitions between components are significant.

3.2. Scenario

We consider the following scenario. The environment is an exhibition centre where we want to reproduce the behaviour of the visitors. We consider one entry door and one exit door. The exposition is composed of two rooms (Figure 5). The room 1 is accessible from the entry and the room 2 is before the exit. Visitors can stay in room 1 as long as they want before leaving the exhibition. As there are two rooms to visit in the exhibition centre, we consider the transition between these rooms. Thus, we have visitors coming from the entry door and going to the first room. There is a transition from the entry to room 1 and also from room 1 to room 2.

These transitions refer to the way chaotic multispiral attractors are analysed. As there are multiple spirals, and since the trajectory evolves from one spiral to another, it is required to consider these transitions. The Poincaré section is composed of several components to handle these multiples transitions. The analogy of transition between spirals is with transitions from one room to another room to generate traces. Moreover, this approach does not prevent to include more components to highlight patterns as it has been done to provide templates of attractors (see Fig. 15 of [42]). For instance, for a Malasoma attractor bounded by a genus one torus, a Poincaré section made of four components has been used to detail torsions or permutations movements [43]. This kind of mechanism can be used to reproduce the movements of the agents.

In this particular scenario, we consider some parts of our model as components of a Poincaré section: the entry and the exit as two components of a Poincaré section and the transition between room 1 and 2 is another component. To build the first return map, we set up the following order: the entry is an initial component, the transition between room 1 and 2 is a transitional component and the exit is a final component.

3.3. Numerical experimentation using the Lorenz system

We perform experimentations using the Lorenz system [26]

$$\begin{cases} \dot{x} = \sigma(y - x), \\ \dot{y} = Rx - y - xz, \\ \dot{z} = -\beta z + xy \end{cases} \tag{3}$$

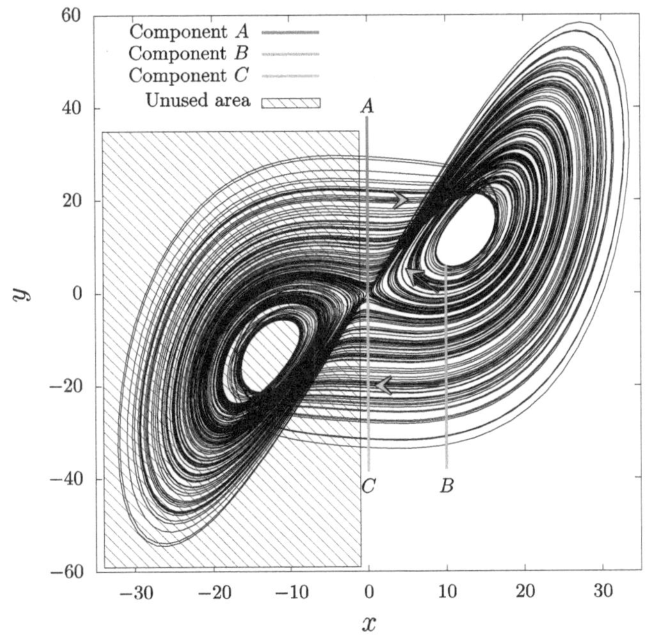

Figure 6. Phase portrait of an attractor solution to the Lorenz system (3) for the parameters values $R = 70$, $\beta = \frac{8}{3}$ and $\sigma = 10$. Arrows highlight the possible transitions from a component to another component (A to B, B to B or C).

in order to illustrate our methodology on a chaotic attractor bounded by a torus with a genus higher than one. This system is solved using a fourth-order Runge–Kutta method. We obtain an attractor solution to this system for the parameters values $R = 70$, $\beta = \frac{8}{3}$ and $\sigma = 10$. Figure 6 represents the projection of the attractor in the plane (x, y). The hatch part is not used to perform the experimentation of our method because we do not consider the whole solution. As a consequence, a partial first return map is obtained from the data (Figure 6) with a list of three components:

- initial component A: Entry
- transitional component B: Transition where an agent decides to stay in room 1 or to proceed to room 2
- final component C: Exit

These components are given by the following equations:

$$\begin{aligned}
\mathcal{P}_A &= \{(y_n, z_n) \mid x_n = 0, \dot{x}_n > 0\}, \\
\mathcal{P}_B &= \{(y_n, z_n) \mid x_n = 10, \dot{x}_n < 0\}, \\
\mathcal{P}_C &= \{(y_n, z_n) \mid x_n = 0, \dot{x}_n < 0\}.
\end{aligned} \tag{4}$$

The three components are represented Figure 6 with arrows showing the flow of the attractor between them. Here we choose arbitrarily 10 for our partition based on bounding torus theory, but we remind that fixed point of the differential equations system can also be used to ensure good partition according to this theory.

As we have done for attractors bounded by high genus torus [42], we build one variable ρ_n with different values to represent the position of the component in the partial first return map depending on its value:

- initial component A: $\rho_n < 1$;
- transitional component B: $1 < \rho_n < 2$ and
- final component C: $2 < \rho_n < 3$.

As introduced in one of our previous work [41], we choose to follow the orientation convention that gives the values of each component from the inside to the outside of the attractor. This convention is mandatory to compare chaotic mechanisms of attractors [41]. Thus, the ρ_n values close to and lower than 1, 2 and 3 are respectively associated to positions in the components close to the letters A, B and C (Figure 6).

We build the partial first return map (Figure 7) based on ρ_n to highlight the possible transitions between all components. It describes the transition between components $X \rightarrow Y$ with X the abscissa an Y the ordinate:

- The absence of points with ordinate value for the component A underlines the fact that it is an initial component. In this map, it is not possible to reach component A.

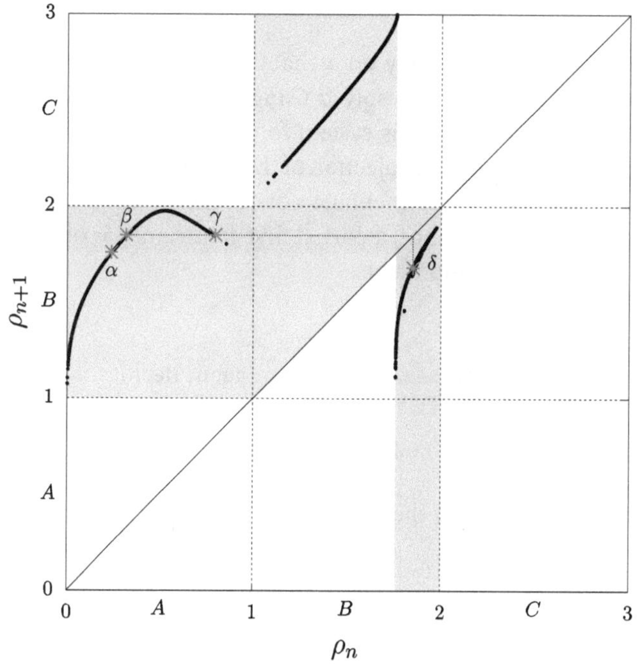

Figure 7. Partial first return map based on ρ_n illustrating the possible transitions between the components. Even if some parts of this figure are empty, we chose to show them to make the partial first return map mechanism understandable. Transitions from A to B are possible and generate congestion; for instance, β and γ have the same image (δ). Conversely, after the component B, there is a tearing mechanism that results in by the separation of closed points (the areas underline this split).

- $A \to B$ in the case (A, B) and $B \to B$ in the case (B, B) indicate that the next choice for the agent depends on the value of component B: stay in room 1 or proceed to room 2.
- $B \to C$ in the case (B, C) indicates that the agent will leave the room 1 to the room 2 and then leave the exhibition
- The absence of points with abscissa value for the component C underlines the fact that it is a final component. From component C there is no successor point.

From the dynamical point of view, even though we do not have the whole dynamical system, such kind of maps gives details on the possible chaotic mechanisms of the system. For instance, if we only consider the (A, B) area, it is an unimodal map with an increasing and a decreasing branch describing the classical 'horseshoe' mechanism (stretching and folding mechanisms). This is illustrated by the positions of the points β and γ in the partial first return map (Figure 7). These two points have the same ordinate but not the same abscissa. Using the map, we obtain only one image from these points: the point δ. In the range $[\beta; \gamma]$ there exist pairs of points in that interval for which this happens because there is only one image for two fibres. In terms of mobility model analysis, it means that even if two agents do not come from the same position in the entry (component A) they can reach the same position in the next component B. Such a mechanism illustrates congestion, i.e. the convergence of agents to the same point. The topological description of a chaotic attractor can be viewed as a series of mechanisms: stretching and folding are enough to generate chaos. Considering the flow of the attractor, a folding mechanism is responsible for gathering trajectories before a stretching mechanism. The congestion is the result of a folding mechanism (continuous map with several branches) because trajectory will collapse to the same area.

Now considering points with abscissa in component B, their ordinate are in components B and C. For an agent coming from component B, there is a split into two different places. The chaotic mechanism associated to such behaviour is the *tearing mechanism*, which is well known for Lorenz attractors (for details about mechanisms in the Lorenz system, see [6]). The coloured areas (Figure 7) underline such a split directly from component A where two points (agents) close to α will move to B but one will stay in component B (green area of Figure 7) while the other will proceed to component C (cyan area of Figure 7). This mechanism is similar in chaotic In the mobility model, it highlights the fact that agents can stay in the same room or proceed to the last room before leaving the exhibition.

Finally, we can use one dynamical system (e.g. Lorenz system) to generate the global pattern of the traces with respects to some conditions based on the transition between areas. This is a proof of concept validated by numerical simulations, a preliminary step before considering experimental data instead of a solution to a dynamical system. From data, we are expecting to find some patterns similar to tearing mechanism or folding mechanism by identify their dynamical signature in partial first return map generated from the data. Between the mechanisms of transitions in experimental real world data we will then look for a dynamical system with similar dynamical properties to use it as a trace generator. The dynamical system will not be directly extracted from the data with an algorithm. One differential equations system, with a set of parameters, has to be found in a database of research articles detailing topology of chaotic attractors (including the list of chaotic mechanisms). The additional search of an appropriate dynamical system including equations and parameters can be achieved using a method we developed providing templates of attractors directly from a bifurcation diagram [40]. This article illustrates the richness of non-equivalent chaotic dynamics that could be found in one dynamical system, and it also provides various chaotic mechanisms with their parameters. To complete, we have to mention that multiple attractors can provide same mechanisms. For instance, the multispiral attractor can be used instead of the Lorenz system because they share common chaotic mechanism. Once a system has been found, the traces can be obtained by solving the dynamical system with initial points (the mobile agent) in the initial components to let them evolve in the environment by visiting the other components and escape via the final components. We thus propose to find a model fitting the data by giving a similar partial first return map. This model could provide a good mobility model, even if there is a suppressed or hidden part that is not used. We consider agents as particles in a flow where the chaotic dynamics provide enough variability despite the deterministic process to be used as a mobility model with our methodology.

4. Conclusion

In systems containing mobile entities, we emphasize the fact that chaos can be built, as well as it can emerge, by synchronization of the entities or by interaction with the environment. In both cases, if a chaotic state is observed, then dedicated tools can be employed to analyse it. For instance, bifurcation diagrams can illustrate transition from limit cycles (periodic solution) to various type of chaotic dynamics. We have seen that, Lyapunov exponents are indicators to distinguish chaos from hyper-chaos. However, for chaotic dynamics, this measure is not well adapted to detail the chaotic mechanism. Thus, others tools (Poincaré section or periodic orbits for instance) leading to topological characterizations can separate non-equivalent chaos, and provide more accurate analysis of these chaotic dynamics. The recent improvements concerning comparison of chaotic attractors and topological characterization method can be used to identify or distinguish chaotic mechanisms and consequently identify and distinguish particular behaviours of dynamical mobile networks.

Consequently, we have focused on the dynamical structure exhibited by the transitions between components of a Poincaré section. These transitions are significant and reliable to design a mobility model with congestion or distribution of agents in a given area. The procedure is applied on a Lorenz system to highlight the advantages of our methodology. The exhibited structure can be used to both generate traces and analyse them. For instance, with the constant increase in the number of connected devices carried by users, there are several ways to collect real traces with their approval. It has been done for the students of the University Politehnica of Bucharest [10] where their social interactions have been studied. In our future work we will apply our methodology to analyse agents traces obtained by such measurements to reproduce and generate traces by finding the most appropriate dynamical system.

Acknowledgments

The authors also would like to thank the reviewers for their useful comments and remarks that have made it possible to improve this article.

Disclosure statement

No potential conflict of interest was reported by the authors.

ORCID

Martin Rosalie http://orcid.org/0000-0003-3676-120X
Serge Chaumette http://orcid.org/0000-0003-4425-2501

References

[1] L.A. Aguirre and C. Letellier, *Modeling nonlinear dynamics and chaos: A review*, Math. Probl. Eng.2009 (2009), pp. 1–35.
[2] M.A. Aziz-Alaoui, *Differential equations with multispiral attractors*, Int. J. Bifurcat. Chaos 9 (1999), pp. 1009–1039.
[3] Y. Bae, *Target searching method in the chaotic UAV*, in *The 23rd Digital Avionics Systems Conference*, IEEE, 2004.
[4] N. Bezzo, P.J.C. Davalos, F. Sorrentino, and R. Fierro, *Decentralized identification and control of networks of coupled mobile platforms through adaptive synchronization of chaos*, Phys. D: Nonlinear Phenom. 267 (2014), pp. 94–103.
[5] G. Boeing, *Visual analysis of nonlinear dynamical systems: Chaos, fractals, self-similarity and the limits of prediction*, Systems 4 (2016), p. 37.
[6] G. Byrne, R. Gilmore, and C. Letellier, *Distinguishing between folding and tearing mechanisms in strange attractors*, Phys. Rev. E 70 (2004), Article ID 056214.
[7] R. Charrier, C. Bourjot, and F. Charpillet, *A nonlinear multi-agent system designed for swarm intelligence: The logistic MAS*, in *First International Conference on Self-Adaptive and Self-Organizing Systems (SASO 2007)*, IEEE, July 2007.
[8] G. Chen and T. Ueta, *Yet another chaotic attractor*, Int. J. Bifurcat. Chaos 9 (1999), pp. 1465–1466.
[9] L. Chua, M. Komuro, and T. Matsumoto, *The double scroll family*, IEEE Trans. Circuits Syst. 33 (1986), pp. 1072–1118.
[10] R.I. Ciobanu and C. Dobre, *Crawdad dataset upb/hyccups (v. 2016-10-17)* (2016).
[11] D.I. Curiac and C. Volosencu, *A 2D chaotic path planning for mobile robots accomplishing boundary surveillance missions in adversarial conditions*, Commun. Nonlinear Sci. Numer. Simul. 19 (2014), pp. 3617–3627.
[12] D.I. Curiac and C. Volosencu, *Path planning algorithm based on arnold cat map for surveillance UAVs*, Def. Sci. J. 65 (2015), p. 483.
[13] S. Das, *Chaotic patterns in the discrete-time dynamics of social foraging swarms with attractant–repellent profiles: An analysis*, Nonlinear Dyn. 82 (2015), pp. 1399–1417.
[14] S. Das, D. Goswami, S. Chatterjee, and S. Mukherjee, *Stability and chaos analysis of a novel swarm dynamics with applications to multi-agent systems*, Eng. Appl. Artif. Intell. 30 (2014), pp. 189–198.
[15] J. Eser, P. Zheng, and J. Triesch, *Nonlinear dynamics analysis of a self-organizing recurrent neural network: Chaos waning*, PLoS ONE 9 (2014), Article ID e86962.
[16] K. Fallahi and H. Leung, *A cooperative mobile robot task assignment and coverage planning based on chaos synchronization*, Int. J. Bifurcat. Chaos 20 (2010), pp. 161–176.
[17] R. Gilmore, *Topological analysis of chaotic dynamical systems*, Rev. Mod. Phys. 70 (1998), p. 1455.
[18] R. Gilmore and M. Lefranc, The Topology of Chaos: Alice in Stretch and Squeezeland, John Wiley & Sons, 2012.
[19] G.A. Gottwald and I. Melbourne, *A new test for chaos in deterministic systems*, in *Proceedings of the Royal Society of London A: Mathematical, Physical and Engineering Sciences*.460.The Royal Society, London, 2004. pp. 603–611.

[20] S. Grossmann and H. Fujisaka, *Diffusion in discrete nonlinear dynamical systems*, Phys. Rev. A 26 (1982), pp. 1779–1782.
[21] J. Harvey, K. Merrick, and H.A. Abbass, *Application of chaos measures to a simplified boids flocking model*, Swarm Intell. 9 (2015), pp. 23–41.
[22] A. Hazan, F. Davesne, V. Vigneron, and H. Maaref, *Topological characterization of mobile robot behavior*, in *Intelligent Robots and Systems*, IEEE, Beijing, 2006, pp. 4157–4162.
[23] R. Klages and J.R. Dorfman, *Simple maps with fractal diffusion coefficients*, Phys. Rev. Lett. 74 (1995), pp. 387–390.
[24] A.J. Lichtenberg and M.A. Lieberman, *Regular and stochastic motion*, ZAMM J. Appl. Math. Mech.64 (1984), pp. 133–134.
[25] V.A. Likhoshvai, V.V. Kogai, S.I. Fadeev, and T.M. Khlebodarova, *Alternative splicing can lead to chaos*, J. Bioinform. Comput. Biol. 13 (2015), Article ID 1540003.
[26] E.N. Lorenz, *Deterministic nonperiodic flow*, J. Atmos. Sci. 20 (1963), pp. 130–141.
[27] S. Mangiarotti, L. Drapeau, and C. Letellier, *Two chaotic global models for cereal crops cycles observed from satellite in northern morocco*, Chaos 24 (2014), Article ID 023130.
[28] L. Martins-Filho and E. Macau, *Trajectory planning for surveillance missions of mobile robots*, in *Autonomous Robots and Agents*, Springer, Berlin, 2007, pp. 109–117.
[29] T. Matsumoto, L.O. Chua, and M. Komuro, *The double scroll*, IEEE Trans. Circuits Syst. 32 (1985), pp. 797–818.
[30] S. Meunier-Guttin-Cluzel, B. Maheu, and G. Gouesbet, *Combined approaches and characterizations of experimental chaotic attractors in thermal lensing*, Phys. D: Nonlinear Phenom. 58 (1992), pp. 423–440.
[31] G.B. Mindlin, X.J. Hou, H.G. Solari, R. Gilmore, and N.B. Tufillaro, *Classification of strange attractors by integers*, Phys. Rev. Lett. 64 (1990), p. 2350.
[32] W.G. Mitchener and M.A. Nowak, *Chaos and language*, Proc. R. Soc. Lond. B 271 (2004), pp. 701–704.
[33] S. Mukhopadhyay and H. Leung, *Cluster synchronization of predator prey robots*, in *Systems, Man, and Cybernetics*, IEEE Conference, Manchester, UK, 2013, pp. 2753–2758.
[34] Y. Nakamura and A. Sekiguchi, *The chaotic mobile robot*, Robot. Autom. 17 (2001), pp. 898–904.
[35] E. Ott, Chaos in Dynamical Systems, Cambridge University Press, 2002.
[36] C.H. Pimentel-Romero, J.M. Munoz-Pacheco, O. Felix-Beltran, L.C. Gomez-Pavon, and C.K. Volos, *Chaotic planning paths generators by using performance surfaces*, in *Fractional Order Control and Synchronization of Chaotic Systems*, Vol. 688, Springer International Publishing, Berlin, 2017, pp. 805–832.
[37] H. Poincaré, Les Méthodes Nouvelles De La Mécanique Céleste: Méthodes De MM. Newcomb, Glydén, Lindstedt Et Bohlin. 1893, Vol. 2, Gauthier-Villars it fils, Paris, 1893.
[38] D. Qu, H. Qu, and Y. Liu, *Emergence in swarming pervasive computing and chaos analysis*, in *Pervasive Computing and Applications*. IEEE Conference, Urumqi, China, 2006, pp. 83–89.
[39] C.W. Reynolds, *Flocks, herds, and schools: A distributed behavioral model*, in *SIGGRAPH '87 Conference Proceedings*, Vol. 21. Computer Graphics, ACM, Anaheim, 1987, pp. 25–34.
[40] M. Rosalie, *Templates and subtemplates of Rössler attractors from a bifurcation diagram*, J. Phys. A: Math. Theor. 49 (2016), Article ID 315101.
[41] M. Rosalie and C. Letellier, *Systematic template extraction from chaotic attractors: I. genus-one attractors with an inversion symmetry*, J. Phys. A: Math. Theor. 46 (2013), Article ID 375101.
[42] M. Rosalie and C. Letellier, *Toward a general procedure for extracting templates from chaotic attractors bounded by high genus torus*, Int. J. Bifurcat. Chaos 24 (2014), Article ID 1450045.
[43] M. Rosalie and C. Letellier, *Systematic template extraction from chaotic attractors: II. genus-one attractors with multiple unimodal folding mechanisms*, J. Phys. A: Math. Theor. 48 (2015), Article ID 235101.
[44] M. Rosalie, G. Danoy, S. Chaumette, and P. Bouvry, *From random process to chaotic behavior in swarms of UAVs*, in *Proc. ACM Symposium on Development and Analysis of Intelligent Vehicular Networks and Applications – DIVANet'16*. ACM, Malte, 2016.

[45] M. Rosalie, M.R. Brust, G. Danoy, S. Chaumette, and P. Bouvry, *Coverage optimization with connectivity preservation for UAV swarms applying chaotic dynamics*, in *Proc. of the IEEE International Conference on Autonomic Computing (ICAC)*, IEEE, 2017.

[46] M. Rosalie, G. Danoy, S. Chaumette, and P. Bouvry, *Chaos-enhanced mobility models for multilevel swarms of UAVs*, Swarm. Evol. Comput. 41 (2018), pp. 36–48.

[47] O.E. Rössler, *An equation for continuous chaos*, Phys. Lett. A 57 (1976), pp. 397–398.

[48] N.R. Rosyid, H. Kikichi, and P. Sooraksa, *Performance evaluation of non-embedded chaotic mobile robots based on minimum complex network*, ASEAN Eng. J. 63 (2012), pp. 65–75.

[49] N. Saiwaki, T. Komatsu, T. Yoshida, and S. Nishida, *Automatic generation of moving crowd using chaos model*, in *IEEE International Conference on Systems, Man, and Cybernetics. Computational Cybernetics and Simulation*, IEEE, 1997.

[50] Y. Sato and J.P. Crutchfield, *Coupled replicator equations for the dynamics of learning in multiagent systems*, Phys. Rev. E 67 (2003), Article ID 015206.

[51] Y. Sato, E. Akiyama, and J.P. Crutchfield, *Stability and diversity in collective adaptation*, Phys. D: Nonlinear Phenom. 210 (2005), pp. 21–57.

[52] M. Shiraishi and Y. Aizawa, *Lyapunov analysis of collective behaviors in self-propelled particle systems*. in *SICE Annual Conference*. IEEE Conference, Sapporo, Japan, 2014, pp. 866–871.

[53] M. Shiraishi and Y. Aizawa, *Collective patterns of swarm dynamics and the lyapunov analysis of individual behaviors*, J. Phys. Soc. Jpn. 84 (2015), Article ID 054002.

[54] F. Sorrentino and E. Ott, *Adaptive synchronization of dynamics on evolving complex networks*, Phys. Rev. Lett. 100 (2008), Article ID 114101.

[55] F. Sorrentino and E. Ott, *Using synchronism of chaos for adaptive learning of time-evolving network topology*, Phys. Rev. E 79 (2009), Article ID 016201.

[56] M. Timme and J. Casadiego, *Revealing networks from dynamics: An introduction*, J. Phys. A: Math. Theor. 47 (2014), Article ID 343001.

[57] T.D. Tsankov and R. Gilmore, *Strange attractors are classified by bounding tori*, Phys. Rev. Lett. 91 (2003), Article ID 134104.

[58] P.F. Verhulst, *Recherches mathématiques sur la loi d'accroissement de la population.*, Nouv. Mém. R. Sci.- 18 (1845), pp. 14–54. http://eudml.org/doc/182533.

[59] C.K. Volos, I.M. Kyprianidis, and I.N. Stouboulos, *A chaotic path planning generator for autonomous mobile robots*, Rob. Auton. Syst. 60 (2012), pp. 651–656.

[60] A. Wolf, J.B. Swift, H.L. Swinney, and J.A. Vastano, *Determining lyapunov exponents from a time series*, Phys. D: Nonlinear Phenom. 16 (1985), pp. 285–317.

[61] Y. Wu, J. Su, H. Tang, and H. Tianfield, *Analysis of the emergence in swarm model based on largest Lyapunov exponent*, Math. Probl. Eng. 2011 (2011). https://doi.org/10.1155/2011/745257.

[62] Y. Yao and W.J. Freeman, *Model of biological pattern recognition with spatially chaotic dynamics*, Neural. Netw. 3 (1990), pp. 153–170.

The SEIR Covid-19 model described by fractional-order difference equations: analysis and application with real data in Brazil

Abdullah Othman Almatroud, Noureddine Djenina, Adel Ouannas and Giuseppe Grassi

ABSTRACT

Several efforts have been recently devoted to the studies on epidemic mathematical models based on fractional-order operators, by virtue of their capability to take into account memory effects and nonlocal features. The aim of this paper is to make a contribution to the topic by introducing a novel Covid-19 model described by non-integer-order difference equations. By conducting a stability analysis, the paper shows that the conceived system has two fixed points at most, i.e. a disease-free fixed point and an endemic fixed point. In particular, a theorem is proved, which assures the global stability of the disease-free fixed point, indicating that the pandemic will disappear when a simple condition on the system parameters is satisfied. Finally, simulation results are carried out with the aim to highlight the capability of the conceived approach.

1. Introduction

The studies of epidemics date back to an Egyptian medical papyrus of herbal knowledge (about 1550 BC) [39]. Over the centuries, several studies have been conducted on epidemics, even though the introduction of mathematical models dates back only to the beginning of the twentieth century [24]. In recent years, a number of epidemic models have been proposed, including those for describing the Covid-19 pandemic [4, 6, 7, 15, 19, 20, 22, 42]. These compartment models, described by ordinary differential equations or difference equations, divide the communities into some classes (for example, susceptible, exposed, infected or removed) and analyse the development of the disease over time. Referring to the Covid-19 pandemic, it is of paramount importance to have knowledge in terms of the time of its peak and of its low points, with the aim to help decision-makers to implement effective actions using the best predictions and forecasts [18]. To this purpose, during the years 2020 and 2021, fractional-order epidemic models described by non-integer-order differential equations or non-integer-order difference equations have been introduced, with the aim to better describe the dynamics of the Covid-19 diffusion [5,

10, 11, 13, 14, 21, 23, 25, 28–36, 40, 41]. Namely, fractional operators play a fundamental role in describing the memory effect of several models associated with biological systems, thus providing more accurate descriptions of the system dynamics [14]. For example, in [25], a fractional-order Covid-19 epidemic model with Mittag–Leffler kernel is illustrated, whereas in [29], stability analyses of a Covid-19 model described by the non-integer Atangana–Baleanu–Caputo derivatives are conducted. In [11], a fractional compartmental model for studying the spread of the Covid-19 pandemic has been proposed. In particular, in [11], the conditions under which the disease-free and endemic equilibrium points are asymptotically stable have been discussed. In [32], the Atangana–Baleanu fractional derivative and the tuberculosis co-infection are exploited to describe the dynamics of the Covid-19 pandemic. In particular, stability analyses of the disease-free equilibrium points are carried out. In [23], a Covid-19 model with fractional stochastic behaviours is studied, whereas in [30], a non-integer order Covid-19 model, which incorporates the reinfection rate in the individuals recovered from the disease, is introduced and studied in detail. In [41], a stability analysis of a Covid-19 model pertaining to fractional operator is conducted. In particular, the stability properties of two different types of equilibrium points are accurately studied. In [31], a fractional-order Covid-19 model involving the Caputo derivative is introduced. Moreover, the stability of the steady-states is discussed, along with the existence and uniqueness of a non-negative solution. In [34], both the Caputo–Fabrizio derivative and the Atangana–Baleanu–Caputo derivative have been used for modelling the spread of the Covid-19 pandemic. Comparisons of these two epidemic modelling approaches have been carried out. Moreover, the local stability of the equilibrium points has been investigated [34]. By including the vaccine rate, in [28], a non-integer order Covid-19 model based on the Atangana–Baleanu derivative is proposed. In [36], the Caputo operator has been used to model the Covid-19 outbreak. The results in [36] have shown that the considered factional-order model provides a better fit to the real data with respect to the corresponding integer-order model. In [13], a fractional model for examining the consequences of adaptive immune responses to the Covid-19 viral mutation is proposed, with the aim to control the disease transmission. In [33], the impact of multiple vaccination strategies on the dynamics of a fractional Covid-19 model is analysed. The existence and uniqueness of the system solution is proved using Banach's fixed point theorem [33]. Moreover, the local and global asymptotic stability of the equilibria is studied. In [14], the dynamics of a novel Covid-19 pandemic model described by commensurate and incommensurate fractional derivatives have been investigated. The paper has analysed the stability of the equilibrium points and has shown a number of chaotic behaviours, obtained by varying the order of the derivative. In [35], a fractional-order Covid-19 model is presented, which enables to compute the number of infections via the chaotic contributions into susceptible, exposed, infectious and removed number of individuals. In [21], an eight-dimensional fractional-order dynamical system is used to model the Covid-19 pandemic. Some optimal control strategies are analysed, along with the existence of stable solutions in the system dynamics. In [40], a non-integer-order Covid-19 model is proposed, according to which the population is divided into susceptible, exposed, infectious, quarantined, recovered and insusceptible individuals. Some conditions are given in order to assure the local asymptotic stability of disease-free and endemic equilibrium points. In light of the above considerations, it must be emphasized that most of the studies mentioned above were constructed on the basis of the time-continuous mode. In fact, this motivated us to propose a new fractional-order discrete model for consideration in the field of epidemiology. Thus, from this perspective, the main objective of this work is to make a contribution to the field of epidemiology by proposing a new fractional-order discrete-time model. The aim of this paper is to make a further contribution to the topic of mathematical modelling of epidemics by introducing a novel Covid-19 model described by non-integer-order difference equations. This is due to the distinctive properties of fractional order operator, which give us more flexibility in modelling, which allows the model to be compatible with any region. The fractional operator is also distinguished by the memory effect, which allows the entry of all previous cases to calculate the next case. The paper shows that the conceived system has two fixed points at most, i.e. a disease-free fixed point and an endemic fixed point. A stability analysis of the equilibria is carried out, which enables a condition for the disappearance of the pandemic to be given. Since this result is rigorous (being proved by a theorem), it represents a remarkable finding of the proposed approach, which may help decision-makers to better understand the epidemiological behaviour of the disease over time. The manuscript is organized as follows. In Section 2, some basic notions on discrete fractional calculus are given. Moreover, a novel fractional Covid-19 model, based on the Caputo discrete operator, is introduced. In Section 3, the basis reproduction number is computed, whereas in Section 4, the stability of the fixed points is analysed in detail. In particular, a theorem is proved, which assures the global stability of the disease-free fixed point, indicating that the pandemic will disappear when a simple condition on the system parameters is satisfied. Finally, In Section 5, simulation results are carried out, with the aim to highlight the capability of the conceived approach.

2. Discrete fractional calculus and COVID-19 pandemic model

In recent times, after coronavirus spreading over the world, a lot of mathematicians have attempted to establish several mathematical models for the purpose of trying to analyse the broadcast of such epidemic. One of these Covid-19 models have been constructed in [39] as follow:

$$\begin{cases} \Delta S(n) = \Lambda - r_1 S(n)E(n) - r_2 S(n)I(n) - \mu S(n) + \tau R(n), \\ \Delta E(n) = r_1 S(n)E(n) + r_2 S(n)I(n) - (\mu + \rho)E(n), \\ \Delta I(n) = \rho E(n) - (\gamma + \mu)I(n), \\ \Delta R(n) = \gamma I(n) - (\mu + \tau)R(n), \end{cases} \qquad n \in \mathbb{N}, \tag{1}$$

where the model's description with its parameters can be outlined in Table 1.

Where $r_i = k_i p_i(1 - v)$, where k is the rate of contact, p is the probability of infection, $1-v$ is the effect of vaccination on the studied population.

The total population N can be gained by adding up all classes announced in model (1), i.e.

$$N = S + E + I + R. \tag{2}$$

During past years, the discrete fractional calculus has demonstrated high capability in describing lots of phenomena by illustrating their dynamics more realistic than that of classical calculus. This, actually, backs to its flexibility in approaching such phenomena better than before. Nevertheless, in the following content, we will recall some fundamental facts associated with discrete fractional calculus. To further obtain more notion insights,

Table 1. Description of the model reported in [39] with real data in Brazil.

Variable/Parameter	Description	Real data
$S(t)$	Susceptible class	17300532 [1]
$E(t)$	Exposed class	169017000 [1]
$I(t)$	Infected class	1027023 [1]
$R(t)$	Recovered class	30921318 [1]
Λ	Recruitment rate into susceptible population	2837456.349 [2]
r_1, r_2	Incidence rates	$4.3726 \times 10^{-11}, 1.3726 \times 10^{-11}$ [1]
μ	Natural mortality rate	0.014 [2]
τ	Relapse rate	0.009 [1]
ρ	Progression rate from exposed to infectious class	0.0005 [1]
γ	Treatment rate for infectious individuals	0.158 [1]

the reader may refer to the works [9, 16, 26, 27, 38]. In this research, we will assume that all functions defined below are defined on the set $\mathbb{N}_a := \{a, a+1, a+2, \ldots\}$, where $a \in \mathbb{R}$.

Definition 2.1 ([3]): The αth-fractional sum of the function $f : \mathbb{N}_a \to \mathbb{R}$ is outlined by:

$$\Delta_a^{-\alpha} f(t) = \frac{1}{\Gamma(\alpha)} \sum_{s=a}^{t-\alpha} (t-s-1)^{(\alpha-1)} f(s), \quad \text{for } t \in \mathbb{N}_{a+\alpha}, \tag{3}$$

where $\alpha > 0$, $\Gamma(\cdot)$ is the gamma function and $t^{(\alpha)} = \frac{\Gamma(t+1)}{\Gamma(t+1-\alpha)}$.

Definition 2.2 ([3]): The αth-order Riemann–Liouville fractional difference operator of the function f is outlined by:

$$\Delta_a^{\alpha} f(t) := \Delta \Delta_a^{-(1-\alpha)} f(t) = \frac{1}{\Gamma(1-\alpha)} \Delta \sum_{s=a}^{t-(1-\alpha)} (t-s-1)^{(-\alpha)} f(s), \quad \text{for } t \in \mathbb{N}_{a+1-\alpha}, \tag{4}$$

where $0 < \alpha < 1$.

Definition 2.3 ([3]): The αth-order Caputo fractional difference operator of the function f defined on $\mathbb{N}_a$ is outlined by:

$$^{C}\Delta_a^{\alpha} f(t) = \begin{cases} \Delta_a^{-(1-\alpha)} \Delta f(t) = \dfrac{1}{\Gamma(1-\alpha)} \\ \quad \times \displaystyle\sum_{s=a}^{t-(1-\alpha)} (t-s-1)^{(-\alpha)} \Delta f(s), & 0 < \alpha < 1 \\ \Delta f(t), & \alpha = 1 \end{cases} \quad \forall\, t \in \mathbb{N}_{a+1-\alpha}, \tag{5}$$

where $0 < \alpha \leq 1$.

In the light of the above notions, we intend to operate the Caputo fractional difference operator on system (1). This would yield the following Fractional-order Discrete System

(FoDS):

$$\begin{cases} {}^C\Delta_a^\alpha S(t+1-\alpha) = \Lambda - r_1 S(t)E(t) - r_2 S(t)I(t) - \mu S(t) + \tau R(t), \\ {}^C\Delta_a^\alpha E(t+1-\alpha) = r_1 S(t)E(t) + r_2 S(t)I(t) - (\mu+\rho)E(t), \\ {}^C\Delta_a^\alpha I(t+1-\alpha) = \rho E(t) - (\gamma+\mu)I(t), \\ {}^C\Delta_a^\alpha R(t+1-\alpha) = \gamma I(t) - (\mu+\tau)R(t), \end{cases} \tag{6}$$

where $0 < \alpha < 1$. For simplicity. When studying the spread of the epidemic in a short period of time (few days), the change in N is very small (because it relates to births and deaths), so we neglect it to facilitate the calculations, so we will consider N as a constant. As reported in [39], we can consider that time of our study to be relatively short. This, however, allows to set $R = N{-}S{-}E{-}I$, and consequently system (6) will be as:

$$\begin{cases} {}^C\Delta_a^\alpha S(t+1-\alpha) = \Lambda + \tau N - r_1 S(t)E(t) - r_2 S(t)I(t) \\ \quad - (\mu+\tau)\,S(t) - \tau E(t) - \tau I(t), \\ {}^C\Delta_a^\alpha E(t+1-\alpha) = r_1 S(t)E(t) + r_2 S(t)I(t) - (\mu+\rho)E(t), \\ {}^C\Delta_a^\alpha I(t+1-\alpha) = \rho E(t) - (\gamma+\mu)I(t), \end{cases} \tag{7}$$

subject to the following initial conditions:

$$S(0), E(0), I(0) \geq 0. \tag{8}$$

Remark 2.1: In [39], the presence of the fixed points was studied well, and at the same time the authors gave a sufficient condition for the disappearance of the disease as well as they also gave a sufficient condition for the positivity of S. However, the fixed point stability has not been comprehensively studied.

Next, we will endeavour to prove the positivity of the solution of system (7) in all its cases, and then we will provide a sufficient condition for the stability of the disease-free stationary fixed point. In other words, we intend to set a proper condition that can guarantee the disappearance of the disease (absence of I) and also guarantee the absence of E. This will be followed by studying a proposed model, which will be very close to the reality, in two cases; commensurate and incommensurate fractional-order cases. After these intentions, we will provide some important notes with the aim of illustrating the fundamental changes between the recent two systems. But before beginning, let us mention the next lemma which would be very useful later on.

Lemma 2.4 ([8]): *Let $\alpha > 0$ such that $\alpha \notin \mathbb{N}$. Suppose the function f is defined on $\mathbb{N}_a$. Then:*

$$f(t) = f(a) + \frac{1}{\Gamma(\alpha)} \sum_{r=a+1-\alpha}^{t-\alpha} (t-r-1)^{(\alpha-1)}\ {}^C\Delta_a^\alpha\ f(r), \quad \forall\, t \in \mathbb{N}_{a+1}. \tag{9}$$

Based on the previous discussion, it can be remarked that:

(1) If ${}^C\Delta_a^\alpha f(t) \geq 0$ then f is non-decreasing for all $t \in \mathbb{N}_a$.
(2) If ${}^C\Delta_a^\alpha f(t) \leq 0$ then f is non-increasing for all $t \in \mathbb{N}_a$.

In this regard and with using Lemma 2.4, we can confirm the following assertions:

$$\begin{cases} {}^{C}\Delta_a^{\alpha} S\big|_{S=0} = \Lambda + \tau N - \tau E(t) - \tau I(t) \geq 0, \\ {}^{C}\Delta_a^{\alpha} E\big|_{E=0} = r_2 S(t) I(t) \geq 0, \\ {}^{C}\Delta_a^{\alpha} I\big|_{I=0} = \rho E(t) \geq 0. \end{cases} \tag{10}$$

This can lead us to conclude that the solution $(S(t), E(t), I(t))^T$ of system (7) belongs to $\mathbb{R}_+^3$. On the other hand, to find the fixed points of such system, we need to put:

$$\begin{cases} \Lambda + \tau N - r_1 S(t)E(t) - r_2 S(t)I(t) - (\mu + \tau)\, S(t) - \tau E(t) - \tau I(t) = 0, \\ r_1 S(t)E(t) + r_2 S(t)I(t) - (\mu + \rho)E(t) = 0, \\ \rho E(t) - (\gamma + \mu)I(t) = 0. \end{cases} \tag{11}$$

This would assert, with some simple calculations, that this system will have two fixed points at most; the first one is called the disease-free fixed point $e_0 = (\frac{\Lambda+\tau N}{(\mu+\tau)}, 0, 0)$, and the second one, if it exists, is called the endemic fixed point.

3. Basic reproduction number

The basic reproduction number R_0 signifies to the average amount of the secondary cases produced from the preliminary infective individuals within a community without resistance to the pandemic [5, 10]. This vital measure has progressively being a fundamental amount utilized for identifying the interventions force required for dominating the pandemics. In epidemiological models, it is widely known that if $R_0 > 1$, then the disease will progressively broadcast in the susceptible community. In contrast, if $R_0 < 1$, then the susceptible community will never suffer from epidemics. The scheme of finding R_0 needs to determine the spectral radius σ of the next-generation matrix. In other words, $R_0 = \sigma(G)$, where G is the next-generation matrix, see [5, 10] to get a full overview about the calculation of such measure. Herein, we briefly intend to describe how R_0 can be calculated. To this aim, we consider the following two equations that are taken out from system (7):

$$\begin{cases} {}^{C}\Delta_a^{\alpha} E(t+1-\alpha) = r_1 S(t)E(t) + r_2 S(t)I(t) - (\mu + \rho)E(t), \\ {}^{C}\Delta_a^{\alpha} I(t+1-\alpha) = \rho E(t) - (\gamma + \mu)I(t). \end{cases} \tag{12}$$

In accordance with the above equations, we can determine the matrix J^* that can be established based on the leaving fluxes from the infected compartment and the newly fluxes to such compartment. That is:

$$J^* = \begin{pmatrix} r_1 \dfrac{\Lambda + \tau N}{(\mu + \tau)} - (\mu + \rho) & r_2 \dfrac{\Lambda + \tau N}{(\mu + \tau)} \\ \rho & -(\gamma + \mu) \end{pmatrix}. \tag{13}$$

Now, we can decompose the matrix J in terms of another two matrices F and V such that $J = F - V$. This, actually, yields:

$$F = \begin{pmatrix} r_1 \dfrac{\Lambda + \tau N}{(\mu + \tau)} & r_2 \dfrac{\Lambda + \tau N}{(\mu + \tau)} \\ 0 & 0 \end{pmatrix} \tag{14}$$

and

$$V = \begin{pmatrix} (\mu + \rho) & 0 \\ -\rho & (\gamma + \mu) \end{pmatrix}. \tag{15}$$

Consequently, we can generate the next generation matrix $G = FV^{-1}$ to be in the following form:

$$G = FV^{-1} = \begin{pmatrix} \frac{r_1}{(\tau + \mu)(\mu + \rho)}(\Lambda + N\tau) + \rho\frac{r_2}{(\tau + \mu)(\gamma + \mu)(\mu + \rho)}(\Lambda + N\tau) & \frac{r_2}{(\tau + \mu)(\gamma + \mu)}(\Lambda + N\tau) \\ 0 & 0 \end{pmatrix}. \tag{16}$$

Therefore, the basic reproduction number R_0 will then be given by:

$$R_0 = \frac{(\gamma r_1 + \mu r_1 + \rho r_2)(\Lambda + N\tau)}{(\mu + \rho)(\tau + \mu)(\gamma + \mu)}. \tag{17}$$

It is worth noting that if $R_0 > 1$, the system has an endemic fixed point of the form $e^* = (S^*, E^*, I^*)$, where:

$$\begin{aligned} S^* &= \frac{(\gamma + \mu)(\mu + \rho)}{\gamma r_1 + \mu r_1 + \rho r_2} \\ E^* &= \frac{(\gamma + \mu)}{\rho} I^* \\ I^* &= \frac{R_0 - 1}{\left(\frac{\tau(\gamma+\mu)+\tau\rho}{\rho(\Lambda+N\tau)}\right) R_0 + \frac{\rho r_2 + r_1(\gamma+\mu)}{\rho(\tau+\mu)}}. \end{aligned} \tag{18}$$

4. Stability analysis of the disease-free fixed point

In this section, we will concern with analysing the stability of the disease-free fixed point of the proposed system (7). This would be achieved by establishing some sufficient conditions for the parameters reported in such system. For this purpose, let us start our investigation by analysing the local stability of system (7) based on the next theorem.

Theorem 4.1 ([12, 37]): *Let $\alpha \in (0, 1)$. If*

$$\lambda \in \left\{ z \in \mathbb{C} : |z| < \left(2\cos\frac{|\arg z| - \pi}{2 - \alpha} \right)^{\alpha} \text{ and } |\arg z| > \frac{\alpha\pi}{2} \right\} \tag{19}$$

for all the eigenvalues λ of J, then the fixed point of system (7) is asymptotically stable, where J is the Jacobian matrix of the second member of the system at the fixed point.

In view of the above result, we state and prove the following stability result that would be very useful to proceed our consideration.

Theorem 4.2: *If* $R_0 < 1$, *then the disease-free fixed point of system* (7) *is locally asymptotically stable if*

$$\max\left\{(\mu+\tau),\ \gamma+2\mu+\rho-\frac{r_1}{\tau+\mu}(\Lambda+N\tau)\right\} < 2^{\alpha}. \tag{20}$$

Proof: In fact, the Jacobian matrix J of system (7) at the fixed point e_0 defined in (18) can be obtained to be as:

$$J(e_0) = \begin{pmatrix} -(\mu+\tau) & -r_1\dfrac{\Lambda+\tau N}{(\mu+\tau)}-\tau & -r_2\dfrac{\Lambda+\tau N}{(\mu+\tau)}-\tau \\ 0 & r_1\dfrac{\Lambda+\tau N}{(\mu+\tau)}-(\mu+\rho) & r_2\dfrac{\Lambda+\tau N}{(\mu+\tau)} \\ 0 & \rho & -(\gamma+\mu) \end{pmatrix}. \tag{21}$$

This would yield:

$$\det(\lambda Id - J(e_0)) = 0 \tag{22}$$

$\Leftrightarrow$

$$((\mu+\tau)+\lambda)\left(\lambda^2+B\lambda+C\right)=0, \tag{23}$$

where $B = (\gamma+2\mu+\rho-\frac{r_1}{\tau+\mu}(\Lambda+N\tau))$, $C = -\rho\frac{r_2}{\tau+\mu}(\Lambda+N\tau)+(\gamma+\mu)(\mu+\rho-\frac{r_1}{\tau+\mu}(\Lambda+N\tau))$. Consequently, we have:

$$(\mu+\tau)+\lambda = 0 \Rightarrow \lambda = -(\mu+\tau) \tag{24}$$

or

$$\left(\lambda^2+B\lambda+C\right)=0. \tag{25}$$

Now, if $R_0 < 1$, then we have:

$$\frac{(\gamma r_1+\mu r_1+\rho r_2)(\Lambda+N\tau)}{(\mu+\rho)(\tau+\mu)(\gamma+\mu)} < 1,$$

$\Leftrightarrow$

$$-B = -(\gamma+2\mu+\rho)+\frac{r_1(\Lambda+N\tau)}{(\tau+\mu)} < -\frac{(\rho r_2)(\Lambda+N\tau)}{(\gamma+\mu)(\tau+\mu)}-\gamma-\mu < 0,$$

and

$$C = -\rho\frac{r_2}{\tau+\mu}(\Lambda+N\tau)+(\gamma+\mu)\left(\mu+\rho-\frac{r_1}{\tau+\mu}(\Lambda+N\tau)\right) > 0,$$

$\Leftrightarrow$

$$B^2+4C > 0,$$

and hence the two roots λ_1 and λ_2 of

$$\left(\lambda^2+B\lambda+C\right)=0 \tag{26}$$

will be negative real numbers. Therefore, if $\lambda_1 + \lambda_2 = -B = -\gamma - 2\mu - \rho + \frac{r_1}{\tau+\mu}(\Lambda + N\tau) > -2^\alpha$, then we have:

$$-2^\alpha < \lambda_1 < 0$$
$$-2^\alpha < \lambda_2 < 0. \tag{27}$$

Hence, we can confirm, according to Theorem 4.1, that the disease-free fixed point is locally asymptotically stable if $R_0 < 1$. ∎

From another point of view, we intend in the following content to present a simple condition that can be employed to ensure the global stability of system (7) at the disease-free fixed point. This would be achieved by using the following theorem.

Theorem 4.3 ([17]): *If there exists a positive definite and decrescent scalar function $V(t, x)$ satisfying*

$$^C\Delta_{t_0}^\alpha V(t, x(t)) \leq 0,$$

for all $t_0 \in \mathbb{N}_a$, then the trivial solution of system (7) is uniformly stable.

Theorem 4.4: *If*

$$\frac{r_2 N}{(\gamma + \mu)} < \frac{(\mu + \rho) - r_1 N}{\rho}, \tag{28}$$

then the pandemic will disappear.

Proof: To prove this result, we intend to consider the Lyapunov function of the following form:

$$L(t) = E(t) + \theta I(t), \tag{29}$$

where $\frac{p_2 N}{(\gamma+\mu)} < \theta < \frac{(\mu+\rho)-p_1 N}{\rho}$. By differentiating L with respect to time, we obtain:

$$\begin{aligned}
^C\Delta_a^\alpha L(t+1-\alpha) &= ^C\Delta_a^\alpha E(t+1-\alpha) + \theta\, ^C\Delta_a^\alpha I(t+1-\alpha), \\
&= (r_1 S(t)E(t) + r_2 S(t)I(t) - (\mu+\rho)E(t)) + \theta\left(\rho E(t) - (\gamma+\mu)I(t)\right), \\
&= (r_1 S(t) - (\mu+\rho) + \theta\rho)\, E + (r_2 S(t) - \theta(\gamma+\mu))\, I, \\
&\leq (r_1 N - (\mu+\rho) + \theta\rho)\, E + (r_2 N - \theta(\gamma+\mu))\, I \\
&< 0.
\end{aligned} \tag{30}$$

Consequently, if condition (28) fulfilled, then $^C\Delta_a^\alpha L(t+1-\alpha)$ will be negative defined, and hence with the help of Theorem 4.3 we can confirm that the pandemic will disappear. ∎

Remark 4.1: It should be noted that each of the two fixed points and the basic reproduction number R_0 are not related to the fractional-order value α. This, actually, makes the fractional-order values of system (7) effective because they are just generalizations of integer-order values of the system and so they do not require to establish new conditions.

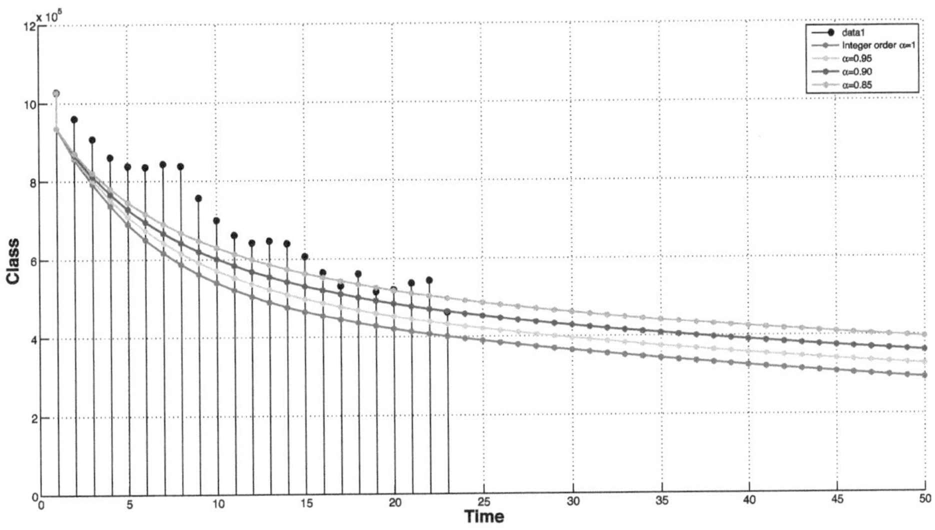

Figure 1. Numerical simulation of infected class and comparison with the real data.

5. Application in Brazil

In this section, we see the effectiveness of this system by applying it directly in Brazil, before that using the website described in [1], we get the number of active infections in Brazil in the period 23 July to 15 August 2022. We have

$$\frac{r_2 N}{(\gamma+\mu)} = 0.0182, \quad \text{and} \quad \frac{(\mu+\rho)-r_1 N}{\rho} = 0.2230.$$

So the condition of the Theorem 4.4 is achieved and the disease disappears with the passage of time.

Figure 1 shows a comparison between the real data and the results obtained using the integer-order model (with order 1) and the fractional-order model with different values of α

Remark 5.1: We note that changing the order allows the system to be improved and made more accurate.

Conclusions

This manuscript has introduced a novel Covid-19 pandemic model described by non-integer order difference equations. A stability analysis has been conducted, which has shown that the system has two fixed points at most, i.e. a disease-free fixed point and an endemic fixed point. In particular, a theorem has been proved, which has assured the global stability of the disease-free fixed point, indicating that the pandemic will disappear when a simple condition on the system parameters is satisfied. This rigorous result represents a remarkable finding of the proposed approach, which may help decision-makers to better understand the epidemiological behaviour of the disease over time. Finally, simulation results have been carried out (for the commensurate/incommensurate cases) with the aim to highlight the capability of the proposed model, which would help decision-makers to select measures for effectively monitoring and controlling the Covid-19 pandemic. In general, we can claim that the model presented in this work can be used to study the COVID-19 pandemic in many different parts of the world, even if this model formulated with the fractional-order backward operator in its commensurate or incommensurate cases.

Disclosure statement

No potential conflict of interest was reported by the author.

ORCID

Noureddine Djenina http://orcid.org/0000-0003-3492-1249

References

[1] Available online: https://www.worldometers.info (accessed on 19 August 2022).
[2] Available online: https://www.data.albankaldawli.org (accessed on 16 April 2023).
[3] T. Abdeljawad, On Riemann and Caputo fractional differences, Comput. Math. Appl. 62 (2011), pp. 1602–1611.
[4] A. Ajbar, R.T. Alqahtani, and M. Boumaza, Dynamics of an SIR-based COVID-19 model with linear incidence rate, nonlinear removal rate, and public awareness, Front. Phys. 9 (2021). DOI:10.3389/fphy.2021.634251.
[5] R. Albadarneh, I. Batiha, A. Ouannas, and S. Momani, Modeling COVID-19 pandemic outbreak using fractional-order systems, Int. J. Math. Comput. Sci. 16 (2021), pp. 1405–1421.
[6] L.J.S. Allen, Y. Lou, and A.L. Nevai, Spatial patterns in a discrete-time SIS patch model, J. Math. Biol. 58 (2009), pp. 339–375.
[7] L.J.S. Allen and P. Driessche, The basic reproduction number in some discrete-time epidemic models, J. Differ. Equ. Appl. 14 (2008), pp. 1127–1147.
[8] G.A. Anastassiou, Discrete fractional calculus and inequalities. 17 (2009), pp. 1–11. Available at arXiv:0911.3370v1.
[9] I.M. Batiha, A. Ouannas, and J. Emwas, A stabilization approach for a novel chaotic fractional-order discrete neural network, J. Math. Comput. Sci. 11 (2021), pp. 5514–5524.
[10] I.M. Batiha, S. Momani, A. Ouannas, Z. Momani, and S.B. Hadid, Fractional-order COVID-19 pandemic outbreak: modelling and stability analysis, Int. J. Biomath. 15 (2021), pp. 2150090. DOI:10.1142/S179352452150090X.
[11] T.A. Biala and A.Q.M. Khaliq, A fractional-order compartmental model for the spread of the COVID-19 pandemic, Commun. Nonlinear Sci. Numer. Simul. 98 (2021), pp. 105764.
[12] J. Čermák, I. Gy*ohuml;ri, and L. Nechvátal, On explicit stability conditions for a linear fractional difference system, Fract. Calc. Appl. Anal. 18 (2015), pp. 651–672.
[13] A.N. Chatterjee and B. Ahmad, A fractional-order differential equation model of COVID-19 infection of epithelial cells, Chaos, Solitons & Fractals. 147 (2021), pp. 110952.
[14] N. Debbouche, A. Ouannas, I.M. Batiha, and G. Grassi, Chaotic dynamics in a novel COVID-19 pandemic model described by commensurate and incommensurate fractional-order derivatives, Nonlinear. Dyn. 109 (2021), pp. 33–45. DOI:10.1007/s11071-021-06867-5.
[15] A. DInnocenzo, F. Paladini, and L. Renna, A numerical investigation of discrete oscillating epidemic models, Physica. A. 364 (2006), pp. 497–512.
[16] N. Djenina, A. Ouannas, I.M. Batiha, G. Grassi, and V.T. Pham, On the stability of linear incommensurate fractional-Order difference systems, Mathematics. 8 (2020), pp. 1754.
[17] J. Fahd, A. Thabet, B. Dumitru, and B. Kübra, On the Stability of Some Discrete Fractional Nonautonomous Systems, Hindawi Publishing Corporation, 2012.

[18] G. Giordano, F. Blanchini, R. Bruno, P. Colaneri, A.D. Filippo, A.D. Matteo, and M. Colaneri, Modelling the COVID-19 epidemic and implementation of population-wide interventions in Italy, Nat. Med. 26 (2020), pp. 855–860.
[19] S. Gounane, Y. Barkouch, A. Atlas, M. Bendahmane, F. Karami, and D. Meskine, An adaptive social distancing SIR model for COVID-19 disease spreading and forecasting, Epidemiol. Methods. 10 (2021), pp. 20200044.
[20] Z.-Y. He, A. Abbes, H. Jahanshahi, N.D. Alotaibi, and Y. Wang, Fractional-order discrete-time SIR epidemic model with vaccination: chaos and complexity, Mathematics. 10(2) (2022), pp. 55–165.
[21] M. Higazy, Novel fractional order SIDARTHE mathematical model of COVID-19 pandemic, Chaos, Solitons & Fractals. 138 (2020), pp. 110007.
[22] H. Jahanshahi, A. Yousefpour, Z. Wei, R. Alcaraz, and S. Bekiros, A financial hyperchaotic system with coexisting attractors: dynamic investigation, entropy analysis. control and synchronization, Chaos, Solitons & Fractals. 126 (2019), pp. 66–77.
[23] S. Jain and Y. El-Khatib, Stochastic covid-19 model with fractional global and classical piecewise derivative, Results Phys. 30 (2021), pp. 104788.
[24] W. Kermack and A. McKendrick, Contributions to the mathematical theory of epidemics, Proc. R. Soc. A. 115 (1927), pp. 700–721.
[25] H. Khan, M. Ibrahim, A.-H. Abdel-Aty, M. Motawi Khashan, F.A. Khan, and A. Khan, A fractional order covid-19 epidemic model with Mittag-Leffler kernel, Chaos, Solitons & Fractals. 148 (2021), pp. 111030.
[26] A.A. Khennaoui, A. Ouannas, S. Momani, I.M. Batiha, Z. Dibi, and G. Grassi, On dynamics of a fractional-Order discrete system with only one nonlinear term and without fixed points, Electronics. 9 (2020), pp. 2179.
[27] A.A. Khennaoui, A.O. Almatroud, A. Ouannas, M.M. Al-sawalha, G. Grassi, V.T. Pham, and I.M. Batiha, An unprecedented 2-dimensional discrete-time fractional-order system and its hidden chaotic attractors, Math. Probl. Eng. 2021 (2021), pp. 1–10.
[28] P. Kumara, V.S. Erturk, and M. Murillo-Arcila, A new fractional mathematical modelling of COVID-19 with the availability of vaccine, Results Phys 24 (2021), pp. 104213.
[29] X.-P. Li, H.A. Bayatti, A. Din, and A. Zeb, A vigorous study of fractional order COVID-19 model via ABC derivatives, Results Phys. 29 (2021), pp. 104737.
[30] J.P.S. Maurício de Carvalho and B. Moreira-Pinto, A fractional-order model for Covid-19 dynamics with reinfection and the importance of quarantine, Chaos, Solitons & Fractals. 151 (2021), pp. 111275.
[31] H. Mohammadi, S. Rezapour, and A. Jajarmi, On the fractional SIRD mathematical model and control for the transmission of COVID-19: the first and the second waves of the disease in Iran and Japan, ISA. Trans. 124 (2021), pp. 103–114. DOI:10.1016/j.isatra.2021.04.012.
[32] A. Omame, M. Abbas, and C.P. Onyenegecha, A fractional-order model for COVID-19 and tuberculosis co-infection using Atangana–Baleanu derivative, Chaos, Solitons & Fractals. 153 (2021), pp. 111486.
[33] A. Omame, D. Okuonghae, U.K. Nwajeri, and C.P. Onyenegecha, A fractional-order multi-vaccination model for COVID-19 with non-singular kernel, Alex. Eng. J. 61 (2022), pp. 6089–6104. DOI:10.1016/j.aej.2021.11.037.
[34] V.S. Panwar, P.S. Sheik Uduman, and J.F. Gómez-Aguilar, Mathematical modeling of coronavirus disease COVID-19 dynamics using CF and ABC non-singular fractional derivatives, Chaos, Solitons & Fractals. 145 (2021), pp. 110757.
[35] O. Postavaru, S.R. Anton, and A. Toma, COVID-19 pandemic and chaos theory, Math. Comput. Simul. 181 (2021), pp. 138–149.
[36] K. Rajagopal, N. Hasanzadeh, F. Parastesh, I.I. Hamarash, S. Jafari, and I. Hussain, A fractional-order model for the novel coronavirus (COVID-19) outbreak, Nonlinear Dyn. 101 (2020), pp. 711–718.
[37] M.T. Shatnawi, N. Djenina, A. Ouannas, I.M. Batiha, and G. Grassi, Novel convenient conditions for the stability of nonlinear incommensurate fractional-order difference systems, Alex. Eng. J. 61 (2022), pp. 1655–1663.

[38] I. Talbi, A. Ouannas, A.A. Khennaoui, A. Berkane, I.M. Batiha, G. Grassi, and V.T. Pham, Different dimensional fractional-order discrete chaotic systems based on the caputo h-difference discrete operator: dynamics, control, and synchronization, Adv. Differ. Equ. 2020 (2020), pp. 1–15.

[39] S. Thanin, Z. Anwar, C. Saowaluck, E. Zohreh, T. Mouhcine, and D. Salih, Analysis of a discrete mathematical COVID-19 model, Results Phys. 28 (2021), pp. 104668.

[40] C. Xu, Y. Yu, Y. Chen, and Z. Lu, Forecast analysis of the epidemics trend of COVID-19 in the USA by a generalized fractional-order SEIR model, Nonlinear. Dyn. 101 (2020), pp. 1621–1634.

[41] S. Yadav, D. Kumar, J. Singh, and D. Baleanu, Analysis and dynamics of fractional order covid-19 model with memory effect, Results Phys. 24 (2021), pp. 104017.

[42] T.H. Zhao, O. Castillo, H. Jahanshahi, A. Yusuf, M.O. Alassafi, F.E. Alsaadi, and Y.M. Chu, A fuzzy-based strategy to suppress the novel coronavirus (2019-NCOV) massive outbreak, Appl. Comput. Math. 20(1) (2021), pp. 160.

Chaotic attractors captured from remote sensing time series for the dynamics of cereal crops

Sylvain Mangiarotti and Flavie Le Jean

ABSTRACT

A chaotic attractor was obtained from remote sensing data, for the first time, in the early 2010s, its Poincaré section revealing a weakly dissipative dynamics. This attractor was captured from a time series of vegetation index, for the cycles of cereal crops in semi-arid region. This attractor was fully unexpected, since it was also the first attractor of a weakly dissipative dynamics directly extracted from observational time series. The generality of this result is questioned here by applying the same modelling approach – the global modelling technique – to four provinces in coastal and inland Morocco: Safi, El Jadida, Khourigba and Khenifra. Several experimentations are considered, applying the analyses at the provinces scale, or to several provinces in association and in aggregation. The obtained models confirm the results obtained in the early 2010s: the dynamics of cereal crops is chaotic and can be approximated by a weakly dissipative three-dimensional dynamics.

1. Introduction

One paradoxical but central property of chaos is to associate determinism with unpredictability: Very simple dynamical relations can give rise to a non-repeating, and unpredictable, time evolution. Chaos is at the edge of complexity. Following Otto E. Rössler, a chaotic evolution can also be seen – aesthetically – as a beautiful dance between multiple variables. Considering the variables of this dance in a corresponding space, by representing the successive states visited by these variables as a continuous trajectory, this dance enables to get a chaotic attractor, that is, a highly non-trivial geometrical structure. The other type of attractor results from chaotic maps and can also be obtained from this structure by considering the continuous trajectory at the crossing of a chosen section of the space: a Poincaré section.

Henri Poincaré, who introduced the original main concepts of chaos at the end of the 19th century [50], could not portray out such a trajectory. He realized that the solution of this problem was so complicated that he did '*not even try to sketch it*'; Amazingly, without computer, he was able to understand, without a figure, that the unpredictability of such systems was the result of the high sensitivity to the initial conditions. The seed of chaos theory was planted, which was about to make a breakthrough in the history of science and give rise to a new way to make science [21].

It is only in the 1960s, with the advent of computers, that the first chaotic attractors could be unveiled. The Lorenz attractor [22] was obtained by reducing to three a seven-equation system introduced by Barry Saltzman [56] for the Rayleigh-Bénard convection. Initially motivated by chemical reactions, Otto Rössler could find chaotic and also hyperchaotic attractors thanks to intuitive considerations involving both mathematics and topology [51–53]. Yoshisuke Ueda discovered the first toroidal chaotic attractor by studying the synchronization between an oscillator and a sinusoidal forcing using an analogue computer [64]. Julian C. Sprott discovered other attractors by a systematic exploration of models of quadratic form using a brute force technique [58]. Jean-Marc Malasoma could identify all the simplest quadratic systems able to produce chaos and provided a demonstration for it [26, 27].

Chaotic attractors resulting from maps also have their history. Some of them were generated by electronic circuits (the Mira map [2]), others by considering physical systems (the Hénon map [8]), or based on the numerical explorations of mathematical recurrence relations [46] (the logistic map introduced by Pierre-François Verhulst in 1838 [65]), or by applying a piece by piece linearization (the Lozi map [24]). The first hyperchaotic map was discovered by Otto Rössler in 1979 [54].

Chaotic attractors and chaotic maps are very closely linked since any continuous chaotic system can produce a chaotic map given a properly chosen Poincaré section. Reversely, chaotic maps can be used to generate a suspension, that is, a continuous system able to reproduce the map [59] or to get topological information about such a suspension [30].

Importantly, René Lozi pointed out the question of non-repeating time evolution specific to chaos by exploring giga-periodic orbits, proving that chaos can be verified in practice up to the limits of computer precision [25].

To extract chaotic attractors from observational data remained a difficult problem. The embedding theorem demonstrated by Floris Takens [60] in the early 1980s and later generalized for various embeddings [57] proved the possibility to reconstruct the state space (or phase space) directly from a single observational time series. Nonetheless, such a reconstruction cannot guaranty the existence of an underlying determinism which is a – necessary condition – for chaos [13].

Among the methodologies introduced to extract deterministic equations from time series [1, 5, 6, 16, 35], few of them enabled to obtain new chaotic attractors directly from observational time series. The first chaotic attractor obtained from an observational times series was for electrochemistry in 1995 [18]. However, to apply it to observations from the real world remained challenging. This difficulty was for instance pointed out in 2006 in Earth science by underlying the fruitless quest for chaotic attractors in hydrology [15]; 12 years later, chaotic attractors were obtained in hydrogeology for the cycles of two karst springs [39]. Previously, a chaotic attractor was obtained in ecology for the cycles of Lynx in Canada [44]. Chaotic attractors were also obtained in epidemiology for the plague in Bombay [28], the epidemic of Ebola Virus Disease in West Africa [37], and the early development of Covid-19 in China [40] and Africa [61]. Chaotic attractors were also obtained from observational time series for the hydro-ecology of soils [42] and for the micro-atmospheres of caves [55].

The present study focuses on the cycles of cereal crops observed from space in semi-arid region. Using the global modelling technique, a chaotic attractor was extracted from time series of vegetation index in 2011 [34]. It was not only the first chaotic attractor directly obtained from remote sensing data. It was also the first weakly dissipative model directly extracted from observational time series (a theoretical one having been discovered by Lorenz in 1984 [23]). The object of the present study is to investigate the reproducibility of this result in other contexts. The global modelling technique is thus applied here to four provinces in Morocco located in the south, namely Safi and El Jadida, located on the coast, and Khourigba and Khenifra located inland.

2. Theoretical background

2.1. Global modeling

The aim of the global modeling technique is to obtain sets of delayed or differential equations able to reproduce the dynamics of an observed system directly from observational time series [20]. Starting from a single time series $X_1^{obs}(t_i)_{i=1...N}$ of variable X_1, with N the number of data points, it is – under proper conditions of observability [19] – possible to rewrite the original n-variable system of ordinary differential equations governing the dynamics, in the canonical form

$$\begin{cases} \dot{X}_1 = X_2 \\ \dot{X}_2 = X_3 \\ \quad\vdots \\ \dot{X}_n = F(X_1, X_2, \ldots, X_n) \end{cases} \tag{1}$$

with $(X_1, X_2, \ldots, X_n)$ the observed variable and its successive derivatives, and $F(X_1, X_2, \ldots, X_n)$ a nonlinear function to be retrieved. In practice, under real conditions, and in particular when considering environmental dynamics, the original system is generally approximated by a low-dimensional system of dimension $n' \leq n$.

2.2. Problem statement

A powerful algebraic structure identification technique is required to obtain successful models of canonical form (Equation (1)), in particular under real observation conditions. Indeed, without structure selection, only divergent and oversimplified models can be expected. The Global Polynomial Modelling (GPoM) package, in R language, was developed at the *Centre d'Etudes Spatiales de la Biosphère* for this purpose [29, 35, 41]. Its aim is to obtain a valid polynomial approximation P of the function F in Equation (1). Three input parameters are required to run the structure identification: The model dimension n', the maximum degree $q^{\max}$ of polynomial P and the integration time length. The aim of the algorithm is to identify a proper approximation of the original dynamics. To be acceptable, a set of equations should be non divergent and should enable a consistent reconstruction of the phase space. It is important to note, here, that the modeling problem is not formulated as a minimization problem because such an objective would foster diverging models and could not guarantee a proper reconstruction of the phase space. Similarly, no constraint is put on the model size (the number of model parameters) since no information is available beforehand on this point.

Of course, the model should also be a good approximation of the original dynamics (precise fitting), and it should be as concise as possible (necessary parameters only). However, these constraints should not be considered in first, but only after considering the quality of the phase space reconstruction.

2.3. Algorithms

Parameters n' and $q^{\max}$ enable to define a general polynomial structure of $n_P = \frac{(n'+q^{\max})!}{n'!\,q^{\max}!}$ terms from which 2^{n_P} particular models can be generated. For instance, for $n' = 3$ and $q^{\max} = 2$, the polynomial will include $n_P = 10$ terms: the constant term, the three linear terms X_1, X_2, X_3, the quadratic terms X_1^2, X_2^2, X_3^2, and the cross terms X_1X_2, X_1X_3 and X_2X_3. And this general structure can generate 2^{10} possible particular models.

To select the best models in this very large set, the following algorithm is applied. First, the model, including all the terms (here $n_P = 10$), is preselected by default. (1) Its parameterization is estimated using a Gram-Schmidt technique and the fitting residual ϵ is estimated. (2) The model parameters are set to zero one by one using a leave-one-out technique and the residuals ϵ_i relating to each polynomial parameters κ_i are calculated. (3) The residuals are compared to the residual obtained when including all the terms. The term κ_j of minimal influence (that is such as $\epsilon_j < \epsilon_i \, \forall j \neq i$) is identified and removed leading to the second selected model (including then $n_P - 1 = 9$ terms in the polynomial). Stages (1) to (3) are repeated until no term remains in the polynomial. The number of possible model has reduced from 2^{n_P} possible models to $n_P + 1$ (here 11) ones, that is, with a polynomial P of size n_P to 0. This selection is drastic since even for $n' = 4$ and $q^{\max} = 5$, it will reduce the model number 2^{126} to 127 models, only.

This new set of models being much smaller, the numeric integration of the whole set of preselected models can now be tested systematically. The integration is performed using the Runge–Kutta technique, starting from the initial conditions $(X_1, X_2, X_3)_0^{obs}$ calculated from the observational time series $X_1^{obs}(t_i)_{i=1...N}$. Integration is tested and models are then classified depending on their dynamical regime after convergence, as: (1) *Diverging* if Not a Number are produced or if the four-sigma distance from the original phase space is overpassed in one of the phase space direction; (2) *Fixed point* if variables do not evolve anymore; (3) *Period-1 attractor* if the trajectory converges to a period-1 cycle; (4) *Period-2 attractor* if the trajectory converges to a period-2 cycle; and otherwise as (5) unclassified. This latter group may include categories of varying levels of complexity, including period-n attractors (with $n \geq 3$), quasi-periodic attractors, and chaotic attractors. A detailed description of the technique and its limitations are provided and discussed in Ref. [29]. The dynamics of cereal crops being obviously neither diverging, nor unvarying, nor periodic either, the best models should be found in the unclassified group. The number of models in this latter group being generally small enough, their detailed characterization can then be performed manually, one by one, using: phase space visualization, Poincaré section extraction, first return map reconstruction, computation of the Lyapunov exponents, estimation of the Kaplan–Yorke dimension. For chaotic models, the robustness to a very long numerical integration can then also be tested (here up to the time length of 1000 years) to distinguish chaos from metastable chaos.

2.4. Protocole

Dimension $n' = 3$ and 4 were tried, with maximum polynomial degree ranging from $q^{max} = 2$ to 5. Indeed, no complexity can be generated with dimensions lower than three (Poincaré-Bendixson theorem) nor with a polynomial degree equal to one (linear). On the other hand, dimensions higher than four and polynomial degree higher than five will generally foster numerical instabilities and should thus be avoided. When no satisfying model could be obtained with the complete time series, the analysis was restarted after truncation (e.g. by excluding the years revealing a high level of non-stationarity such as years 1992, 1993 and 1995). This is a common situation under real environmental conditions that may result from temporary non stationary conditions (e.g. special meteorological events). The exact parameters under which the models were obtained are all provided in Appendix 1 (Table A2). To have an estimate of the error associated with parameters values, once a model could be obtained on a selected window, several estimates were performed on the same selected window but with a moving delay varying from 1 to 7 days. The variation coefficients associated with the obtained sets of parameter values are provided in Appendix 1 (Table A3) where their variability is briefly discussed.

3. Area of study and data

3.1. Area of study

The analyses are applied to four Morocco provinces (see Figure 1) characterized by semi-arid climate. Located on the Atlantic coast, Safi and El Jadida provinces are dominated by plain lands with annual precipitations varying from 200 mm to 300 mm. The other two provinces, Khourigba and Khenifra, are located inland and characterized by a continental climate with marked seasonal variability of temperature and precipitations, and an annual precipitation of 500 mm on average. Despite their proximity, the two latter provinces are quite different one to another since Khouribga lies on the phosphated plateau of the Medium Atlas, whereas Khenifra gathers different reliefs ranging from the Medium to the High Atlas.

3.2. Data and preprocessing

The Normalized Difference Vegetation Index (NDVI), hereafter noted γ, from the *Global Inventory Modeling and Mapping Study* of the *Global Land Cover Facility* [63] is used from 1982 to 2008, period from which the first cereal crops attractor was obtained in 2011 [35]. This product is based on the *Advanced Very High-Resolution Radiometer* (AVHRR) sensor onboard *National Oceanic Atmospheric Administration* satellites (NOAA-7, -9, -11, -14, -16 and -17). This is an 8×8–km^2 product that bimonthly sampling results from a compositing algorithm (the maximum NDVI value is extracted every 15 days from the daily daytime values) aiming to reduce atmospheric interferences and view angle effects on the signal [9]. This data set is corrected from sensor degradation over time and from the atmospheric aerosols resulting from the volcanic eruptions of El Chichon (in April 1982) and Mt Pinatubo (in June 1991).

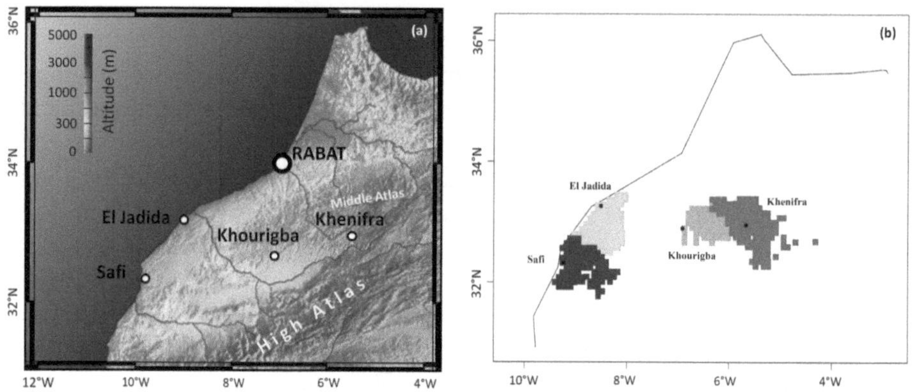

Figure 1. Area of study (a) and data location (b) in the Northern part of Morocco. Four Morocco provinces considered in the study are located in different geographic contexts (a): Safi and El Jadida are coastal provinces of low altitude (0–250 m) extending on the plains of Abda and Doukkala; Khourigba and Khenifra provinces are located inland. Khourigba is located on the phosphated plateau of the Medium Atlas (≥500 m.), Khenifra includes reliefs of both Medium and High Atlas (≥750 m.). The pixels of the NDVI product, selected based on the GLC2000 cover land map, used to obtain time series of mean values (b) for Safi, El Jadida, Khourigba and Khenifra.

Vegetation index relates to the fraction of photosynthetically active radiation absorbed by vegetation in the canopy and on the leaf area index (LAI) [48]. It is defined as:

$$\gamma = \frac{\rho_{NIR} - \rho_{Red}}{\rho_{NIR} + \rho_{Red}} \tag{2}$$

with ρ_{NIR} and ρ_{Red} the reflectances in the near infrared and red bands, respectively. The higher γ, the denser and more active the vegetation is. For areas with vegetation, it ranges between 0.1 and 0.9. This index has proven to be efficient for the monitoring of open grasslands and grain crops in semiarid regions [62], the characterization of vegetation cycles [10, 33], the control of vegetation models [11, 32], and the forecasts of vegetation development [36, 45, 49], primary production [3] or yields [12].

The following preprocessing was applied to the time series. First, a pixel selection based on the GLC2000 cover land map [47] was applied. Pixels presenting more than 50% of crops were selected, the others were rejected (see Figure 1(b)). For each province, a time series was calculated by simply averaging the data of the selected zone in each image. The time series were then resampled at a higher sampling rate using cubic splines. Time series of the four scenes of study are presented in Figure 2. The aggregated surfaces correspond to 6592 km^2, 6848 km^2, 4288 km^2 and 9408 km^2 for Safi, El Jadida, Khourigba and Khenifra, respectively. These scenes of study are mostly dominated by *durum* wheat, bread wheat and barley. A Savitzky-Golai filter was used to compute the derivatives required to apply the global modelling technique (see Equation (1)).

Crop signal is clearly dominated by an annual cycle in each case. The time series corresponding to the first and third quartiles on the areas of study are also superimposed to the plots (in orange). It exhibits a weak dispersion of the signal for the provinces of Safi, El Jadida and Khourigba which reflects the relatively high homogeneity of the scenes of study (Figure 2(a–c). A much larger dispersion is observed in Khenifra (Figure 2(d)), due

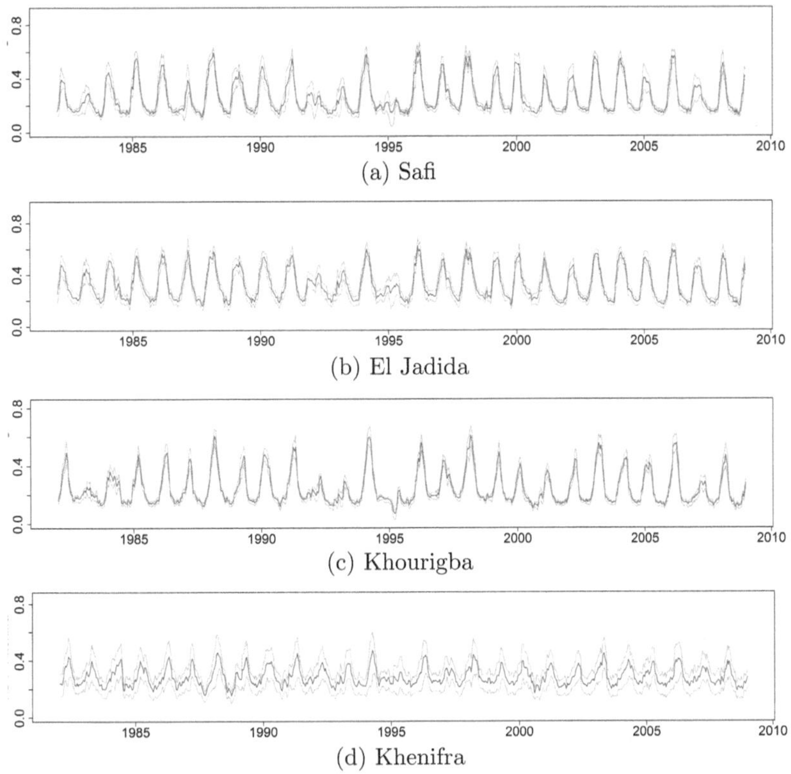

Figure 2. Time series (in green) of vegetation index γ averaged spatially over the provinces of (a) Safi, (b) El Jadida, (c) Khourigba and (d) Khenifra. The time series corresponding to the first and third quartiles are also provided (in orange).

to the contrasted influence of the Moyen Atlas plain lumped out with the lower part of the High Atlas.

Differences in the dynamics between the four provinces are illustrated by the cross-phase portraits (Figure 3) and their two-by-two correlations are presented in Table 1. The time series of Safi and El Jadida (Figure 3(a)) are highly correlated ($\rho = 0.96$), almost fully synchronized, in a relation close to linear that results from very similar climatic conditions on the coastline. Despite their proximity, the cross-plot between Khourigba and Khenifra (Figure 3(c)) exhibits a lower correlation ($\rho = 0.83$) with an obvious hysteresis, a consequence of the delay resulting from the climatic gradients between Moyen and High Atlas. This hysteresis is reinforced by the different climatic conditions between coast and plain when considering the cross-plots between Khenifra and El Jadida ($\rho = 0.64$, Figure 3(e)) and Khenifra and Safi ($\rho = 0.69$, Figure 3(f)). Despite the increased delay, the time evolution does never become completely asynchronous between the provinces.

4. Experiments & characterization

4.1. Experiments

Three experiments of global modeling analysis have been performed: simple, associated and aggregated analyses. The simple analysis consists of directly applying the global modeling technique to the single time series presented in Figure 2.

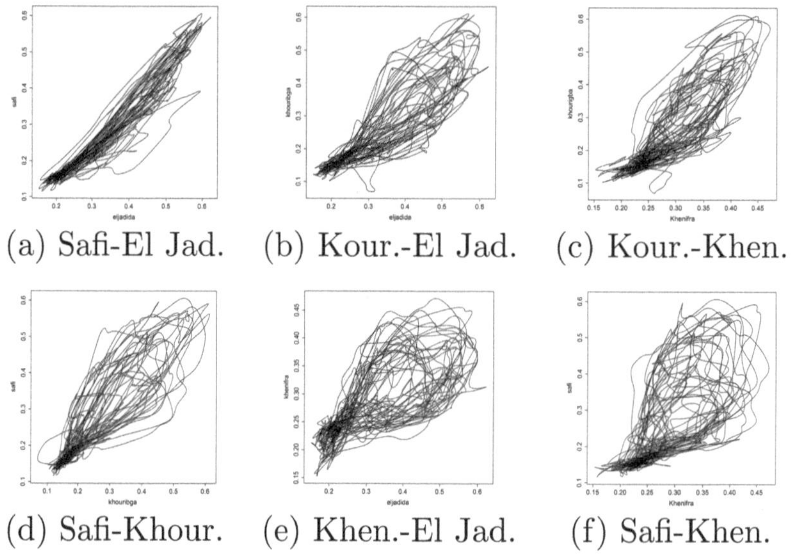

Figure 3. Cross-portraits between the four provinces considered in the study (Safi, El Jadida, Khourigba and Khenifra). (a) Safi–El Jad. (b) Kour.–El Jad. (c) Kour.–Khen. (d) Safi–Khour. (e) Khen.–El Jad and (f) Safi–Khen.

Table 1. Correlation ρ between the time series of the four provinces.

	El Jadida	Khourigba	Khenifra
Safi	0.96	0.87	0.69
El Jadida		0.88	0.64
Khourigba			0.83

The associated analysis consists of considering concatenated time series (carefully avoiding any overlapping effect at the junction). Considering time series in the association can provide various advantages, in particular when several short time series are available due to gaps in the recording, or when time series are spatially distributed with possible overlapping periods. One important interest of the associated analysis is its insensitiveness to phase.

Aggregated analysis aims to analyse time series built by an aggregating process. Its application requires a sufficient level of synchronization between the aggregated time series.

Associated and aggregated analyses were applied considering the provinces two by two and then the four ones altogether. Their conditions of applicability were investigated based on both algebraic and numerical considerations in a previous work, taking the Rössler system as a benchmark [38].

4.2. Characterization

The models performances were characterized based on their forecasts error growth. To do so, the error distribution $F_\tau(e)$ for a forecasting time τ is defined as

$$F_\tau(e) = \mathbb{P}(|\gamma(t+\tau) - \gamma^{obs}(t+\tau)| \leq e). \tag{3}$$

The error growth $e_p(\tau)$ after time τ is defined as

$$e_p(\tau) = F_\tau^{-1}(p), \tag{4}$$

with p the percentile corresponding to the level of confidence. In practice, error growth were estimated for $p = 75\%$, 90% and 95% based on forecast ensembles of 1000 runs with initial conditions regularly distributed along the observational time series.

The attractors reached at the convergence were characterized by estimating their Lyapunov spectrum and the Kaplan–Yorke dimension. To calculate the Lyapunov exponents, the method considering the tangential direction separately was used [7]. For each model, the exponents were estimated from an ensemble of 40,000 simulations, run along the attractor after the convergence was reached, with time steps of 10^{-3}. Error bars were deduced from the dispersions based on the last 1000 iterations.

To estimate the dimension of the attractor reached at the convergence, the Kaplan–Yorke dimension [14]

$$D_{KY} = 2 + \frac{\lambda_1}{|\lambda_3|} \tag{5}$$

was used since well adapted when algebraic equations are available, and because it can be deduced straightforward from the Lyapunov spectrum.

5. Results and discussion

5.1. *Simple analysis*

To start, each province has been analysed separately. Models for the Khenifra province were all rejected. Several models could be sometimes obtained for the same province. In this case, models presenting (1) similar amplitudes compared to the data portraits and (2) presenting the more parsimonious formulation were systematically preferred. Single analyses are presented in Figure 4. Model performances are presented and compared in terms of horizon of predictability H_P, based on a maximum error taken equal to 25%, with a 90% confidence level. These results are all synthesized in Table 2. Note that the curves corresponding to confidence levels of 75%, 90% and 95% of error growth are systematically presented on all the figures of error growth (Figures 4–6, last column) so that any other convention can be used for comparison.

The geometrical and dynamical characteristics of the attractors were also calculated (these are also made available in Table 2). Phase portraits reconstructed from the data all exhibit a rather jittery dynamics, generally characterized by large loops corresponding to the annual cycle presenting more or less dispersed variability at the maximum; small loops around the low values of NDVI ($0.1 \leq y \leq 0.25$) often relating to growth of adventices before ploughing.

The model obtained for the Safi province is presented in Figure 4(a) (in orange) after the transient was removed. Note that the details of all the models equations are provided in Appendix 2 together with the initial conditions. In comparison to the original phase portrait (in black in the background), the resulting attractor is smoother and exhibits smaller amplitudes of annual maxima. Despite these differences, the model is quite interesting since it can catch several aspects of the original behaviour: the annual cycles are

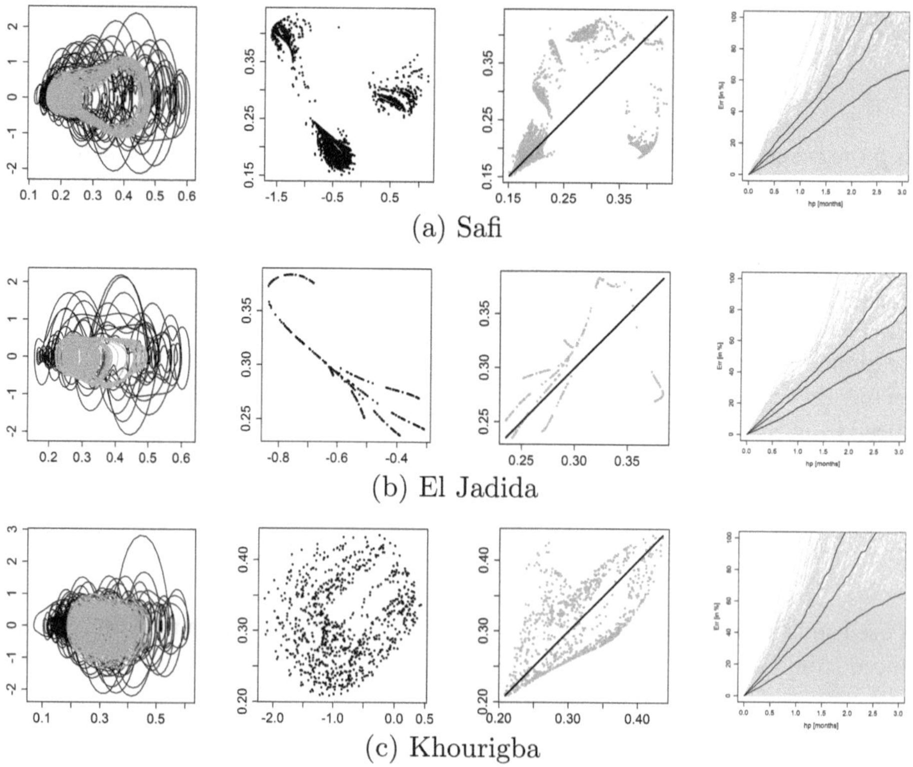

Figure 4. Global models obtained at the province scale. The model phase portraits are provided in the first column (in orange) with the original phase portrait in background (in black); A Poincaré section to the model is provided in the second column and the first-return map in the third column. The model error growth (in black) is presented in the last column for probabilities of 75%, 90% and 95%, respectively; Forecasts error growth resulting from individual trajectories (in green) are also provided. (a) Safi. (b) El Jadida and (c) Khourigba.

clearly distinguishable with some modulations of the annual cycle, although not presenting marked interannual variability; and the cyclic but quite dense oscillations for lower values of NDVI. Note that, adapting the initial conditions, the model can also perform valid forecasts out of the attractor region. In practice, the model performances can reach a horizon of predictability of 24 days. The computation of the Lyapunov exponents demonstrates the average divergence of the flow over the attractor ($\lambda_1 > 0$). The Poincaré section and the first return map are also presented in Figure 4(a). The first return map reveals the presence of folding and the thickness of the Poincaré section is characteristic of a weakly dissipative chaos as confirmed by the Kaplan–Yorke dimension $D_{KY} = 2.64$. All the properties are thus gathered to have a strong argument for chaos in a single and consistent object (the model and its attractor): deterministic equations + trajectory divergence + fractal dimension + folding.

The global model obtained for the El Jadida province (Figure 4(b)) is also smoother and with small maximal amplitudes in comparison to the original phase portrait. Despite this limitation, it exhibits better forecasting performances than the Safi model, with a horizon of predictability significantly larger ($H_P = 28$ days). Oscillations of low amplitudes are

Table 2. Characteristics of the obtained models in terms of forecasting performance (horizon of predictability H_P [in day], for a relative error of 25% with 90% probability), Lyapunov exponents λ_i [in year^{-1}] and fractal dimension of the attractor obtained after convergence (Kaplan–Yorke dimension D_{KY}).

Simple analysis	Safi	El Jadida	Khourigba
H_P [in day]	24. ± 0.5	28. ± 0.5	24. ± 0.5
λ_1 [in yr^{-1}]	+0.60 ± 0.01	+0.26 ± 0.01	+0.72 ± 0.02
λ_2 [in yr^{-1}]	−0.01 ± 0.02	+0.00 ± 0.01	+0.00 ± 0.01
λ_3 [in yr^{-1}]	−0.94 ± 0.01	−2.54 ± 0.01	−0.90 ± 0.02
D_{KY}	2.64 ±0.01	2.10 ± 0.01	2.80 ± 0.04
Associated analysis	Safi–El Jadida	Khourigba–Khenifra	Four provinces
H_P [in day]	29. ± 0.5	34. ± 0.5	29. ± 0.5
λ_1 [in yr^{-1}]	+0.52 ± 0.02	+0.34 ± 0.01	+0.81 ± 0.01
λ_2 [in yr^{-1}]	−0.02 ± 0.02	+0.00 ± 0.01	−0.00 ± 0.01
λ_3 [in yr^{-1}]	−1.86 ± 0.01	−1.18 ± 0.01	−1.99 ± 0.01
D_{KY}	2.28 ± 0.01	2.28 ± 0.01	2.40 ± 0.01
Aggregated analysis	Safi–El Jadida	Khourigba–Khenifra	Four provinces
H_P [in day]	30. ± 0.5	38. ± 0.5	35. ± 0.5
λ_1 [in yr^{-1}]	+0.40 ± 0.01	+0.71 ± 0.01	+0.42 ± 0.01
λ_2 [in yr^{-1}]	−0.00 ± 0.01	−0.02 ± 0.01	−0.06 ± 0.01
λ_3 [in yr^{-1}]	−1.36 ± 0.01	−1.40 ± 0.01	−0.08 ± 0.01
D_{KY}	2.30 ± 0.01	2.51 ± 0.01	2.52 ± 0.01

Note: In each case, 40,000 points were used to compute the λ_i and D_{KY}. Details are provided in Section 4.2.

captured from the dense and agitated oscillations of the original phase portrait; the annual cycle of large amplitude is captured and made visible thanks to a lower density of trajectories reaching a maximum in the region $0.38 \le y \le 0.44$, as observed in the original phase portrait. The first Lyapunov exponent demonstrates the average divergence of the flow over the attractor ($\lambda_1 > 0$). The Poincaré section and the first return map (Figure 4(b)) reveal the presence of folding. Contrarily to Safi, the very thin Poincaré section is characteristic of a more dissipative chaos confirmed by the computation of the Kaplan–Yorke dimension $D_{KY} = 2.10$. Here again, all the properties are gathered for chaos.

The global model captured for the Khourigba province appears quite dense and does not exhibit any directly observable hole in its centre. The first Lyapunov exponent is clearly positive and the first-return map is characterized by an annular structure of higher density compared to its centre filled more parsimoniously. This structure clearly evokes a weakly dissipative (and probably toroidal) chaos as confirmed by its Kaplan–Yorke dimension $D_{KY} = 2.80$.

5.2. *Associated analysis*

The objective of the associated analysis, here, is to obtain a global model common to several provinces. The analysis was applied to all the provinces two by two and then to the four provinces taken together. Global models could be obtained only in a few situations presented hereafter.

A global model was obtained for the coastal provinces of Safi and El Jadida, taken together. The attractor reached at the convergence exhibits very similar amplitudes compared to the original phase portrait, except for the lowest values of NDVI mainly relating to the Safi province. The Poincaré section of the model is characterized by the superposition of two similar patterns presenting a v shape. We were not able to get any simpler pattern of the first-return map by chosing alternative Poincaré section. The model

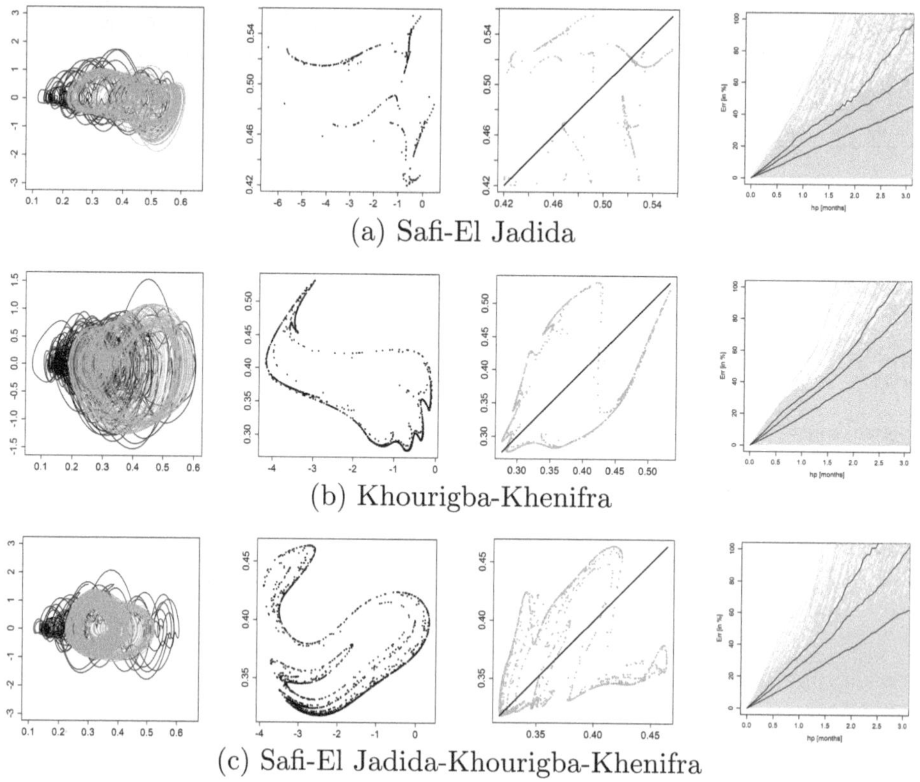

Figure 5. Provinces associated analysis. Phase portraits from the data (black) and from the model (grey) in the first column; Poincaré section to the model (second column) and its corresponding first-return map (third column); Forecasting error growth (last column), see Figure 4 caption for details. (a) Safi–El Jadida. (b) Khourigba–Khenifra and (c) Safi–El Jadida–Khourigba–Khenifra.

exhibits a relatively long horizon of predictability $H_P = 29$ days. Here again, all the necessary conditions and properties of chaos are retrieved (determinism, $\lambda_1 > 0$, $D_{KY} = 2.28$, folding).

A model was also obtained for the inland provinces (Khourigba with Khenifra) which also presents a high coherency in amplitude in comparison to the original phase portrait (lowest values excepted). Its forecasting performances are even higher than the previous models, with a horizon of predictability reaching $H_P = 34$ days. Its flow has a diverging component ($\lambda_1 > 0$) and its annular structure with multiple folding is very characteristic of toroidal chaos. A toroidal attractor can be defined as an attractor both bounding a torus and bounded by another torus. Toroidal chaos is rare in dimension three [31]; the first ones were published in the early 1980s: one by William F. Langford [17], another one by Edouard N. Lorenz [23]. Here, the obtained attractor is also weakly dissipative ($D_{KY} = 2.28$) as for the associated analysis of the coastal provinces. All the conditions for chaos are thus gathered here again.

Finally, a global model was obtained too from the four provinces considered in association. The range of the attractor reached at the convergence is rather reduced ([0.23; 0.52]) in comparison to the original phase portrait ([0.12; 0.61]). The model performances lead to a horizon of predictability of $H_P = 29$ days. The flow is characterized by a positive first Lyapunov exponent and a fractal geometry ($D_{KY} = 2.40$). The resulting flow is characterized by a very foliated bi-dimensional structure corresponding to a weakly dissipative (probably toroidal) dynamics presenting multiple folds easily observed in the Poincaré section. None simple first-return map could be found. All the conditions and properties for chaos are here again found.

The results here obtained by associated analyses of the provinces provide arguments for the existence of a common or similar system underlying the dynamics of cereal crops in the four provinces.

5.3. *Aggregated analysis*

The objective of the aggregated analysis is not to capture a dynamical model common to several spatialized dynamics, but to capture the dynamics of the whole (aggregated) system, taken as a single one. Here also, the analysis was applied to all the provinces aggregated two by two, and then to the four aggregated provinces. Global models were obtained for the same situations as with the associated analysis.

A global model was obtained for the aggregated signal of Safi and El Jadida provinces (Figure 6(a)). The attractor reached after convergence has a quite small range ([0.22; 0.44]) in comparison to the original phase portrait ([0.14; 0.60]). Despite this limitation, it shows good forecasting abilities ($H_P = 30$ days). This means that it is a good local model rather than a global one. Its Poincaré section is characterized by the superposition of two main foliated layers which may correspond to a moderately dissipative system ($D_{KY} = 2.30$). The first-return map clearly exhibits multiple branches characteristic of chaos.

A global model was captured from the signal aggregated from Khourigba and Khenifra (Figure 6(b)). It presents the largest horizon of predictability of all the models obtained in this study ($H_P = 38$ days). The obtained attractor covers most of the range of the original portrait except the lowest values of NDVI ($\gamma < 0.23$). The flow is characterized by a positive first Lyapunov exponent and the Poincaré section by a fully bi-dimensional foliated pattern, corresponding to a weakly dissipative (probably toroidal) dynamics as confirmed by the Kaplan–Yorke dimension ($D_{KY} = 2.51$). Although obviously less dissipative, the pattern of first-return map exhibits striking similarities with the one obtained from the associated analysis (Figure 5(b)). Contrarily, no salient similarity appears with the patterns obtained from any of the simple analysis (Figure 4(c)). All the conditions for chaos are here again retrieved in a single object and the resulting attractor is characteristic of a well developed weakly dissipative chaos.

Finally, a global model was also obtained from the aggregation of the four provinces. The range of the model is in particularly good agreement with the original phase portrait (Figure 6(c)) and the model performances are also high ($H_P = 35$ days). The flow is characterized by a positive first Lyapunov exponent and a fractal geometry ($D_{KY} = 2.52$). The Poincaré section is also characteristic of a weakly dissipative toroidal chaos. Interestingly, the first-return map exhibits a toroidal structure for the lower values (in the range 0.15–0.25) likely to correspond to a beginning of the growth of fallow resulting from earlier rains in the season before sowing. This behaviour could not be captured by other models.

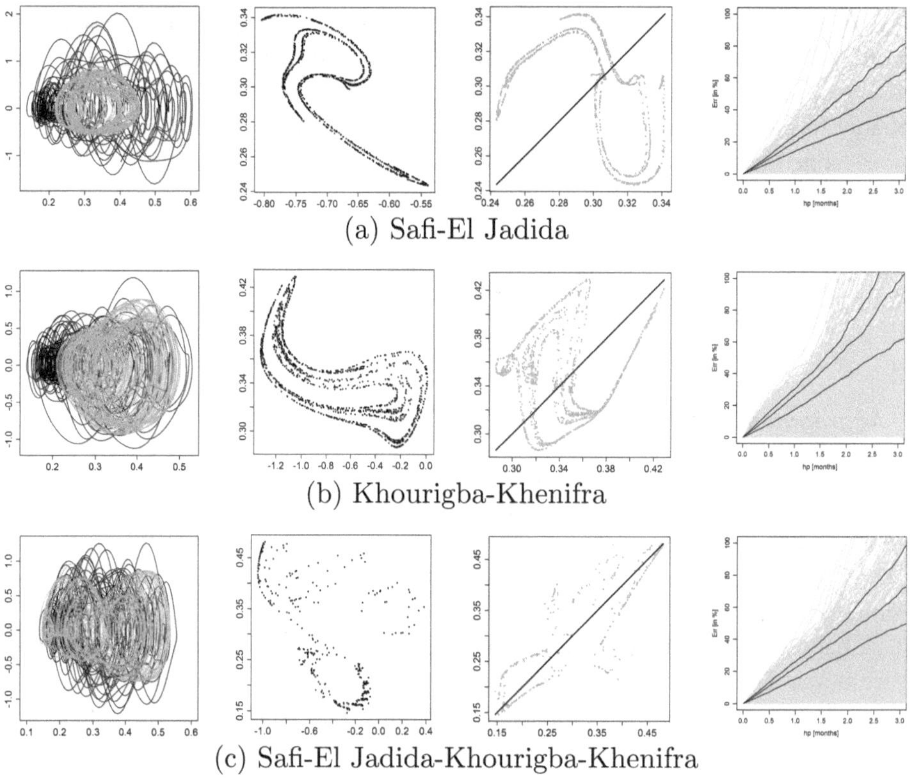

Figure 6. Provinces aggregated analysis. Phase portraits from the data (black) and from the model (grey) in the first column; Poincaré section to the model (second column) and its corresponding first-return map (third column); Forecasting error growth (last columnn), see Figure 4 caption for details. (a) Safi–El Jadida. (b) Khourigba–Khenifra and (c) Safi–El Jadida–Khourigba–Khenifra.

6. Conclusive discussion

By applying the global modeling approach to the cycles of cereal crops on four provinces of Morocco, Safi, El Jadida, Khourigba and Khenifra, at the province scale, and at larger scales (by association or by aggregation), it was possible to obtain three-dimensional global models in many cases.

Although the algebraic formulation of the models cannot be interpreted term by term, which is a situation common to models of canonical form, obtaining these low-dimensional models shows the possibility of approximating the cycles of cereal crops by very few variables. In a forecasting perspective, the obtained models enable horizon of predictability varying from $H_P = 24$ to 38 days. This is less than what was previously found for North Morocco ($H_P = 48$ days [43]) where irrigation is more developed, making the cycles more predictable.

In terms of phase space, from one model to another, the representativeness of the attractors can be very variable in comparison to the original phase portrait. It is generally found that the early stages of growth, from sowing to tillering, which correspond to low values of vegetation index ($0.1 \leq \gamma \leq 0.25$), are more difficult to model, mainly because the fraction of active biomass is particularly low at these stages (the signal is dominated by bare soil).

It is also found in some cases, that the attractors generated by the models cannot reach the maximum values of the observational datasets. It is particularly the case for the attractor obtained for the aggregation of Safi and El Jadida provinces that poorly covers the original phase space. Despite this, the forecasting skills of the corresponding model are good (30 days), and the maximum values can be reached by the model when updating the initial conditions of the model during the growing stage. This is therefore a good local model although its capacity to reproduce the global dynamics is limited.

This limitation is far not systematic, as illustrated by the model obtained for the aggregation of the four provinces. In this case, both the minima and maxima are particularly well covered by the attractor at the convergence. The global dynamics is particularly well reproduced. Moreover, this model also allows for high forecasting performances (35 days).

All these models enable us to have, at the same time, the necessary conditions and all the properties of chaos concentrated in a single object, that is, a set of ordinary differential equations with its corresponding attractor in the phase space. These conditions and properties are the following: (1) determinism guaranteed by a set of ordinary differential equations of canonical form, (2) flow divergence characterized by a positive first Lyapunov exponent ($\lambda_1 > 0$), (3) fractal dimension characterized by a non-integer dimension (here $2.1 \leq D_{KY} \leq 2.8$), (4) folding of the flow revealed by the Poincaré section and/or its first return map. Following the recommendations by René Lozi [25], the persistence of all the attractors was tested on a long numerical integration period, here 1000 years, which is quite significant in comparison to the data length (27 years) and to the timescales of changes to be expected for climate in this region during the coming decades in relation to anthropogenic activities [4]. These global models thus provide strong and consistent arguments for chaos.

Moreover, almost all the models have a weakly dissipative dynamics (8 over 9 models of dimension such as $2.28 \leq D_{KY} \leq 2.80$). These values are fully consistent with the dimensions estimated for the first chaotic attractors obtained for cereal crops in the early 2010s ($D_{KY} = 2.68$ and 2.75 [43]). The weakly dissipative property for cereal crops attractor is thus largely confirmed here. It can also be compared to the first weakly dissipative system of Ordinary Differential Equations obtained by Edouard N. Lorenz in 1984 for the dynamics of the Earth atmosphere [23], which dimension was estimated equal to 2.39.

The lower dimension of the attractor obtained for El Jadida ($D_{KY} = 2.10$) may very likely reveal an oversimplification of the cereal crops dynamics for this province. The resulting dimension is, however, significantly larger than the dimension of the most iconic dissipative system, namely the Lorenz-1963 ($D_{KY} = 2.05$) and Rössler-1976 ($D_{KY} = 2.06$), under usual parameterization.

Although toroidal chaos may be suspected for several attractors, it could be identified surely only in a single case here, for the associated analysis of provinces Khourigba with Khenifra, thanks to the very characteristic central hole in the Poincaré section. Toroidal chaos as a general property of cereal crops thus remains an open question that will require further topological analyses and investigations.

It was found possible to obtain models by associated analysis considering either the coastal provinces or the inland ones. A model for both coastal and inland provinces was also obtained when considering the four provinces at the same time. These results provide an argument for the common underlying dynamics of the four provinces, despite the geographical delay observed between them in practice. All the models obtained by associated analysis have longer horizons of predictability than the ones obtained at the province scale which reveals the complementary of the information provided by the time series and the efficacy of the associated analysis.

It was found possible to obtain models by aggregated analysis for the same situations as by associated analysis. This confirms the determinism of the dynamics at larger spatial scales (13,440 km^2, 17,984 km^2 and 27,136 km^2), despite the delays observed between the four studied provinces. In other words, the delays taking place between the provinces are not sufficient to lose the annual synchronization of the ensemble. These models offer the largest horizons of predictability which is very likely due to the smoothing resulting from the spatial averaging.

After the first results obtained in the early 2010s [35], weakly dissipative regime and, in some cases toroidal chaos, are confirmed for the dynamics of cereal crops in semi-arid region. What can generate such specific dynamics? This detection is questioning.

Toroidal structure relies on the superposition of dominant periodicities. Although no strict periods can be expected in the present case, processes of characteristic time scales can be put forward. In north Morocco, cereal crops sowing is scheduled by the end of the year (between November and December) to avoid the risks of excessive drought immediately after the summer at the earlier stage of growth and to benefit from a maximum of rainfall during winter. Crops are harvested by April and May, generating one main seasonal cycle. This behaviour generates one dominant cycle of annual period clearly visible in the time series. Superimposed to this main cycle, the irregular distribution of rainfall will generate periods of varying growth speed which characteristic time will depend on the water made available in the soil by the rainfall events and on the crops response to water availability and water stress. Typically, these fluctuations will give rise to oscillations of 1–3 weeks characteristic time scale in the phase space. After harvesting, other cycles relying on regrowth of weeds and remaining crops may take place, also depending on rainfall events, that will maintain this oscillating effect until the next crop seeding. The entanglement of this pseudo-periodic process with the seasonal cycles may very likely be at the origin of toroidal chaos.

To interpret the weakly dissipative behaviour is more challenging. Indeed, important amounts of energy, water and nutrients are known to be necessary to grow plants. Such a result is thus puzzling. The weakly dissipative models here obtained for crops cycles do not take into account the input and the output as separate components. Rather, the obtained models focus on the global dynamics of the crops as it can be unveiled from the single variable used (here, the vegetation index observed from space), that is, as a balance between the input and the output. For the same reason, it cannot either be expected to separate the numerous processes at work (photosynthesis, evapotranspiration, etc.). The weakly dissipative models here obtained thus simply suggest that, despite the important amount of energy, water, carbon and nutrients involved in the processes at work, the part used in practice by the plants, in the present ecosystem context, is close to optimal – at the crop level as it can be seen from the vegetation index point of view.

However, this optimality has very likely nothing to do with optimal productivity. Yields may probably be increased by using further irrigation and fertilizers, but this may actually reduce the optimality of the processes at work, in the plant: a more dissipative dynamics may be expected to artificially increase the grain production, that is, by driving the plant production out of the natural equilibrium of the ecosystem. In other words, an increase in the inputs may not be proportional – but suboptimal only – in comparison to its impact on the grain production.

Acknowledgements

S. M. would like to thank Lionel Jarlan and Christophe Letellier for stimulating discussions, respectively, about land cover maps and weakly dissipative chaos. SM would like to thank Pr. Yves Couder (1941–2019) for the stimulating discussions we had at the 19e *Rencontres du Non Linéaires* in 2016, and for his warm encouragements to publish the present results. We would like to dedicate this work to his memory.

Disclosure statement

No potential conflict of interest was reported by the author(s).

Funding

This work was supported by the French programs *Les Enveloppes Fluides et l'Environnement (LEFE)* of the Institut National des Sciences de l'Univers (INSU), *Programme National de Télédétection Spatiale (PNTS)* of the Centre National de la Recherhce Scientifique, and *Spirales* of the Institut de Recherche pour le Développement (IRD); by the Centre National d'Etudes Spatiales (CNES), and by Université Paul Sabatier Toulouse III.

ORCID

Sylvain Mangiarotti http://orcid.org/0000-0002-5959-9770

References

[1] L.A. Aguirre and S.A. Billings, *Dynamical effects of overparametrization in nonlinear models*, Physica D 80(1–2) (1995), pp. 26–40.

[2] J. Bernussou, L. Hsu, and C. Mira, *Quelques exemples du second ordre*, Collected Preprints of Colloque International du CNRS, 229, Transformations ponctuelles et applications (Toulouse, September 1973), Proceedings Editions du CNRS (Paris), 1976, pp. 195–226.

[3] L.J. Blanco, J.M. Paruelo, M. Oesterheld, and F.N. Biurrun, *Spatial and temporal patterns of herbaceous primary production in semi-arid shrublands: A remote sensing approach*, J. Veg. Sci. 27(4) (2016), pp. 716–727.

[4] E. Bouras, L. Jarlan, S. Khabba, S. Er-Raki, A. Dezetter, F. Sghir, and Y. Tramblay, *Assessing the impact of global climate changes on irrigated wheat yields and water requirements in a semi-arid environment of Morocco*, Sci. Rep. 9(1) (2019), Article ID 19142.

[5] D. Coca and S.A. Billings, *Continuous-time system identification for linear and nonlinear systems using wavelet decompositions*, Int. J. Bifur. Chaos 7(1) (1997), pp. 87–96.

[6] G. Gouesbet and C. Letellier, *Global vector field reconstruction by using a multivariate polynomial L_2-approximation on nets*, Phys. Rev. E 49(6) (1994), pp. 4955–4972.

[7] F. Grond, H.H. Diebner, S. Sahle, A. Mathias, S. Fischer, and O.E. Rössler, *A robust locally interpretable algorithm for Lyapunov exponents*, Chaos Solitons Fractals 16(5) (2003), pp. 841–852.

[8] M. Hénon, *A two-dimensional mapping with a strange attractor*, Commun. Math. Phys. 50(1) (1976), pp. 69–77.

[9] B.N. Holben, *Characteristics of maximum-value composite images for temporal AVHRR data*, Int. J. Remote Sens. 7(11) (1986), pp. 1417–1434.

[10] L. Jarlan, Y.M. Tourre, E. Mougin, N. Philippon, and P. Mazzega, *Dominant patterns of AVHRR NDVI interannual variability over the Sahel and linkages with key climate signals (1982–2003)*, Geophys. Res. Lett. 32(4) (2005), Article ID L04701.

[11] L. Jarlan, S. Mangiarotti, E. Mougin, P. Mazzega, and P. Hiernaux, *Assimilation of SPOT/VEGETATION NDVI data into a Sahelian vegetation dynamics model*, Remote Sens. Environ.112(4) (2008), pp. 1381–1394.

[12] L. Jarlan, J. Abaoui, B. Duchemin, A. Ouldbba, Y.M. Tourre, S. Khabba, M. Le Page, R. Balaghi, A. Mokssit, and G. Chehbouni, *Linkages between common wheat yields and climate in Morocco (1982–2008)*, Int. J. Biometeorol. 58 (2014), pp. 1489–1502.

[13] D.T. Kaplan and L. Glass, *Direct test for determinism in a time series*, Phys. Rev. Lett. 68(4) (1992), pp. 427–430.

[14] J.L. Kaplan and J.A. Yorke, *Chaotic behavior of multidimensional difference equations*, Lect. Notes Math. 730 (1979), pp. 204–227.

[15] D. Koutsoyiannis, *On the quest for chaotic attractors in hydrological processes*, Hydrolog. Sci. J. 51(6) (2006), pp. 1065–1091.

[16] C. Lainscsek, C. Letellier, and I. Gorodnitsky, *Global modeling of the Rössler system from the z-variable*, Phys. Lett. A 314(5–6) (2003), pp. 409–427.

[17] W.F. Langford, Numerical studies of torus bifurcations, in *International Series of Numerical Mathematics*, Vol. 70, 1985, pp. 285–295.

[18] C. Letellier, L. Le Sceller, E. Maréchal, P. Dutertre, B. Maheu, G. Gouesbet, Z. Fei, and J.L. Hudson, *Global vector field reconstruction from a chaotic experimental signal in copper electrodissolution*, Phys. Rev. E 51(5) (1995), pp. 4262–4266.

[19] C. Letellier, L.A. Aguirre, and J. Maquet, *Relation between observability and differential embeddings for nonlinear dynamics*, Phys. Rev. E 71(6) (2005), Article ID 066213.

[20] C. Letellier, L.A. Aguirre, and U.S. Freitas, *Frequently asked questions about global modeling*, Chaos: Interdiscip. J Nonlinear Sci. 19(2) (2009), Article ID 023103.

[21] C. Letellier, R. Abraham, D.L. Shepelyansky, O.E. Rössler, P. Holmes, R. Lozi, L. Glass, A. Pikovsky, L.F. Olsen, I. Tsuda, C. Grebogi, U. Parlitz, R. Gilmore, L.M. Pecora, and T.L. Carroll, *Some elements for a history of the dynamical systems theory*, Chaos: Interdiscip. J. Nonlinear Sci.31(5) (2021), Article ID 053110. https://doi.org/10.1063/5.0047851.

[22] E.N. Lorenz, *Deterministic nonperiodic flow*, J. Atmos. Sci. 20 (1963), pp. 130–141.

[23] E.N. Lorenz, *Irregularity: A fundamental property of the atmosphere*, Tellus 36(A) (1984), pp. 98–110.

[24] R. Lozi, *Un attracteur étrange (?) du type attracteur de Hénon*, J. Phys. 39(C5) (1978), pp. 9–10.

[25] R. Lozi, *Giga-periodic orbits for weakly coupled tent and logistic discretized maps*, in *Modern Mathematical Models: Methods and Algorithms for Real World Systems*, A.H. Siddiqi, I.S. Duff, O. Christensen, eds., Proc. Conf. Intern. on Industrial and Appl. Math., New Delhi, Anamaya, New Delhi, 4–6 Dec. 2004.

[26] J.M. Malasoma, *What is the simplest dissipative chaotic jerk equation which is parity invariant?*, Phys. Lett. A 264(5) (2000), pp. 383–389.

[27] J.M. Malasoma, *Minimal smooth chaotic flows*, in *World Scientific Series on Nonlinear Science, Series A*, Vol. 84, C. Letellier and R. Gilmore, eds., World Scientific, Singapore, 2013, pp. 281–301.

[28] S. Mangiarotti, *Low dimensional chaotic models for the plague epidemic in Bombay (1896–1911)*, Chaos Solitons Fractals 81 (2015), pp. 184–196.

[29] S. Mangiarotti and M. Huc, *Can the original equations of a dynamical system be retrieved from observational time series?*, Chaos: Interdiscip. J Nonlinear Sci. 29(2) (2019), Article ID 023133.

[30] S. Mangiarotti and C. Letellier, *Topological analysis for designing a suspension of the Hénon map*, Phys. Lett. A 379(47–48) (2015), pp. 3069–3074.

[31] S. Mangiarotti and C. Letellier, *Topological characterization of toroidal chaos: A branched manifold for the Deng toroidal attractor*, Chaos: Interdiscip. J. Nonlinear Sci. 31(1) (2021), Article ID 013129.

[32] S. Mangiarotti, P. Mazzega, L. Jarlan, E. Mougin, F. Baup, and J. Demarty, *Evolutionary bi-objective optimization of a semi-arid vegetation dynamics model with NDVI and 0 satellite data*, Remote Sens. Environ. 112(4) (2008), pp. 1365–1380.

[33] S. Mangiarotti, P. Mazzega, E. Mougin, and P. Hiernaux, *The vegetation cycle in West Africa from AVHRR-NDVI data: Horizons of predictability versus spatial scales*, Remote Sens. Environ. 114(9) (2010), pp. 2036–2047.

[34] S. Mangiarotti, R. Coudret, L. Drapeau, and L. Jarlan, *Modélisation par approche globale de la dynamique du blé pluvial observée par télédétection spatiale en zone semi-aride*, Rencontre Du Non Linéaire 14 (2011), pp. 103–108.
[35] S. Mangiarotti, R. Coudret, L. Drapeau, and L. Jarlan, *Polynomial search and global modeling two algorithms for modeling chaos*, Phys. Rev. E 86(4) (2012), Article ID 046205.
[36] S. Mangiarotti, P. Mazzega, P. Hiernaux, and E. Mougin, *Predictability of vegetation cycles over the semi-arid region of Gourma (Mali) from forecasts of AVHRR-NDVI signals*, Remote Sens. Environ.123 (2012), pp. 246–257.
[37] S. Mangiarotti, M. Peyre, and M. Huc, *A chaotic model for the epidemic of Ebola virus disease in West Africa (2013–2016)*, Chaos: Interdiscip. J. Nonlinear Sci. 26 (2016), Article ID 113112.
[38] S. Mangiarotti, F. Le Jean, M. Huc, and C. Letellier, *Global modeling of aggregated and associated chaotic dynamics*, Chaos Solitons Fractals 83 (2016), pp. 82–96.
[39] S. Mangiarotti, Y. Zhang, and M. Leblanc, *Chaos theory applied to the modeling of karst springs: First results from univariate time series*, Hydrogeol. J. 27(6) (2019), pp. 2027–2043.
[40] S. Mangiarotti, M. Peyre, Y. Zhang, M. Huc, F. Roger, and Y. Kerr, *Chaos theory applied to the outbreak of COVID-19: An ancillary approach to decision-making in pandemic context*, Epidemiol. Infect. 148 (2020), Article ID e95.
[41] S. Mangiarotti, M. Huc, F. Le Jean, M. Chassan, and L. Drapeau, *GPoM (Generalized polynomial modelling)*, Comprehensive R Archive Network, version 1.3, 2020-02-18. Available at https://CRAN.R-project.org/package = GPoM.
[42] S. Mangiarotti, E. Fu, P. Jouquet, M.T. Tran, M. Huc, and N. Bottinelli, *Earthworm activity and its coupling to soil hydrology: A deterministic analysis*, Chaos: Interdiscip. J. Nonlinear Sci. 31 (2021), Article ID 013134.
[43] S. Mangiarotti, L. Drapeau, and C. Letellier, *Two chaotic global models for cereal crops cycles observed from satellite in northern Morocco*, Chaos: Interdiscip. J. Nonlinear Sci. 24(2) (2021), Article ID 023130.
[44] J. Maquet, C. Letellier, and L.A. Aguirre, *Global models from the Canadian lynx cycles as a direct evidence for chaos in real ecosystems*, J. Math. Biol. 55(1) (2007), pp. 21–39.
[45] N. Martiny, N. Philippon, Y. Richard, C. Camberlin, and C.J.C. Reason, *Predictability of NDVI in semi-arid African regions*, Theor. Appl. Climatol. 100(3–4) (2010), pp. 467–484.
[46] R. May, *Simple mathematical models with very complicated dynamics*, Nature 261(5560) (1976), pp. 45–67.
[47] P. Mayaux, E. Bartholome, S. Fritz, and A. Belward, *A new land-cover map of Africa for the year 2000*, J. Biogeogr. 31(6) (2004), pp. 861–877.
[48] R.B. Myneni and D.L. Williams, *On the relationships between FAPAR and NDVI*, Remote Sens. Environ. 49(3) (1994), pp. 200–211.
[49] N. Philippon, N. Martiny, and P. Camberlin, *Forecasting the vegetation photosynthetic activity over the Sahel: A model output statistics approach*, Int. J. Climatol. 29(10) (2009), pp. 1463–1477.
[50] H. Poincaré, *Les Méthodes Nouvelles De La Mécanique Céleste, Tome iii*, Gauthier-Vilard, Paris, 1899.
[51] O.E. Rössler, *An equation for continuous chaos*, Phys. Lett. A 57(5) (1976), pp. 397–398.
[52] O.E. Rössler, *Chaotic behavior in simple reaction system*, Z. Naturforsch. A 31(3–4) (1976), pp. 259–264.
[53] O.E. Rössler, *Different types of chaos in two simple differential equations*, Z. Naturforsch. A 31(12) (1976), pp. 1664–1670.
[54] O.E. Rössler, *An equation for hyperchaos*, Phys. Lett. A 71(2–3) (1979), pp. 155–157.
[55] M. Sáez, S. Mangiarotti, S. Cuezva, A. Fernandezé Cortés, B. Molero, S. Sanchez-Moral, and D. Benavente, *Global models for CO2 and 222Rn concentrations in the cave of Altamira*, Theor. Appl. Climatol. 143(1–2) (2021), pp. 603–626.
[56] B. Saltzman, *Finite amplitude free convection as an initial value problem – I*, J. Atmos. Sci. 19 (1962), pp. 329–341.
[57] T. Sauer, J.A. Yorke, and M. Casdagli, *Embedology*, J. Stat. Phys. 65(3–4) (1991), pp. 579–616.
[58] J.C. Sprott, *Some simple chaotic flows*, Phys. Rev. E 50(2) (1994), pp. R647–R650.

[59] J. Starrett and C. Nicholas, *A suspension of the Hénon map by periodic orbits*, Chaos Solitons Fractals 45(12) (2012), pp. 1486–1493.
[60] F. Takens, *Detecting strange attractors in turbulence*, Lect. Notes Math. 898 (1981), pp. 366–381.
[61] N. Thenon, M. Peyre, M. Huc, A. Touré, F. Roger, and S. Mangiarotti, *COVID-19: Underreporting, demographic effect, chaotic dynamics, and mitigation strategy impact*, PLoS Negl. Trop. Dis. 16(9) (2022), Article ID e0010735.
[62] C.J. Tucker, *Satellite remote sensing of total herbaceous biomass production in the Senegalese Sahel: 1980–1984*, Remote Sens. Environ. 17(3) (1985), pp. 223–249.
[63] C.J. Tucker, J.E. Pinzon, M.E. Brown, D.A. Slayback, E.W. Pak, R. Mahoney, E.F. Vermote, and N.E. Saleous, *An extended AVHRR 8 km NDVI dataset compatible with MODIS and SPOT vegetation NDVI data*, Int. J. Remote Sens. 26(20) (2005), pp. 4485–4498.
[64] Y. Ueda, *The Road to Chaos*, Aerial Press, Santa Cruz, CA, 1993.
[65] P.-F. Verhulst, *Notice sur la loi que la population suit dans son accroissement*, in A. Quetelet, ed., Correspondance mathématique et physique de l'Observatoie de Bruxelles, Ghent, 1838, pp. 113–121.

Appendices

Appendix 1. Preprocessing and global modelling parameterization

The preprocessing parameters, that is, the time span and the filtering window length from which, and the algorithm parameters for which, the models were obtained (n' and q) are provided in Table A1.

Appendix 2. Algebraic formulation and initial conditions of the global models

The global models obtained in this study were all built on a polynomial structure of 4th degree

$$\begin{aligned}
P(X_1, X_2, X_3) = {}& \kappa_0 + \kappa_1 X_3 + \kappa_2 X_3^2 + \kappa_3 X_3^3 + \kappa_4 X_3^4 + \kappa_5 X_2 + \kappa_6 X_2 X_3 \\
& + \kappa_7 X_2 X_3^2 + \kappa_8 X_2 X_3^3 + \kappa_9 X_2^2 + \kappa_{10} X_2^2 X_3 + \kappa_{11} X_2^2 X_3^2 + \kappa_{12} X_2^3 + \kappa_{13} X_2^3 X_3 \\
& + \kappa_{14} X_2^4 + \kappa_{15} X_1 + \kappa_{16} X_1 X_3 + \kappa_{17} X_1 X_3^2 + \kappa_{18} X_1 X_3^3 + \kappa_{19} X_1 X_2 \\
& + \kappa_{20} X_1 X_2 X_3 + \kappa_{21} X_1 X_2 X_3^2 + \kappa_{22} X_1 X_2^2 + \kappa_{23} X_1 X_2^2 X_3 + \kappa_{24} X_1 X_2^3 \\
& + \kappa_{25} X_1^2 + \kappa_{26} X_1^2 X_3 + \kappa_{27} X_1^2 X_3^2 + \kappa_{28} X_1^2 X_2 + \kappa_{29} X_1^2 X_2 X_3 + \kappa_{30} X_1^2 X_2^2 \\
& + \kappa_{31} X_1^3 + \kappa_{32} X_1^3 X_3 + \kappa_{33} X_1^3 X_2 + \kappa_{34} X_1^4, \qquad \text{(A1)}
\end{aligned}$$

used as a low-dimensional approximation ($n' = 3$ instead of n the genuine dimension of the observed system) of function F in Equation (1).

Table A1. Preprocessing and modelling parameters: model dimension n', polynomial degree q, time series time period and Savitzky-Golai filter window length [in day].

	Model	n'	q	Time span	Filter length
Simple analysis	Safi	3	4	1985.5–2009.0	±5.
	El Jadida	3	4	1995.9–2006.3	±5.
	Khourigba	3	4	1982.0–2009.0	±5.
Associated analysis	Safi-El Jadida	3	4	1985.5–1992.4	±18.
	Khourigba–Khenifra	3	4	1983.7–2009.0	±24.
	Four provinces	3	4	1985.5–1989.0	±11.
Aggregated analysis	Safi-El Jadida	3	4	1982.3–1999.4	±13.
	Khourigba–Khenifra	3	4	1983.7–2009.0	±24.
	Four provinces	3	4	1983.7–2009.0	±18.

Table A2. Coefficients $\kappa_{.}$ of polynomial $P(X_1, X_2, X_3)$ in Equation (A1) with their relating algebraic term.

	Simple			Aggregation			Association		
Term	Khourigba	Safi	El Jadida	Khourigba–Khenifra	Safi–El Jadida	4 provinces	Khourigba–Khenifra	Safi–El Jadida	4 provinces
Ct	0	0	−457.579478241111	−49.0900087072149	0	−47.2340911289641	−49.3030759738641	0	310.345652020337
X_3	37.1446280303406	42.0924613158961	73.2749415970538	74.8549016801808	17.0549477108904	58.4102863055197	32.3363313823215	1.96234201117236	0
X_3^2	0	0	−1.24685195828636	−0.830432625038667	0	0	−0.0696039338744489	0	1.88687410005021
X_3^3	0	0	0.0149102336007161	−0.0540813680546796	0	0	−0.0529839738897092	0	−0.0505650060609916
X_3^4	0	0	−2.19917423634495e−05	0	0	0	0	0	0
X_2	−1897.82079118271	−3180.9145288928	−1553.22133994658	−824.278061327948	−1183.89155078371	−1486.42036696856	−96.4138334361615	−778.256467301351	−779.544280093871
X_2X_3	0	0	22.0132098798609	−15.5318271482612	0	44.4285562202986	0	3.81772953112116	0
$X_2X_3^2$	0	0	1.09604563820986	8.43961360829089	0	0	4.65442049331681	0	0.587249513291798
$X_2X_3^3$	0	0	0	−0.366903949156773	0	0	−0.219062961104181	0	0
X_2^2	271.811903630662	721.451319030473	289.928925420794	−44.0426376025801	530.803795844138	373.9444569262	0	254.168425897352	265.222069492646
$X_2^2X_3$	0	0	0.918237628419584	33.7702902463926	0	22.4212049680297	15.9829156233081	0	0
$X_2^2X_3^2$	0	0	−0.142876127405803	0	0	−1.97475390771277	0	0	0
X_2^3	−555.174951561873	−780.378072769522	−680.402726484078	−282.434830946912	−378.069506292303	−435.058727585872	−45.7227328753026	−497.309539383868	−191.950771234462
$X_2^3X_3$	0	0	1.49341951983493	0	0	−10.5551057335364	0	0	0
X_2^4	0	0	0	0	0	33.0487373748508	0	73.3942866019287	58.3823567826858
X_1	0	0	6432.82888172374	542.770378220312	−171.664050038886	461.161211682532	687.454482673793	−263.713579521342	−4749.98132545213
X_1X_3	−399.433056459286	−378.352273063614	−558.471829085978	−754.262357241434	−93.3944028481972	−535.191192259003	−346.286036903516	145.351862812883	38.5313319612348
$X_1X_3^2$	0	1.00167624179017	6.36448723093675	3.49746323009847	4.09714821914238	5.53490781681206	0.852300907336551	0	−7.44062101490793
$X_1X_3^3$	0	0	−0.0338832891111151	0	0	−0.147801656703085	0	0	0.0959880206768607
X_1X_2	14350.9114027582	23976.9953797272	4891.84490764222	5680.55652089658	6685.9051900618	10966.9585195001	−789.61534234004	4873.48157749048	3049.58021868494
$X_1X_2X_3$	0	80.9389195140528	−97.8827600713231	250.982838505221	0	−138.326709795728	106.824226846663	0	28.7184106097492
$X_1X_2X_3^2$	0	0	−2.19741665739779	−25.2458791903576	0	0	−12.4865235318369	0	0
$X_1X_2^2$	−2364.75917541058	−2829.65470291957	1159.98925896369	0	−1867.44016859388	−1888.80497048555	−83.9366578355843	−875.518354073242	−1951.48494180805
$X_1X_2^2X_3$	0	0	−3.1299411113225	−89.0633719953311	0	−58.3740845634334	−43.2228542951278	0	0
$X_1X_2^3$	1002.76135726358	1434.2531400107	1023.64539941861	498.241099798343	535.450942849229	608.793912115731	0	772.307570053207	0
X_1^2	579.970500809391	−1471.71854528292	−31923.5022280187	−2206.88674479295	392.456892270235	−2026.52896040491	−3495.10022248718	1932.03078242585	23923.6564274458
$X_1^2X_3$	1265.09764372305	1009.58093338788	1291.39731854554	2303.01239597873	112.602195454777	1464.86548140725	1075.11773333463	−823.916967751904	−214.106887028165
$X_1^2X_3^2$	0	−1.48373396312514	−7.05481154502859	0	−7.80421077319243	−8.64981343241282	0	0	8.46649064377366
$X_1^2X_2$	−32709.7021150445	−51256.3672267725	6716.4164830191	−12318.3574836012	−9253.84107999825	−24449.6219435027	4686.08259009234	−6962.07841618702	750.957483726637
$X_1^2X_2X_3$	0	−176.541077496132	83.7233391128399	−510.418976469506	0	90.7380456991754	−219.377375001981	0	−53.5652112640876
$X_1^2X_2^2$	3745.67765660233	2435.71756698492	−3733.50807669325	0	1396.83647842247	1948.40156054604	0	0	2008.83209997655
X_1^3	−1259.57061844765	6185.09270796082	62419.4564091972	4443.09372373432	0	4516.66146900253	7832.36690990651	−4410.41929939361	−47264.0982620063
$X_1^3X_3$	−1207.96307205208	−831.274708133515	−930.521430652595	−2108.99317845018	0	−1195.15367683992	−944.195731566203	1033.2395281691	297.263730034574
$X_1^3X_2$	20491.4252437999	28742.0371474332	−21687.2680330135	7509.44797464508	0	15613.8166855079	−5860.50922308738	0	−8450.88508914614
X_1^4	0	−6198.57307134981	−41134.7244634171	−3442.71541837163	0	−3665.89684745479	−6075.77266057362	3676.11624533309	32347.9871875151
N_{terms}	15+2	20+2	33+2	27+2	15+2	29+2	25+2	17+2	25+2

Table A3. Coefficient of variation and median value of the $\kappa_{\cdot}$ parameters presented in Table A2.

	Simple			Aggregation			Association		
Term	Khourigba	Safi	El Jadida	Khourigba–Khenifra	Safi-El Jadida	4 provinces	Khourigba–Khenifra	Safi–El Jadida	4 provinces
Ct	–	–	0.2025	0.0851	–	0.0607	0.1434	–	0.1952
X_3	0.0373	0.0852	0.1046	0.0416	0.0295	0.0271	0.013	0.0056	–
X_3^2	–	–	0.4614	0.0691	–	–	0.0165	–	0.0476
X_3^3	–	0.0502	0.2924	0.0276	–	–	0.0053	–	0.4132
X_3^4	–	–	0.424	–	–	–	–	–	–
X_2	0.0201	0.0175	0.2017	0.0134	0.0022	0.0089	0.0209	0.0059	0.0086
X_2X_3	–	0.0046	0.2029	0.0284	–	0.0505	–	0.0105	–
$X_2X_3^2$	–	–	0.2644	0.0223	–	–	0.0095	–	0.0138
$X_2X_3^3$	–	–	–	0.0119	–	–	0.0021	–	–
X_2^2	0.0466	0.0438	0.0755	0.0118	0.0187	0.0189	–	0.0198	0.0082
$X_2^2X_3$	–	–	1.1379	0.0217	–	0.0251	0.0087	–	–
$X_2^2X_3^2$	–	–	0.9928	–	–	0.0293	–	–	–
X_2^3	0.0198	0.0004	0.0189	0.0042	0.0228	0.0102	0.0017	0.0194	0.0033
$X_2^3X_3$	–	–	0.0814	–	–	0.0217	–	–	–
X_2^4	–	–	–	–	–	0.0236	–	0.0153	0.0219
X_1	–	–	0.1825	0.0796	0.011	0.0437	0.1049	0.0153	0.1887
X_1X_3	0.0358	0.1447	0.1275	0.0367	0.0295	0.0242	0.0111	0.0069	0.0413
$X_1X_3^2$	–	–	0.5721	0.0622	0.0469	0.0147	0.0155	–	0.0561
$X_1X_3^3$	–	0.0243	0.3172	–	–	0.0821	–	–	0.2534
X_1X_2	0.0205	0.022	0.593	0.0152	0.0013	0.0112	0.021	0.0047	0.0258
$X_1X_2X_3$	–	0.0063	0.2478	0.0546	–	0.1084	0.0006	–	0.0079
$X_1X_2X_3^2$	–	–	0.3671	0.0206	–	–	0.0188	–	–
$X_1X_2^2$	0.0385	0.0813	1.4511	–	0.0266	0.0186	0.0046	0.0124	0.0092
$X_1X_2^2X_3$	–	–	1.2718	0.0188	–	0.0251	0.005	–	–
$X_1X_2^3$	0.019	0.0006	0.0471	0.0039	0.0321	0.0389	–	0.0937	–
X_1^2	0.0192	0.1458	0.168	0.0773	0.0252	0.0386	0.0877	0.0182	0.2028
$X_1^2X_3$	0.0349	1.6899	0.1588	0.0333	0.0288	0.0242	0.0103	0.0073	0.0329
$X_1^2X_3^2$	–	–	0.6646	–	0.0448	0.019	–	–	0.0699
$X_1^2X_2$	0.0211	0.0323	1.1175	0.0167	0.0014	0.0176	0.0093	0.0038	0.0155
$X_1^2X_2X_3$	–	–	0.3437	0.1069	–	0.0611	0.0007	–	0.0075
$X_1^2X_2^2$	0.0378	0.361	0.1103	–	0.0376	0.0204	–	–	0.0125
X_1^3	0.0239	0.2185	0.1684	0.0788	–	0.0393	0.0844	0.0186	0.2550
$X_1^3X_3$	0.0345	0.1193	0.1957	0.0319	–	0.0278	0.0106	0.0074	0.0284
$X_1^3X_2$	0.024	0.0705	0.3278	0.0187	–	0.04	0.006	–	0.0085
X_1^4	–	0.4993	0.181	0.0826	–	0.0432	0.0859	0.0185	0.3642
Median value	0.0240	0.0603	0.2478	0.0284	0.0266	0.0251	0.0106	0.0124	0.0284

The $\kappa_{\cdot}$ parameters of all the models are provided in Table A2. An estimate of their associated error is also provided in Table A3. The parameters can exhibit very different sensitivity levels from one parameter to another. This sensitivity is however difficult to interpret individually because they can be linked one to another. This variation in sensitivity is easier to understand when considering the relation between the original equations and the canonical reformulation which can be done for some theoretical systems as it is the case for the Lorenz or the Rössler systems [6]. This sensivity is difficult to interpret parameter by parameter when the original equations are unknown as it is the case when starting from observational time series from the real world. Moreover, small variations of some parameters may have a large impact whereas large variations of other ones may be of marginal effect. For this reason, the median value is preferred to estimate the error associated with the coefficients. Except for the model obtained for the El Jadida province, for which the median value is particulary high (24.8% of error), it is relatively low for the other models (6.0% for Safi province, <2.9% for all the other models).

The initial conditions $(X_1, X_2, X_3)_0$ are also provided in Table A4.

Table A4. Initial conditions of the global models presented in the study.

	Models	$(X_1, X_2, X_3)_0$
Simple analysis	Safi	(0.2201067, 0.5456500, −0.5666293)
	El Jadida	(0.3871700, −0.6916930, −0.8473171)
	Khourigba	(0.3588249, 0.2364964, −7.0099925)
Associated analysis	Safi-El Jadida	(0.4489411, −0.2451920, 1.4257687)
	Khourigba–Khenifra	(0.29475997, −0.05883727, 0.78191018)
	Four provinces	(0.3820839, −0.1365062, −6.1009947)
Aggregated analysis	Safi-El Jadida	(0.4411259, −0.1653706, −3.7228289)
	Khourigba–Khenifra	(0.4354163, 0.8340796, −3.2927661)
	Four provinces	(0.4669276, 0.3643090, −2.5227233)

Nonlinear dynamic and chaos in a remanufacturing duopoly game with heterogeneous players and nonlinear inverse demand functions

Habiba Meskine, Mohammed-Salah Abdelouahab and René Lozi

ABSTRACT

The purpose of this paper is to investigate the dynamics of a remanufacturing duopoly game model with heterogeneous players and different competition strategies, assuming that the original equipment manufacturer has bounded rationality and aims to maximize profit, and the third-party remanufacturer has an adaptive nature and aims to maximize share in order to get a certain profit. The model behaviours are discussed mainly from the point of view of stability, bifurcations and chaos. By utilizing Jury's stability criteria, the Cournot-Nash equilibrium's asymptotic stability is examined. We found that under certain conditions, when consumer willingness to pay was taken as a bifurcation parameter, the system undergoes a flip bifurcation and a Neimark-Sacker bifurcation. Lyapunov exponents are used to demonstrate that the system becomes chaotic through each of the previous bifurcations.

1. Introduction

Remanufacturing operations are considered environmentally friendly due to the fact that they involve restoring used products to new condition and reselling them [7]. Recently there has been an increase in remanufacturing promotion initiatives in various countries and a well-known companies (such as IBM, Kodak, HP, Xerox, etc.) are actively researching efficient reverse supply chain management models [6,23]. But some original equipment manufacturers (OEMs) prefer not to pursue the remanufacturing plan because doing so could hurt their profits and cannibalize sales of their original products. This may encourage third-party competitors to gather and remanufacture the used products that have first sold by the OEM. For instance, the USA's remanufacturing business is valued at around 53 billion \$ in the last few years, which indicates that OEMs must deal with the competitive threat from TPRs [14–16,22,24]. Such competition can be analysed using oligopoly game models, in which, firms must take into account the consumers' behaviour as well as the competitors' actions and reactions regarding either output or price, leading to complex behaviours [26].

In 1838 the French economist Cournot has introduced the first well-known mathematical description of competition in a duopoly market and a the formal theory of oligopoly game [12]. The theory was significantly updated after one century later, by H. von Stackelberg [32]. Recently, several studies have shown that the Cournot model can lead to complex dynamics such as bifurcation and chaotic behaviours [1–5,9,10,19,25,27–31,33].

Puu [27,28] was among the first researchers who highlighted that Cournot's duopoly model may encompass a variety of complex dynamics, including chaotic attractors.

Realistic expectations are an important factor for the sake of modelling economic events. Expectations strategies, were proposed in various Dynamical Cournot games (with homogeneous players), including those with boundedly rational [4,5,10] and with adaptive expectations [9,25,30].

There has also been extensive research on dynamic Cournot games with heterogeneous players. In their study [2], Agiza and Elsadany analysed the dynamics of a Cournot duopoly game with a naïve player and a boundedly rational player. They also considered a duopoly game with identical demand and cost functions [3] in which one player had bounded rational expectations while the other had adaptive expectations.

Chaos theory has been effectively used in manufacturing and has generated considerable attention [13,18]. However, its application in remanufacturing markets is restricted, and further research and practical implementations are needed to understand its potential benefits and implications, since manufacturers, suppliers, consumers, and policymakers often interact with each other in complex ways.

In this paper, we investigate the dynamics of a nonlinear repeated remanufacturing duopoly game with different competition strategies and heterogeneous players, where the OEM is boundedly rational and pursues profit maximization as his primary business goal, and the TPR is adaptive and chooses share maximization based on achieving a certain profit as his business aim. By mean of the stability theory one determines some stability conditions of the fixed points and validate the appearance of chaos by using bifurcation diagrams and Lyapunov exponent.

The paper is structured as follows. In Section 2 a remanufacturing duopoly game model with heterogeneous players and different competition strategies is proposed. A discussion of equilibrium points and their stability conditions is provided in Section 3. Numerical simulations are carried out in Section 4 to confirm the complexity of the proposed model. Finally, Section 5 concludes the paper with a brief summarization.

2. Problem description and duopoly model building

In this paper, We consider a $2D$-discrete map describing the behaviour of two firms. The first is an original equipment manufacturer (OEM) producing only original products and sells them, whereas the third-party independent firm, also called the third-party remanufacturer, remanufactures only returned goods to make differentiated products.
We consider the following assumptions.

- OEM adopts a bounded rational strategy and maximizes its own profit, while TPR adopts an adaptive strategy and maximizes market share in order to achieve a certain profit. Therefore, both the remanufacturer's profitability status and its market share affect the assessment of its operating effect.

- The OEM and the TPR continually compete against each other, and there was only partial information about the market available to the two manufacturers.
- The decision variables representing the production quantities are carried out in discrete time periods, $t = 0, 1, 2, \ldots$, and restricted to the phase space $S = \{(q_n, q_r) : q_n, q_r > 0\}$, which has an economic meaning.
- The first firm produces new (original) products, q_n at period time t and the second firm may take returned products q_r in order to remanufacture and sell them again in the market at period time $t+1$.
- The consumers' willingness to pay (CWP) is heterogeneous and uniformly distributed in the interval $[0, 1]$, and each consumer's willingness to pay for a remanufactured product is a fraction δ of his willingness to pay for an original product.
- If $\delta = 0$, then the consumer is only interested in original products, and will not purchase returned remanufactured products, but if $\delta = 1$ then the consumer is willing to pay the same price for both original and returned remanufactured products.

The marginal costs of OEM and TPR are c_n and c_r, respectively, restricted in this study to $1 \geq c_n > c_r \geq 0$.
Considering the above assumptions one can derive the following nonlinear inverse demand functions

$$P_n = 1 - \sqrt{q_n + \delta q_r}, \tag{1}$$

$$P_r = \delta(1 - \sqrt{q_n + q_r}). \tag{2}$$

The two firms are supposed having linear cost functions of the form

$$C_i(q_i) = c_i q_i, i = n, r. \tag{3}$$

Therefore, their profit functions are given by

$$\Pi_n = P_n q_n - C_n(q_n) = (1 - c_n - \sqrt{q_n + \delta q_r})q_n, \tag{4}$$

$$\Pi_r = P_r q_r - C_r(q_r) = (\delta(1 - \sqrt{q_n + q_r}) - c_r)q_r. \tag{5}$$

Thus, their marginal profit functions are

$$\frac{\partial \Pi_n}{\partial q_n} = 1 - c_n - \sqrt{q_n + \delta q_r} - \frac{q_n}{2\sqrt{q_n + \delta q_r}}, \tag{6}$$

$$\frac{\partial \Pi_r}{\partial q_r} = \delta(1 - \sqrt{q_n + q_r}) - c_r - \frac{\delta q_r}{2\sqrt{q_n + q_r}}. \tag{7}$$

Hence, OEM's output adjustment mechanism can be modelled as follows:

$$q_n(t+1) = q_n(t) + \alpha q_n(t)\frac{\partial \Pi_n(t)}{\partial q_n(t)}, \tag{8}$$

where $\alpha \geq 0$ is the relative speed of the output adjustment.

When the TPR seeks market share maximization (MSM) completely it loses its profit (i.e. $\Pi_r = 0$). Thus, from (5), one gets the TPR's optimum output as

$$\tilde{q}_{r_{\text{MSM}}} = \left(1 - \frac{c_r}{\delta}\right)^2 - q_n. \tag{9}$$

Moreover, when the TPR seeks profit maximization (PM) completely its marginal profit vanishes. (i.e. $\frac{\partial \Pi_r}{\partial q_r} = 0$). Thus, from (7), one gets the TPR's optimum output as

$$\tilde{q}_{r_{\text{PM}}} = 2\frac{(\delta - c_r)^2 - 3\delta^2 q_n + (\delta - c_r)\sqrt{3\delta^2 q_n + (\delta - c_r)^2}}{9\delta^2}. \tag{10}$$

The TPR takes into account market share expansion and profit maximization as a weighted average and its attitude in the market is a trade-off between profit and market share, which can be represented using a coefficient $\nu \in [0, 1]$, that is

$$\tilde{q}_r = \nu \tilde{q}_{r_{\text{MSM}}} + (1 - \nu)\tilde{q}_{r_{\text{PM}}}.$$

Where $\nu = 0,\ \nu = 1$ and $0 < \nu < 1$ indicate TPR's profit maximization strategy, market share maximization strategy and the mixed mod strategy respectively.

Hence, from (9) and (10) one gets the TPR's reaction function

$$\tilde{q}_r = \nu((1 - \frac{c_r}{\delta})^2 - q_n) + 2(1 - \nu)\frac{(\delta - c_r)^2 - 3\delta^2 q_n + (\delta - c_r)\sqrt{3\delta^2 q_n + (\delta - c_r)^2}}{9\delta^2}. \tag{11}$$

In order to keep $\tilde{q}_r$ positive the CWP δ will be restricted to the interval $[\delta_0, 1]$, with $\delta_0 > 2c_r$. In this study one takes $\delta_0 = \frac{11}{5}c_r$ (This choice has been done based on the obtained numerical results).

As an adaptive player, the TPR thinks in terms of adaptive expectations and calculates its output for the period t+1 by weighting its previous period's output $q_r(t)$ and its reaction function $\tilde{q}_r(t)$. Now introducing a coefficient $\beta \in [0, 1]$ representing the TPR's adjustment speed, one can get its output adjustment process as follows:

$$q_r(t + 1) = (1 - \beta)q_r(t) + \beta\tilde{q}_r(t). \tag{12}$$

Taking into account the constraint that the TPR's output in period $t + 1$ do not exceed OEM's output in period t, i.e. $q_r(t + 1) \leq q_n(t)$ the TPR's output adjustment mechanism can be rewritten as follows

$$q_r(t + 1) = \min\{(1 - \beta)q_r(t) + \beta\tilde{q}_r(t), q_n(t)\}. \tag{13}$$

Now one can formulate a remanufacturing duopoly game model using (6), (11), and combining (8) with (13), as follows:

$$\begin{cases} q_n(t+1) = q_n(t) + \alpha q_n(t)\left[1 - c_n - \sqrt{q_n(t) + \delta q_r(t)}\right. \\ \left. - \frac{q_n(t)}{2\sqrt{q_n(t)+\delta q_r(t)}}\right], \\ q_r(t+1) = \min\{(1-\beta)q_r(t) + \beta\left[\nu((1 - \frac{c_r}{\delta})^2 - q_n(t))\right. \\ \left. +2(1-\nu)\frac{(\delta-c_r)^2 - 3\delta^2 q_n(t) + (\delta-c_r)\sqrt{3\delta^2 q_n(t) + (\delta-c_r)^2}}{9\delta^2}\right], q_n(t)\} \end{cases} \tag{14}$$

Which is a two dimension discrete dynamical system of remanufacturing duopoly game model with different competition strategies and heterogeneous players.

3. Local stability analysis

This section analyses the local stability by looking for possible steady states and analyzing their stability.
System (14) has a singularity at $(0,0)$, therefore only no-boundary Nash equilibria $E = (q_n^*, q_r^*)$, with $q_n^* \geq q_r^* > 0$ will be considered. For this purpose one can distinguish the two cases $f_r > q_n^*$ and $f_r \leq q_n^*$, where

$$f_r = (1-\beta)q_r^* + \beta\left[\nu((1-\frac{c_r}{\delta})^2 - q_n^*) + 2(1-\nu)\frac{(\delta-c_r)^2 - 3\delta^2 q_n^* + (\delta-c_r)\sqrt{3\delta^2 q_n^* + (\delta-c_r)^2}}{9\delta^2}\right].$$

3.1. $f_r > q_n^*$

By imposing the steady state conditions $q_i(t+1) = q_i(t) = q_i^*, i = n, r$, in the map (14), one can see that (14) has the fixed point

$$E = (q_n^*, q_r^*).$$

Where

$$q_n^* = q_r^* = \frac{4(1-c_n)^2(\delta+1)}{(2\delta+3)^2}. \tag{15}$$

Which is a Nash equilibrium of the game providing that $c_n \neq 1$ and $f_r > \frac{4(1-c_n)^2(\delta+1)}{(2\delta+3)^2}$.

The Jacobian matrix of system (14) for this considered case has the form

$$J(q_n, q_r) = \begin{bmatrix} 1+\alpha\varphi & -\alpha\delta\phi \\ 1 & 0 \end{bmatrix} \tag{16}$$

Where

$$\varphi(q_n, q_r) = 1 - c_n - \sqrt{q_n + \delta q_r} - \frac{q_n}{\sqrt{q_n + \delta q_r}}\left(1 + \frac{q_n + 2\delta q_r}{4(q_n + \delta q_r)}\right), \tag{17}$$

and

$$\phi(q_n, q_r) = \frac{q_n(q_n + 2\delta q_r)}{4(q_n + \delta q_r)^{\frac{3}{2}}} \geq 0. \tag{18}$$

Particularly

$\varphi(q_n^*, q_r^*) = 1 - c_n - \sqrt{(1+\delta)q_n^*} - \frac{q_n^*}{\sqrt{(1+\delta)q_n^*}}(1 + \frac{1+2\delta}{4(1+\delta)})$ and $\phi(q_n^*, q_r^*) = \frac{(1+2\delta)\sqrt{q_n^*}}{(1+\delta)^{\frac{3}{2}}}$.

The characteristic equation is

$$\lambda^2 - Tr(J)\lambda + Det(J) = 0.$$

Where $Tr(J) = 1 + \varphi = 1 - \alpha\frac{(1-c_n)(4\delta+3)}{2(2\delta+3)(1+\delta)}$ and $Det(J) = \alpha\delta\frac{(1-c_n)(2\delta+1)}{2(2\delta+3)(1+\delta)}$.
Now one introduces the two quantities

$$\alpha_n^{NS} = \frac{2(2\delta+3)(1+\delta)}{\delta(1-c_n)(1+2\delta)} \quad \text{and} \quad \alpha_n^F = \frac{4(2\delta+3)(1+\delta)}{(1-c_n)(-2\delta^2+3\delta+3)},$$

Hence, the stability of E is summarized in the following proposition

Proposition 3.1: *If the fixed point E exists then it will be locally asymptotically stable if the following condition holds*

$$\alpha < \min\left\{\alpha_n^F, \alpha_n^{NS}\right\}. \tag{19}$$

Proof: Suppose that the fixed point E exists, then it will be locally asymptotically stable as long as the eigenvalues of the Jacobian matrix $J(E)$ lie within the unit circle of the complex plane.

According to the Jury's stability criteria, this holds if:

$$\begin{cases} 1 - Det(J) > 0, \\ 1 - Tr(J) + Det(J) > 0, \\ 1 + Tr(J) + Det(J) > 0. \end{cases} \tag{20}$$

The first condition can be write as $1 - \alpha\delta\frac{(1-c_n)(2\delta+1)}{2(2\delta+3)(1+\delta)} > 0$, which implies that

$$\alpha < \frac{2(2\delta+3)(1+\delta)}{\delta(1-c_n)(1+2\delta)} = \alpha_n^{NS}. \tag{21}$$

The second condition is $1 - Tr(J) + Det(J) = \alpha\frac{(1-c_n)}{2} > 0$, which is always satisfied.

Then the third condition is

$$1 + Tr(J) + Det(J) = 2 + \alpha\frac{1-c_n}{2(2\delta+3)(1+\delta)}(2\delta^2 - 3\delta - 3)) > 0,$$

which can be written as

$$\alpha < \frac{4(2\delta+3)(1+\delta)}{(1-c_n)(-2\delta^2+3\delta+3)} = \alpha_n^F, \tag{22}$$

One has

$$\alpha_n^{NS} - \alpha_n^F = \frac{2(-6\delta^2+\delta+3)(2\delta+3)(\delta+1)}{\delta(2\delta+1)(-2\delta^2+3\delta+3)(1-c_n)},$$

Clearly the denominator is positive for all $\delta \in [0,1]$. Thus, $\alpha_n^{NS} - \alpha_n^F$ has the same sign as $(-6\delta^2+\delta+3)$, which is a second degree polynomial of δ having the two roots $\delta_1 = \frac{1+\sqrt{73}}{12}$ and $\delta_2 = \frac{1-\sqrt{73}}{12} < 0$.

Therefore, $\alpha_n^{NS} > \alpha_n^F$ for $0 \leq \delta < \frac{1+\sqrt{73}}{12}$, but $\alpha_n^{NS} < \alpha_n^F$ for $\frac{1+\sqrt{73}}{12} < \delta \leq 1$.
It follows that for $\delta \in [0, \frac{1+\sqrt{73}}{12}]$ the three Jury's conditions are satisfied if $\alpha < \alpha_n^F$, and for $\delta \in [\frac{1+\sqrt{73}}{12}, 1]$ the three Jury's conditions are satisfied if $\alpha < \alpha_n^{NS}$. ■

In the case $f_r > q_n^*$ the local stability region of the Cournot-Nash equilibrium E is determined by (19).
Therefore, the condition (19) defines surfaces in the parameter space on which a flip bifurcation and a Neimark-Sacker bifurcation take places. The threshold for the flip bifurcation is

$$\alpha = \alpha_n^F = \frac{4(2\delta+3)(1+\delta)}{(1-c_n)(-2\delta^2+3\delta+3)}.$$

Moreover, the threshold for the Neimark–Sacker bifurcation is given by

$$\alpha = \alpha_n^{Ns} = \frac{2(2\delta+3)(1+\delta)}{\delta(1-c_n)(1+2\delta)}.$$

By differentiating α_n^F with respect to δ one gets

$$\frac{\partial \alpha_n^F}{\partial \delta} = \frac{8(8\delta^2+12\delta+3)}{(1-c_n)(-2\delta^2+3\delta+3)^2} > 0.$$

Moreover, by differentiating α_n^{NS} with respect to δ one gets

$$\frac{\partial \alpha_n^{NS}}{\partial \delta} = \frac{-2(8\delta^2+12\delta+3)}{(1-c_n)(2\delta^2+\delta)^2} < 0.$$

Hence, for $\delta \in [0, \frac{1+\sqrt{73}}{12}]$ the stability region of the equilibrium E becomes large with increasing δ, and it losses its stability via flip bifurcation. While, for $\delta \in [\frac{1+\sqrt{73}}{12}, 1]$ the stability region of the equilibrium E becomes smaller with increasing δ, and it losses its stability via Neimark-Sacker bifurcation as shown in Figure 1.

3.2. $f_r < q_n^*$

In this case, the Nash equilibrium $E = (q_n^*, q_r^*)$ satisfies $q_n^* > q_r^*$, where

$$q_n^* = 2\frac{(1-c_n)^2 - 3\delta q_r^* + (1-c_n)\sqrt{3\delta q_r^* + (1-c_n)^2}}{9}$$

and

$$q_r^* = \nu(1-\frac{c_r}{\delta})^2 - q_n^* + 2(1-\nu)\frac{(\delta-c_r)^2 - 3\delta^2 q_n^* + (\delta-c_r)\sqrt{3\delta q_n^* + (\delta-c_r)^2}}{9\delta^2}$$

The Jacobian matrix of system (14) at the equilibrium E has the form

$$J(q_n^*, q_r^*) = \begin{bmatrix} 1+\alpha\varphi & -\alpha\delta\phi \\ -\beta\psi & 1-\beta \end{bmatrix} \tag{23}$$

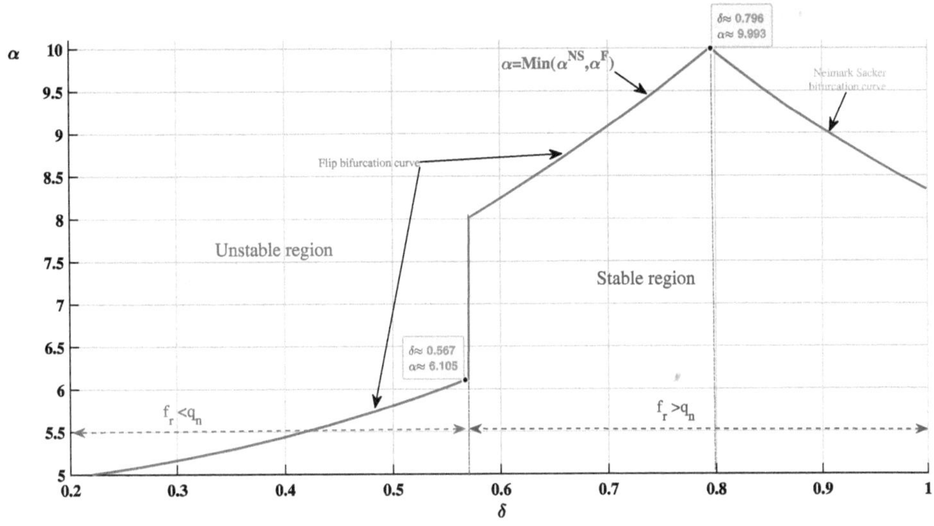

Figure 1. Stability region in the (δ, α) plane, for $\beta = 0.1, c_n = 0.2, c_r = 0.1$ and $\nu = 0.1$.

Where, φ, ϕ are given by (17), (18) respectively, and

$$\psi(q_n, q_r) = \frac{1}{3}\left(2 + \nu - \frac{(1-\nu)(\delta - c_r)}{\sqrt{3\delta^2 q_n^* + (\delta - c_r)^2}}\right) \geq 0. \tag{24}$$

The characteristic equation can be written as

$$\lambda^2 - Tr(J)\lambda + Det(J) = 0,$$

where, $Tr(J) = \varphi - \beta + 2$ and $Det(J) = (1+\varphi)(1-\beta) - \psi\phi$.

Therefore, one can summarize the stability of E as in the proposition below

Proposition 3.2: *Suppose that* $f_r < q_n^*$.
If $\varphi < -\delta\phi\psi$ *then the fixed point E will be locally asymptotically stable if the following condition holds*

$$\alpha < \frac{2(2-\beta)}{\beta\delta\phi\psi + (\beta - 2)\varphi} = \alpha^F, \tag{25}$$

and it loses its stability via a flip bifurcation when α *exceeds the threshold value* α^F.

Proof: According to the Jury stability criterion the equilibrium E will be locally asymptotically stable if

$$\begin{cases} 1 - Det(J) > 0, \\ 1 - Tr(J) + Det(J) > 0, \\ 1 + Tr(J) + Det(J) > 0, \end{cases} \tag{26}$$

which can be written after some simplifications as follows

$$\begin{cases} \alpha\,[\beta\delta\phi\psi - \varphi(1-\beta)] > -\beta, \\ \alpha\beta[\varphi + \delta\phi\psi] < 0, \\ -\alpha[\beta\delta\phi\psi + (\beta-2)\varphi] + 2(2-\beta) > 0. \end{cases} \tag{27}$$

Taking into-account the assumption $\varphi < -\delta\phi\psi$, which imply that $\varphi < 0$, one can observe that the first and the second conditions are satisfied for all positive values of α and β, and the third condition responsible of the flip bifurcation is equivalent to condition (25). ■

The threshold α^F for the flip bifurcation can be represented in the parameter space by a surface (or a curve in the plane). For example, it is represented by a curve in the (δ, α) plane as shown in Figure 1. From this figure, one can observe that in the case $f_r < q_n^*$ the stability region of the Nash equilibrium E becomes larger by increasing δ, and the curve $\alpha = \alpha^F$ jumps when $f_r = q_n^*$ due to the discontinuity of the "*Min*" function.

4. Numerical analysis

Our main goal in this section is to emphasize the complex dynamics of the proposed duopoly game with heterogeneous players described by the dynamical system (14)). Here are the values of the parameters, which can be conveniently used to study the local stability properties of the equilibrium points: $\beta = 0.1, c_n = 0.2, c_r = 0.1, \nu = 0.1$, then $\delta_0 = 0.22$ and α, δ are taken as bifurcation parameters.
Based on the stability conditions (19) and (25), one draws Figure 1, showing the stability region of the equilibrium point E in the (δ, α) parameter plane. As seen in this

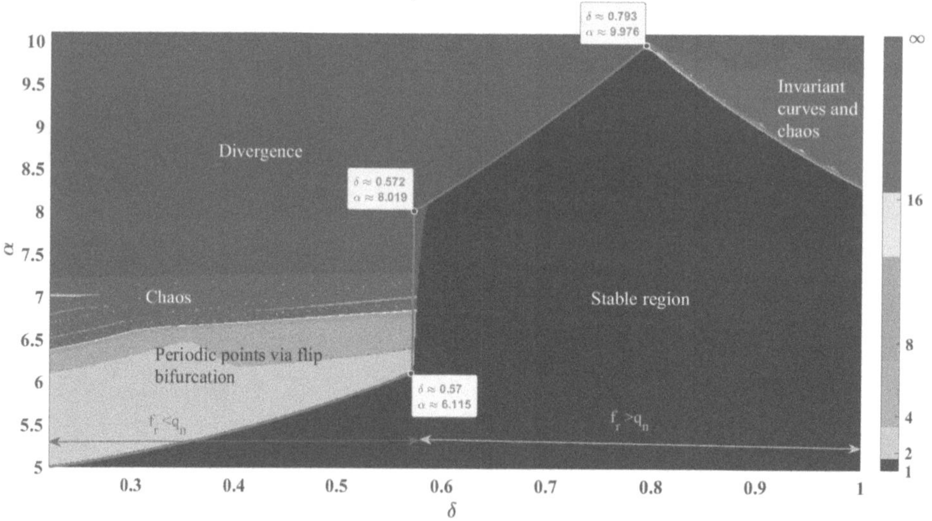

Figure 2. 2-D Bifurcation diagram in (δ, α)-plane for $\beta = 0.1, c_n = 0.2, c_r = 0.1$ and $\nu = 0.1$. The blue colour is used to indicate a stable fixed point. The cyan, green, green dark and yellow colours correspond to period-2, period-4, period-8, and period-16 cycles, respectively, while the red and brown, colours correspond to chaos and divergence respectively.

figure, the stability region of the Nash equilibrium increases with an increase of δ in the interval $(0.22, 0.796)$. Therefore, in this range, a higher CWP would enhance the stability of the game. Furthermore, the equilibrium can lose its stability via flip bifurcation when the pair (δ, α) crosses the stability boundary curve. Particularly in the sub-interval $\delta \in (0.22, 0.567)$, one has $f_r < q_n$, and the emerging periodic points through the flip bifurcation are stable (supercritical flip bifurcation). Furthermore a period doubling scenario to chaos is observed when α increases, as shown in Figure 2. While, in the sub-interval $\delta \in (0.567, 0.796)$, one has $f_r > q_n$, and the emerging periodic points through the flip bifurcation are unstable (sub-critical flip bifurcation). Above the bifurcation curve no attractor is observed and the trajectories diverge. For $(0.7953, 1)$, the stability region decreases with an increase of δ and becomes smaller, meaning that in this range a higher WTP would destabilize the game, and when the pair (δ, α) crosses the stability boundary curve a Neimark-Sacker bifurcation occurs and invariant curve emerges.

Using the initial conditions $(q_{n0}, q_{r0}) = (q_n^* + 0.0001, q_r^*)$, one can examine the 2-d bifurcation and the existence of various attractors in the parameter plane (δ, α). Different results are shown in Figure 2, where the blue colour is used to indicate a stable fixed point. The cyan, green, green dark and yellow colours correspond to period-2, period-4,

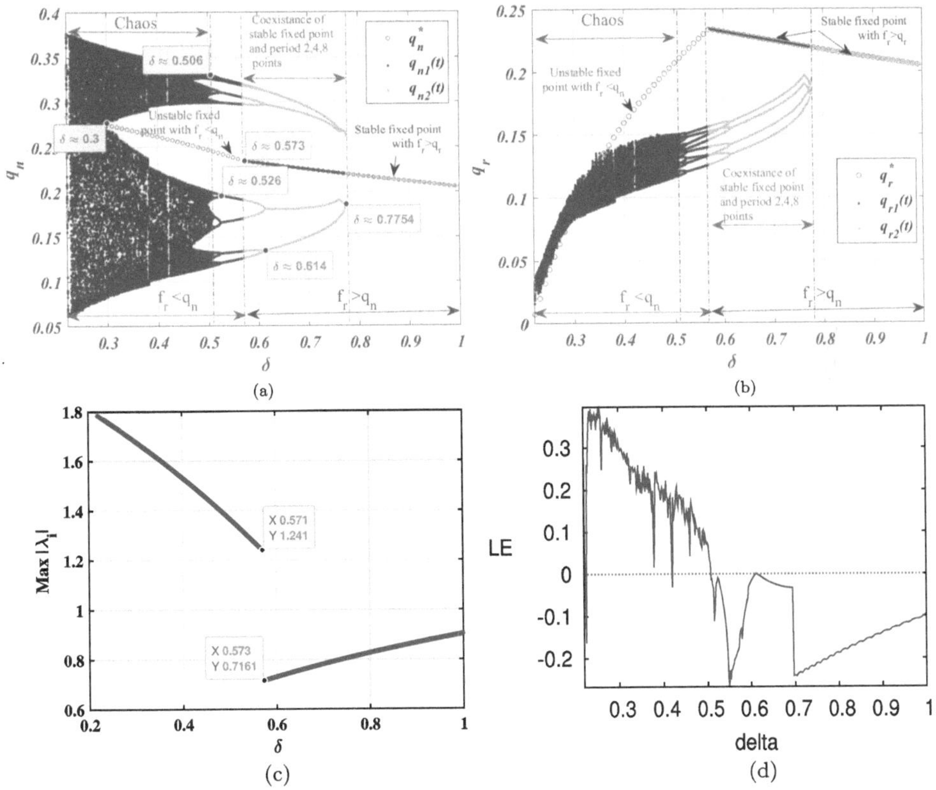

Figure 3. Bifurcation diagrams versus δ for $\alpha = 6.68, \beta = 0.1, c_n = 0.2, c_r = 0.1$ and $\nu = 0.1$. (a) The Bifurcation diagram of the solution q_n. (b) The Bifurcation diagram of the solution q_r. (c) Radius of eigenvalues corresponding to the Nash equilibrium E. (d) The Maximum Lyapunov exponents.

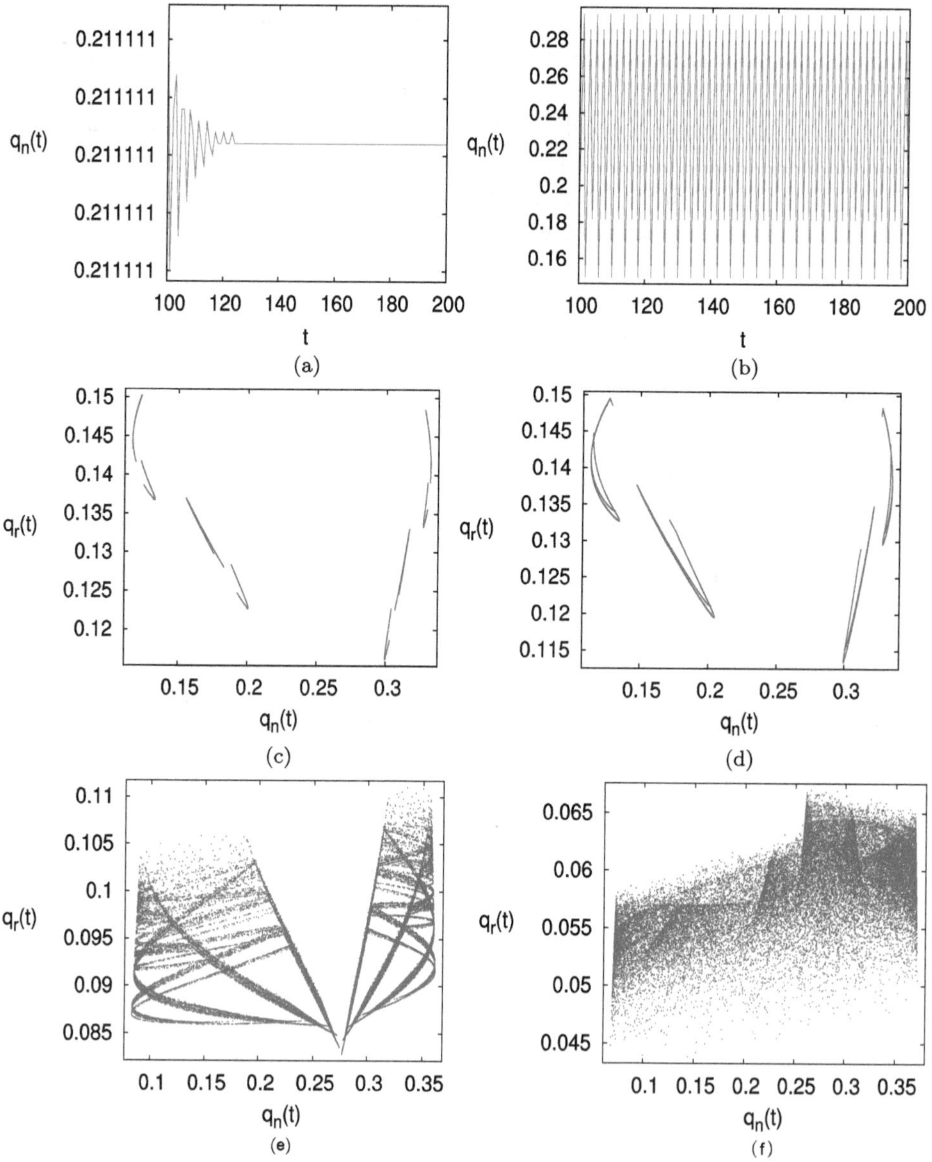

Figure 4. Time series and phase portrait of system (14) for some values of δ with $\alpha = 6.68, \beta = 0.1, c_n = 0.2, c_r = 0.1$ and $\nu = 0.1$. (a) Stable fixed point for $\delta = 0.9$. (b) Period-4 cycle for $\delta = 0.7$. (c) 8 pieces chaotic attractor for $\delta = 0.5$. (d) 4 pieces chaotic attractor for $\delta = 0.48$. (e) 2 pieces chaotic attractor for $\delta = 0.3$. (f) 1 piece chaotic attractor for $\delta = 0.25$.

period-8, and period-16 cycles, respectively, while the red and brown, colours correspond to chaos and divergence (the trajectories do not converge to any attractor) respectively.

Figure 3 presents some bifurcation diagrams versus δ for $\alpha = 6.68, \beta = 0.1, c_n = 0.2, c_r = 0.1$ and $\nu = 0.1$. From this figure, by the decreasing values of δ one can see that for $\delta \in (0.573, 1)$ the Nash equilibrium E is locally asymptotically stable, whereas a bifurcation occurs when $\delta \approx 0.7754$ where a period-2 cycle appears and bifurcates rapidly to a period-4 cycle as shown in Figure 4(b) which co-exists with the Nash equilibrium E until $\delta \approx 0.614$ where it bifurcates in-tour to a period-8 cycle that co-exists with the fixed point which loses its stability at $\delta \approx 0.573$, and the period-8 cycle bifurcates to a period-16 when $\delta \approx 0.526$ then the period doubling bifurcation posses increase dramatically and the chaotic behaviour appears at $\delta \approx 0.506$ where multi-pieces attractors manifest and the number of pieces decrease with decreasing of δ until it becomes a one piece chaotic attractor when $\delta < 0.3$. Chaotic attractors for two-dimensional mappings have been revealed forty years ago by the initial model of Hénon [17], Lozi [11,20], Belykh [?]. Since then they have been extensively studied, however many other kinds of attractors have to be explored like those shown in Figure 4(c)–(f). Of course, one should be careful with numerical simulations [21], and different tools (such as bifurcation diagram, Lyapunov exponents and phase portrait) must be used to overcome wrong results.

5. Conclusion

This paper defines and analyses an industry duopoly game model involving two heterogeneous firms: an original product manufacturer and a third-party remanufacturer, which enters the market and competes with the original manufacturer, whenever an original product has reached the end of its useful life.

The obtained results show that the stability of the Nash equilibrium point can be lost through flip (period doubling) or Neimark-Sacker bifurcations. The coexistence of the Nash equilibrium with multi-period cycles is also highlighted. Chaotic behaviour is investigated via bifurcation diagrams, Lyapunov exponents and phase portraits, showing the existence of different attractors when varying CWP, including 8-pieces, 4-pieces, 2-pieces and one-piece chaotic attractors. In the forthcoming paper, we will investigate the control of chaos in the proposed model by stabilizing the Nash equilibrium in the range of parameters where it is unstable and making it more profitable for the market.

Disclosure statement

No potential conflict of interest was reported by the author(s).

References

[1] H.N. Agiza, *On the stability, bifurcations, chaos and chaos control of Kopel map*, Chaos Soliton Fractals 10(11) (1999), pp. 1909–1916.
[2] H.N. Agiza and A.A. Elsadany, *Nonlinear dynamics in the Cournot duopoly game with heterogeneous players*, Phys. A 320 (2003), pp. 512–524.
[3] H.N. Agiza and A.A. Elsadany, *Chaotic dynamics in nonlinear duopoly game with heterogeneous players*, Appl. Math. Comput. 149(3) (2004), pp. 843–860.
[4] H.N. Agiza, A.S. Hegazi, and A.A. Elsadany, *The dynamics of Bowley's model with bounded rationality*, Chaos Soliton Fractals 12(9) (2001), pp. 1705–1717.
[5] H.N. Agiza, A.S. Hegazi, and A.A. Elsadany, *Complex dynamics and synchronization of a duopoly game with bounded rationality*, Math. Comput. Simul 58(2) (2002), pp. 133–146.
[6] A. Atasu, M. Sarvary, and L.N. Van Wassenhove, *Remanufacturing as a marketing strategy*, Manag. Sci. 54(10) (2008), pp. 1731–1746.
[7] A. Atasu, V.D.R. Guide, and L.N. Van Wassenhove, *So what if remanufacturing cannibalizes new product sales?*, Calif. Manag. Rev. 52(2) (2010), pp. 56–76.

[8] V.N. Belykh, *On Models of Phase Synchronization Systems and Their Study. Dinamika System*, Gorky State University Press, 1976, vol. 11, pp. 23–32 (in Russian).
[9] G.I. Bischi and M. Kopel, *Equilibrium selection in a nonlinear duopoly game with adaptive expectations*, J. Econ. Behav. Organ 46(1) (2001), pp. 73–100.
[10] G.I. Bischi and A. Naimzada, *Global analysis of a dynamic duopoly game with bounded rationality*, in *Advanced in Dynamics Games and Application*, J.A. Filar, V. Gaitsgory, K. Mizukami, eds., Birkhauser, Switzerland, 1999.
[11] R. Bououden, M.-S. Abdelouahab, F. Jarad, and Z. Hammouch, *A novel fractional piecewise linear map: regular and chaotic dynamics*, Int. J. Gen. Syst. 50(5) (2021), pp. 501–526. doi: 10.1080/03081079.2021.1919102
[12] A. Cournot, *Recherches Sur Les Principes Mathématiques De La Théoris Des Richesses*, Hachette, Paris, 1838.
[13] P.E. Dessert and S.D. James, *Applying chaos to manufacturing process optimization*, J. Adv. Manuf. Syst. 01(02) (2002), pp. 201–210.
[14] M.E. Ferguson and L.B. Toktay, *The effect of competition on recovery strategies*, Prod. Oper. Manag. 15(3) (2006), pp. 351–368.
[15] G. Ferrer and J.M. Swaminathan, *Managing new and remanufactured products*, Manag. Sci. 52(1) (2006), pp. 15–26.
[16] G. Ferrer and J.M. Swaminathan, *Managing new and differentiated remanufactured products*, Eur. J. Oper. Res. 203(2) (2010), pp. 370–379.
[17] M. Hénon, *A two-dimensional mapping with a strange attractor*, Commun. Math. Phys. 50 (1) (1976), pp. 69–77. doi: 10.1007/BF01608556
[18] Y. Hu, F. Zhu, L. Zhang, Y. Lui, and Z. Wang, *Scheduling of manufacturers based on chaos optimization algorithm in cloud manufacturing*, Robot. Comput. Integr. Manuf. 58 (2019), pp. 13–20.
[19] M. Kopel, *Simple and complex adjustment dynamics in Cournot duopoly models*, Chaos Soliton Fractals 7(12) (1996), pp. 2031–2048.
[20] R. Lozi, *Un attracteur étrange (?) du type attracteur de Hénon*, J. Phys. Colloq. 39 (C5) (1978), pp. C5-9–C5-10. doi: 10.1051/jphyscol:1978505
[21] R. Lozi, *Can we trust in numerical computations of chaotic solutions of dynamical systems ?*, World Scientific Series on Nonlinear Science Series A Topology and Dynamics of Chaos, 2013, pp. 63–98.
[22] P. Majumder and H. Groenevelt, *Competition in remanufacturing*, Prod. Oper. Manag. 10(2) (2001), pp. 125–141.
[23] S. Mitra and S. Webster, *Competition in remanufacturing and the effects of government subsidies*, Int. J. Prod. Econ. 111(2) (2008), pp. 287–298.
[24] A. ñrsdemir, E. Kemahlıoğlu-Ziya, and A.K Parlaktürk, *Competitive quality choice and remanufacturing*, Prod. Oper. Manag. 23(1) (2014), pp. 48–64.
[25] T. Onazaki, G. Sieg, and M. Yokoo, *Stability, Chaos and multiple attractors: a single agent makes a difference*, J. Econ. Dyn. Control 27(10) (2003), pp. 1917–1938.
[26] Y. Peng and Q. Lu, *Complex dynamics analysis for a duopoly stackelberg game model with bounded rationality*, Appl. Math. Comput. 271 (2015), pp. 259–268.
[27] T Puu, *Chaos in duopoly pricing*, Chaos Solitons Fractals 1 (6) (1991), pp. 573–581.
[28] T Puu, *The chaotic duopolists revisited*, J. Econom. Behav. Organ. 33 (3-4) (1998), pp. 385–394.
[29] D. Rand, *Exotic phenomena in games duopoly models*, J. Math. Econom. 5 (2) (1978), pp. 173–184.
[30] S. Rassenti, S.S. Reynolds, V.L. Smith, and F. Szidarovszky, *Adaptation and convergence of behavior in repeated experimental Cournot games*, J. Econ. Behav. Organ 41(2) (2000), pp. 117–146.
[31] L. Shi, Z. Sheng, and F. Xu, *Complexity analysis of remanufacturing duopoly game with different competition strategies and heterogeneous players*, Nonlinear Dyn. 82 (3) (2015), pp. 1081–1092.
[32] H. Von Stackelberg, *Probleme der unvollkommenen konkurrenz*, Weltwirtsch. Arch. 48 (1938), pp. 95–141.
[33] J.X. Zhang, Q.L. Da, and Y.H. Wang, *Analysis of nonlinear duopoly game with heterogeneous players*, Econ. Model 24(1) (2007), pp. 138–148.

Index

Note: **Bold** page numbers refer to tables and *italic* page numbers refer to figures.

Kolibri 360 GmbH
Ferdinand-Jühlke-Straße 7
99095 Erfurt, Deutschland
produktsicherheit@kolibri360.de